STUDENT SOLUTIONS MANUAL TO ACCOMPANY

STATISTICS IN PRACTICE

by ERNEST A. BLAISDELL

RONALD L. SHUBERT
Elizabethtown College

SAUNDERS COLLEGE PUBLISHING
Harcourt Brace Jovanovich College Publishers
Fort Worth • Philadelphia
Boston • New York • Chicago • Orlando • San Francisco • Atlanta
Dallas • London • Toronto • Austin • San Antonio

Printed in the United States of America.

Shubert: Student Solutions Manual to accompany STATISTICS IN PRACTICE
by Ernest Blaisdell

ISBN 0-03-032227-8

5 021 9876543

PREFACE

This solutions manual accompanies the text Statistics in Practice by Ernest A. Blaisdell. It contains solutions to all of the odd Exercises in the text. In addition, solutions are provided for all Review Exercises .

Users of the manual may occasionally obtain answers that differ slightly due to rounding. Although we have checked my work carefully, errors, most likely, remain. I would appreciate these being called to my attention.

All MINITAB* solutions were done on a 386 personal computer using the most current version (Release 8.2) of the MINITAB Statistical Software. I wish to acknowledge the cooperation of MINITAB Inc. for providing me with needed support.

I wish to acknowledge the contributions of Elizabethtown College students Beth Batcho, Renee Hatter, and Richard Graybill who were helpful in the early stages of the project and especially my daughter, Erika Shubert, who was extremely helpful in the final editing. I also appreciate the many suggestions for improvement made by Ernest Blaisdell as a result of his careful proofreading of the manuscript. Finally, I thank my entire family for their support during the past several months while I worked on this project.

TABLE OF CONTENTS

CHAPTER 1
A FIRST LOOK AT STATISTICS AND MINITAB

CHAPTER ONE EXERCISES

1.1 The 100 volunteers who participated in the study constitute a sample from a population of coffee drinkers.

1.3 **a.** We note that, with 202 cases, Arizona did have the largest number of recorded cases of tuberculosis in 1988. Therefore, the statement is descriptive.
b. With only 5 recorded cases of tuberculosis we may conclude there is little risk of contracting tuberculosis in Wyoming so the statement is inferential. However, since the population of Wyoming is smaller than that of the other states in the study, the data do not justify this conclusion.
c. We note that Colorado and New Mexico each record 89 cases and are tied for the second largest number of recorded cases. Therefore, the statement is descriptive.
d. In 1988, there were more recorded cases occurring in Nevada (48) than in Utah (18). Therefore, the statement is descriptive.
e. From the data, we might conjecture that a person living in Arizona (202 cases) is more likely to contract tuberculosis than one living in Colorado (89 cases) so this statement is inferential. However, since Arizona has a larger and possibly older population of people, the data do not justify the conclusion.

1.5 **a.** The data would be regarded as a sample if one were interested in making inferences about all college faculty in the United States, for example, if we used it to estimate the difference, among all college faculty, in the proportion of men versus women who possess the doctorate. (other answers possible)
b. If the study were restricted to describing faculty at this one particular college, then it would be regarded as a population.

1.7 **a.** The population of interest to the candidate is the set of people who will actually vote in the mayoral election.
b. There is no way to determine in advance who will vote in an election.

1.9 **a.** The population of interest is all sardines (i.e. their weights) in the fishing trawler's catch.
b. The parameter of interest is the population mean which would be the average weight of all sardines in the catch.
c. The statistic used to estimate this parameter is the sample mean, which is 38.9.
d. These would probably have been caught in the same location and may contain either mostly small or large sardines.

1.11 **a.** The population of interest is all homes served by the sewage treatment facility.
b. The parameter of interest is the percentage (or proportion) of homes served by the sewage treatment facility which have an in-sink garbage disposal.
c. The statistic used to estimate the parameter is the percentage of the sample of 880 homes in the survey with an in-sink garbage disposal, and its value was 35%.

MINITAB LAB ASSIGNMENTS

1.13(M)

```
MTB > # Exercise 1.13(M)
MTB > SET C3
DATA > 762 498 489 457 439 421 410 364 360 335
DATA > END
MTB > SUM C3
   SUM      =        4535.0
```

1.15(M)

```
MTB > # Exercise 1.15(M)
MTB > SET C1
DATA> 0:100/10
DATA> END
MTB > LET C2 = 5*(C1-32)/9
MTB > NAME C1 'F' C2 'C'
MTB > PRINT C1 C2

 ROW      F          C

   1      0   -17.7778
   2     10   -12.2222
   3     20    -6.6667
   4     30    -1.1111
   5     40     4.4444
   6     50    10.0000
   7     60    15.5556
   8     70    21.1111
   9     80    26.6667
  10     90    32.2222
  11    100    37.7778
```

1.17(M)

```
MTB > # Exercise 1.17(M)
MTB > SET C1
DATA> 1:25
DATA> END
MTB > LET C2 = C1**3
MTB > LET C3 = C1**(1/3)
MTB > NAME C1 'N' C2 'CUBE' C3 'CUBE RT'
MTB > PRINT C1-C3

 ROW        N     CUBE    CUBE RT

   1        1        1    1.00000
   2        2        8    1.25992
   3        3       27    1.44225
   4        4       64    1.58740
   5        5      125    1.70998
   6        6      216    1.81712
   7        7      343    1.91293
   8        8      512    2.00000
   9        9      729    2.08008
  10       10     1000    2.15443
```

```
  11        11     1331   2.22398
  12        12     1728   2.28943
  13        13     2197   2.35133
  14        14     2744   2.41014
  15        15     3375   2.46621
  16        16     4096   2.51984
  17        17     4913   2.57128
  18        18     5832   2.62074
  19        19     6859   2.66840
  20        20     8000   2.71442
  21        21     9261   2.75892
  22        22    10648   2.80204
  23        23    12167   2.84387
  24        24    13824   2.88450
  25        25    15625   2.92402
```

1.19(M)

```
MTB > # Exercise 1.19(M)
MTB > READ C1 C2
DATA> 1 73
DATA> 2 85
DATA> 3 72
DATA> 4 69
DATA> 5 87
DATA> 6 94
DATA> 7 86
DATA> 8 81
DATA> 9 58
DATA> 10 98
DATA> END
     10 ROWS READ
MTB > NAME C1 'STUDENT' C2 'FE'
MTB > PRINT C1 C2

 ROW   STUDENT     FE

   1         1     73
   2         2     85
   3         3     72
   4         4     69
   5         5     87
   6         6     94
   7         7     86
   8         8     81
   9         9     58
  10        10     98
```

1.21(M)

```
MTB > # Exercise 1.21(M)
MTB > # Student Numbers are in C1 and Final Exam Scores are in C2
MTB > SORT C2, carry along C1, place results in C2 and C1
MTB > PRINT C1 C2

 ROW   STUDENT     FE

   1         9     58
   2         4     69
   3         3     72
   4         1     73
   5         8     81
   6         2     85
   7         7     86
   8         5     87
   9         6     94
  10        10     98
```

1.23(M)

```
MTB > # Exercise 1.23(M)
MTB > READ C1 - C5
DATA> 1 73 69 78 54
DATA> 2 85 89 87 72
DATA> 3 69 75 81 47
DATA> 4 72 91 86 58
DATA> 5 87 93 91 69
DATA> 6 94 99 95 83
DATA> 7 86 72 82 68
DATA> 8 81 84 89 70
DATA> 9 58 65 69 39
DATA> 10 98 94 94 82
DATA> END
     10 ROWS READ
MTB > NAME C1 'STUDENT' C2 'FE' C3 'E1' C4 'E2' C5 'E3' C6 'AVG'
MTB > LET C6 = (2*C2+C3+C4+C5)/5
MTB > PRINT C1 C3-C5 C2 C6

 ROW   STUDENT     E1     E2     E3     FE     AVG

   1         1     69     78     54     73    69.4
   2         2     89     87     72     85    83.6
   3         3     75     81     47     69    68.2
   4         4     91     86     58     72    75.8
   5         5     93     91     69     87    85.4
   6         6     99     95     83     94    93.0
   7         7     72     82     68     86    78.8
   8         8     84     89     70     81    81.0
   9         9     65     69     39     58    57.8
  10        10     94     94     82     98    93.2
```

1.25(M)

```
MTB > # Exercise 1.25(M)
MTB > SET C1
DATA> 8.13 23.64 14.53 5.98 15.89 18.01 22.10 8.22 11.93 10.06
DATA> 17.23 9.84 19.25 17.95 11.85 15.85 7.04 11.53 12.66 24.45
DATA> 19.18 13.83 18.62 6.58 13.37 7.56 13.16 9.65 14.46 21.16
DATA> 11.34 29.48 10.55 8.01 23.34 10.24 14.74 10.99 11.71 8.72
DATA> 13.09 7.70 8.94 16.64 12.37 16.29 19.81 7.46 16.53 19.83
DATA> END
MTB > RANK C1 C2
MTB > NAME C1 'AMOUNT' C2 'RANK'
MTB > PRINT C1 C2
```

ROW	AMOUNT	RANK
1	8.13	8
2	23.64	48
3	14.53	30
4	5.98	1
5	15.89	33
6	18.01	39
7	22.10	46
8	8.22	9
9	11.93	22
10	10.06	14
11	17.23	37
12	9.84	13
13	19.25	42
14	17.95	38
15	11.85	21
16	15.85	32
17	7.04	3
18	11.53	19
19	12.66	24
20	24.45	49
21	19.18	41
22	13.83	28
23	18.62	40
24	6.58	2
25	13.37	27
26	7.56	5
27	13.16	26
28	9.65	12
29	14.46	29
30	21.16	45
31	11.34	18
32	29.48	50
33	10.55	16
34	8.01	7
35	23.34	47
36	10.24	15
37	14.74	31
38	10.99	17
39	11.71	20
40	8.72	10
41	13.09	25
42	7.70	6
43	8.94	11
44	16.64	36
45	12.37	23
46	16.29	34
47	19.81	43
48	7.46	4
49	16.53	35
50	19.83	44

Note: The ranking shown is from low to high. In order to rank the expenditures from high to low, simply type in the command **LET C2 = 51 - C2** before the PRINT command.

CHAPTER 2
DESCRIBING DATA: GRAPHICAL AND NUMERICAL METHODS

EXERCISES for Section 2.1
Graphical Methods for Quantitative Data: Frequency Distributions and Histograms

2.1 We use classes of equal width beginning with the class 30 - 39 and count the percentage which fall within each interval. The results follow:

Percentage	Frequency
30 - 39	1
40 - 49	2
50 - 59	4
60 - 69	2
70 - 79	6
80 - 89	1

2.3 We first obtain a relative frequency distribution by dividing each frequency by 16, the total frequency:

Percentage	Frequency	Relative Frequency
30 - 39	1	0.0625
40 - 49	2	0.1250
50 - 59	4	0.2500
60 - 69	2	0.1250
70 - 79	6	0.3750
80 - 89	1	0.0625

The relative frequency histogram is constructed by plotting the class boundaries along the horizontal axis and bars with heights corresponding to the relative frequencies above the intervals which correspond to the appropriate classes. The result follows:

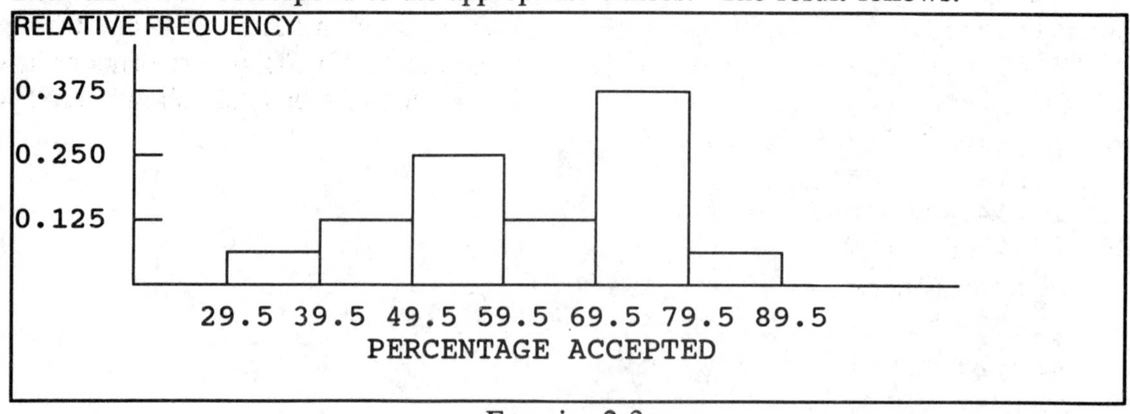

Exercise 2.3

2.5 Since the lower limit of the first class is 20, and the class width is 8, we see that the lower limit of the second class is 28, so the upper limit of the first class is 27. The limits of the five classes are as follows:

Class
20 - 27
28 - 35
36 - 43
44 - 51
52 - 59.

2.7 We note that the lower limit of the second class is 140 and that the upper limit of the first class is 140 - 1 = 139. Therefore, the upper boundary of the first class is (139 + 140)/2 = 139.5. We note that the class width is 40, so the lower boundary of the first class would be 139.5 - 40 = 99.5. Therefore, the class boundaries are as follows:

Class
99.5 - 139.5
139.5 - 179.5
179.5 - 219.5
219.5 - 259.5
259.5 - 299.5
299.5 - 339.5

2.9 The class width is 9 so the lower boundary of the first class is 74 - 9/2 = 69.5 and the upper boundary is 69.5 + 9 = 78.5. We continue to obtain other class boundaries as follows:

Class
69.5 - 78.5
78.5 - 87.5
87.5 - 96.5
96.5 - 105.5
105.5 - 114.5
114.5 - 123.5
123.5 - 132.5
132.5 - 141.5

2.11 The class boundaries of the first class are 7.5 to 8.5 and therefore represent children of age 8. The relative frequency of this class is indicated by the height of the bar which is 0.50. Since there are 20 participants in the special education program, this means that there are (0.50)(20) = 10 students of age 8. Similarly, there are (0.30)(20) = 6 students of age 9 and (0.10)(20) = 2 students each of ages 10 and 11. The list of ages is as follows: 8, 8, 8, 8, 8, 8, 8, 8, 8, 8, 9, 9, 9, 9, 9, 9, 10, 10, 11, 11.

2.13 **a.** The class width is (30 - 25) = 5.
b. The boundaries of the fifth class are 44.5 - 49.5.
c. The lower boundary of the first class is (24 + 25)/2 = 24.5.
d. The upper boundary of the last class is (69 + 70)/2 = 69.5.
e. The lower limit of the seventh class is 55.
f. The upper limit of the third class is 39.
g. The number of faculty younger than 40 is (4 + 14 + 19) = 37, which is the sum of the frequencies of the first three classes.

h. The number of faculty older than 59 is the sum of the frequencies of the last two classes which is (16 + 8) = 24.
i. The number of faculty older than 39 but younger than 60 is (28 + 26 + 23 + 20) = 97.

2.15 We use classes of equal width beginning with the class 2.0 - 2.9 and count the number of states whose 1987 unemployment rates fall within each interval. We obtain relative frequencies by dividing each frequency by 50. The results follow: (other answers are possible)

Unemployment Rate	Frequency	Relative Frequency
2.0 - 2.9	1	1/50 = 0.02
3.0 - 3.9	6	6/50 = 0.12
4.0 - 4.9	8	8/50 = 0.16
5.0 - 5.9	8	8/50 = 0.16
6.0 - 6.9	8	8/50 = 0.16
7.0 - 7.9	8	8/50 = 0.16
8.0 - 8.9	7	7/50 = 0.14
9.0 - 9.9	0	0/50 = 0.00
10.0 - 10.9	3	3/50 = 0.06
11.0 - 11.9	0	0/50 = 0.00
12.0 - 12.9	1	1/50 = 0.02
Total =	50	

2.17 The relative frequency histogram for the 50 unemployment rates that appear in Exercise 2.15 is constructed by plotting the class boundaries along the horizontal axis and bars with heights corresponding to the relative frequencies above the intervals which correspond to the appropriate classes. The result follows: (other answers are possible)

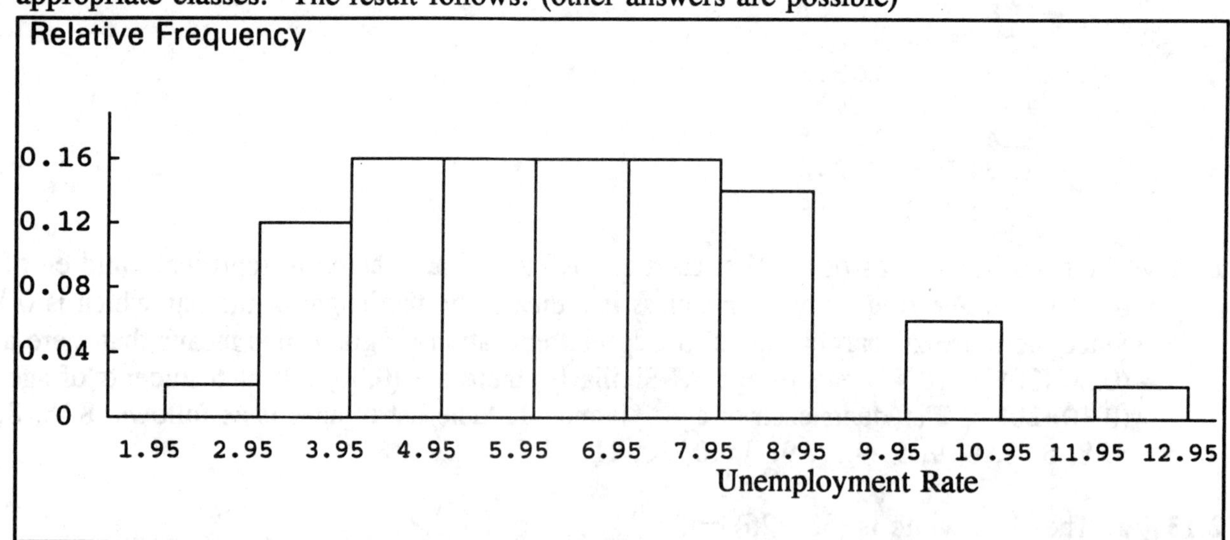

Exercise 2.17

MINITAB LAB ASSIGNMENTS

2.19(M)

```
MTB > # Exercise 2.19(M)
MTB > # We assume the data are in C1
MTB > HISTOGRAM C1;
SUBC> START 470;
SUBC> INCREMENT 20.


Histogram of C1   N = 79

Midpoint   Count
   470.0       2  **
   490.0       6  ******
   510.0       4  ****
   530.0       7  *******
   550.0      13  *************
   570.0      13  *************
   590.0      16  ****************
   610.0       7  *******
   630.0       1  *
   650.0       1  *
   670.0       2  **
   690.0       5  *****
   710.0       1  *
   730.0       1  *
```

2.21(M)

```
MTB > # Exercise 2.21(M)
MTB > # We assume the data have been READ into C1 and C2
MTB > NAME C1 '1980' C2 '1987'
MTB > HIST C1 C2

Histogram of 1980   N = 32

Midpoint   Count
      60      10  **********
      70       5  *****
      80       8  ********
      90       4  ****
     100       1  *
     110       2  **
     120       2  **


Histogram of 1987   N = 32

Midpoint   Count
      80       2  **
      90       6  ******
     100       3  ***
     110       5  *****
     120       2  **
     130       5  *****
     140       2  **
     150       0
     160       2  **
     170       1  *
     180       3  ***
     190       1  *
```

EXERCISES for Section 2.2

Graphical Methods for Quantitative Data: Dotplots and Stem-and-Leaf Displays

2.23 From the dot plot we read the times required by 18 laboratory animals to perform a complex task as follows: 20, 20.5, 21.7, 21.8, 21.8, 21.8, 21.8, 22, 22, 22, 22, 22.2, 22.2, 22.5, 22.5, 22.7, 23, 23.6.

2.25 From the stem-and-leaf display we read the hourly wage rates of 21 part-time employees of a fast food restaurant as follows: $4.25, $4.25, $4.25, $4.40, $4.40, $4.40, $4.45, $4.50, $4.50, $4.50, $4.50, $4.55, $4.60, $4.65, $4.75, $4.75, $4.75, $5.00, $5.00, $5.00, $5.00.

Graphical Methods for Quantitative Data: Dotplots and Stem and Leaf Displays

2.27 For numbers between 1 and 99 we use the tens digit as the stem, while for numbers larger than 99 both the tens and the one-hundreds digit are used for the stem. The units digit is used for the leaf. The stem-and-leaf display for the students-per-crime ratio is as follows:

Stem	Leaf
0	5 9 8 6
1	7 3 9 1 0 6 0 5 7 4 2 1 7 4 3
2	6 6 6 9 7
3	9 5 1 7
4	5 9 7 7
5	0
6	6
7	
8	
9	
10	4
11	
12	
13	
14	
15	
16	
17	7
18	
19	4

2.29 We note that the number of gold medals range from 1 to 12. We draw a number line to cover this range and construct a dotplot as follows:

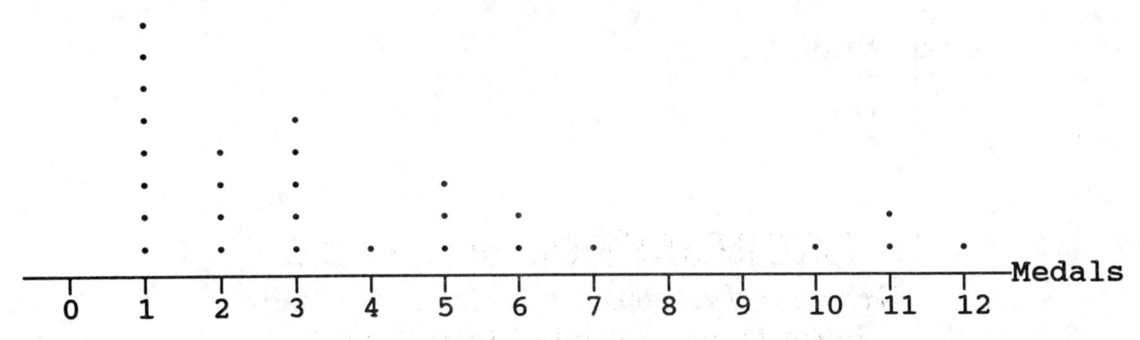

2.31 We note that the ages of 40 blood donors which were given in Exercise 2.2 range from 21 to 64. We draw a number line to cover this range and a dotplot as follows:

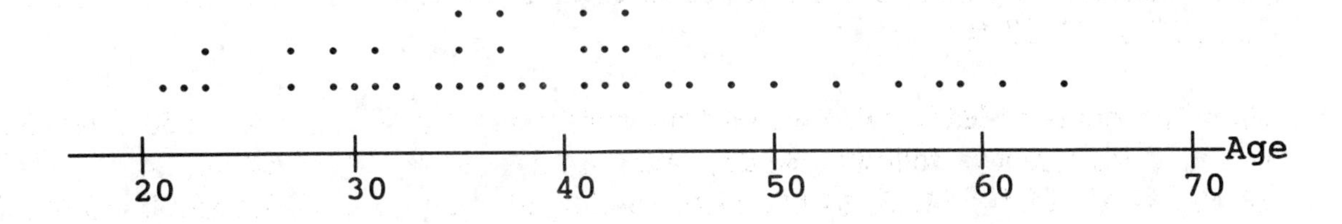

2.33 We use the tens digit for the stem and the units digit for the leaf. The stem-and-leaf display for the 40 ages that appear in Exercise 2.31 follows:

Stem	Leaf
2	1 3 9 7 7 2 9 3
3	5 5 7 9 1 6 7 5 2 4 8 1 0 7
4	3 2 8 1 2 3 3 1 5 6 1
5	3 8 9 0 6
6	1 4

2.35 **a.** We use the whole number and the one-tenths part of the numbers for the stem and the one-hundreds digit for the leaf. The stem-and-leaf display for the cumulative grade point averages of the 18 students in an honors section of Probability and Statistics follows:

Stem	Leaf
3.2	9
3.3	9
3.4	5 1
3.5	3 0
3.6	7 0 4 2
3.7	1 4 5 3
3.8	8
3.9	0
4.0	0 0

b. With the leaves arranged in order of magnitude, we get the following stem-and-leaf display for the cumulative grade point averages:

Stem	Leaf
3.2	9
3.3	9
3.4	1 5
3.5	0 3
3.6	0 2 4 7
3.7	1 3 4 5
3.8	8
3.9	0
4.0	0 0

MINITAB LAB ASSIGNMENTS

2.37(M)

```
MTB > # Exercise 2.37(M)
MTB > # The grade point averages are in C2
MTB > NAME C2 'gpa'
MTB > STEM C2

Stem-and-leaf of gpa        N  = 47
Leaf Unit = 0.010


    4    28 0479
    9    29 44459
   15    30 112689
   22    31 4567899
  (5)    32 14689
   20    33 1266799
   13    34 11
   11    35 147
    8    36 167
    5    37 257
    2    38 7
    1    39 7
```

2.39(M)

```
MTB > # Exercise 2.39(M)
MTB > # The SAT math scores were read into C3 in Exercise 2.38(M)
MTB > STEM C3;
SUBC> INCREMENT 100.

Stem-and-leaf of SAT math  N  = 79
Leaf Unit = 10

    8    4 77889999
 (53)    5 01112223333444444444455566666666777778888888888999999
   18    6 0000011247788889
    2    7 12
```

EXERCISES for Section 2.3

Graphical Methods for Qualitative Data: Bar and Pie Charts

2.41 A bar is drawn for each grade with heights corresponding to the number of students at each grade. The result follows:

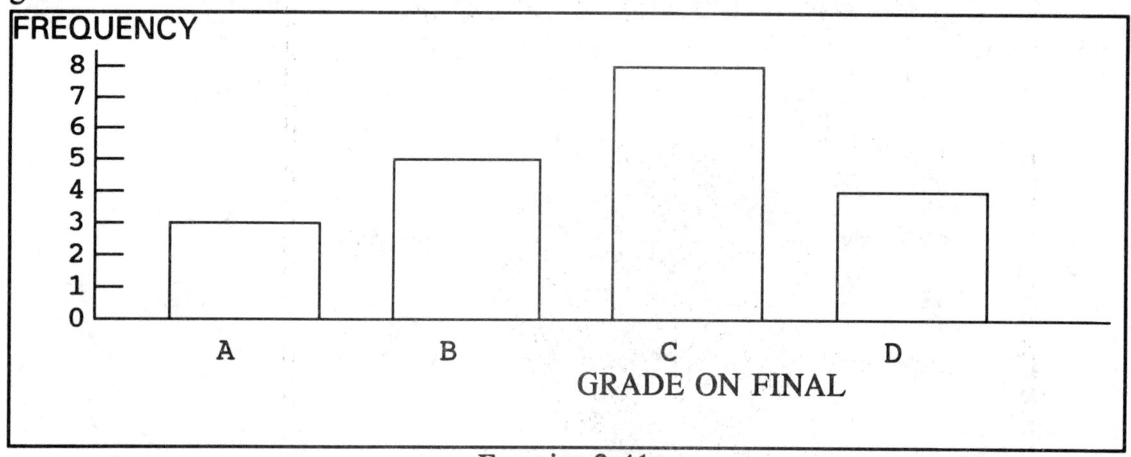

Exercise 2.41

2.43 A slice of the circle, with central angle corresponding to the number of students, is allocated for each final examination grade given by the professor. The relative frequency for "A" is 3/20 = 0.15, so we allocate (0.15)(360°) = 54° to the central angle corresponding to "A". Similarly, the others are: "B", 90°; "C", 144°; and "D", 72°. The pie chart is shown below.

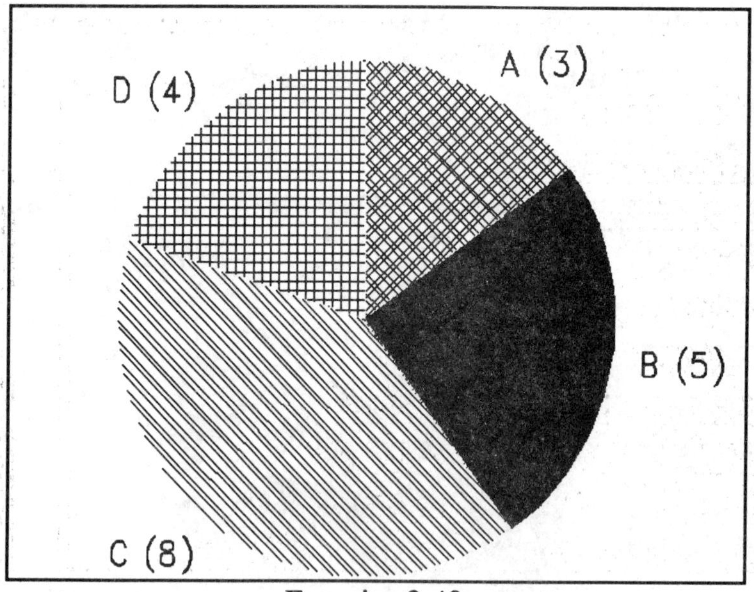

Exexcise 2.43

Graphical Methods for Qualitative Data: Bar and Pie Charts

2.45 A slice of the circle, with central angle corresponding to expenditures, is allocated for each of the various categories. The relative frequency for "Housing" is 41.36/100 = 0.4136, so we allocate (0.4136)(360°) = 149° to the central angle corresponding to "Housing". Similarly, the others are: "Miscellaneous", 23°; "Transportation", 64°; "Clothing", 22°; "Food", 64°; "Entertainment",16°; and "Health Care", 23°. The pie chart is shown.

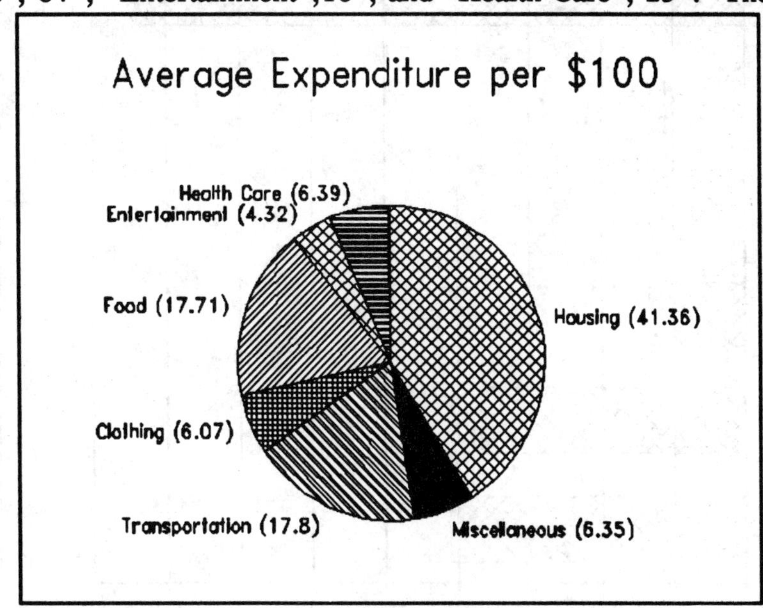

Exercise 2.45

2.47 A horizontal bar is drawn for each player with length corresponding to his number of rushing yards in a Super Bowl. The result follows:

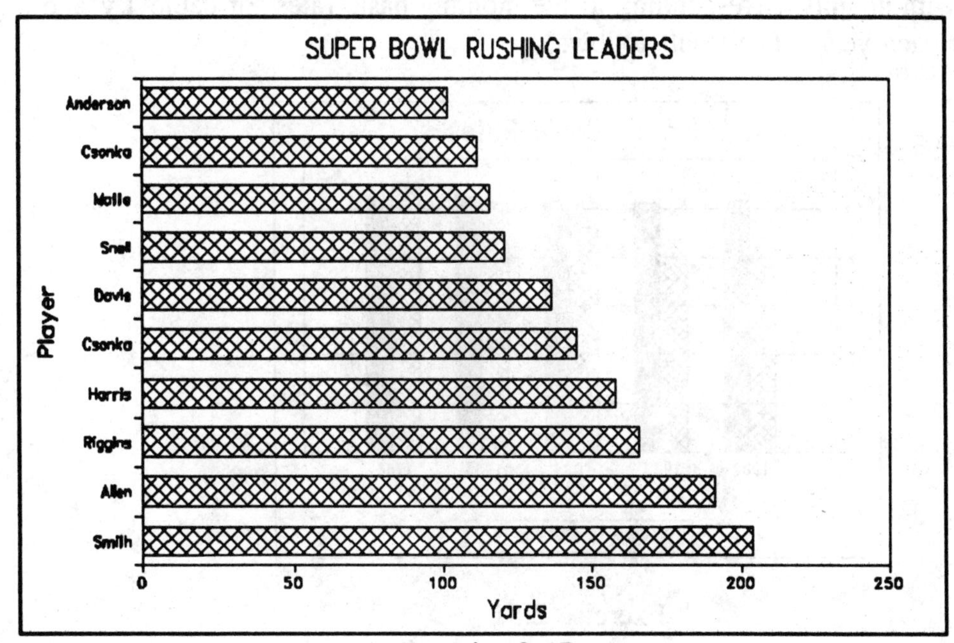

Exercise 2.47

2.49 A horizontal bar is drawn for each company with length corresponding to the percentage market share for each company. The result follows:

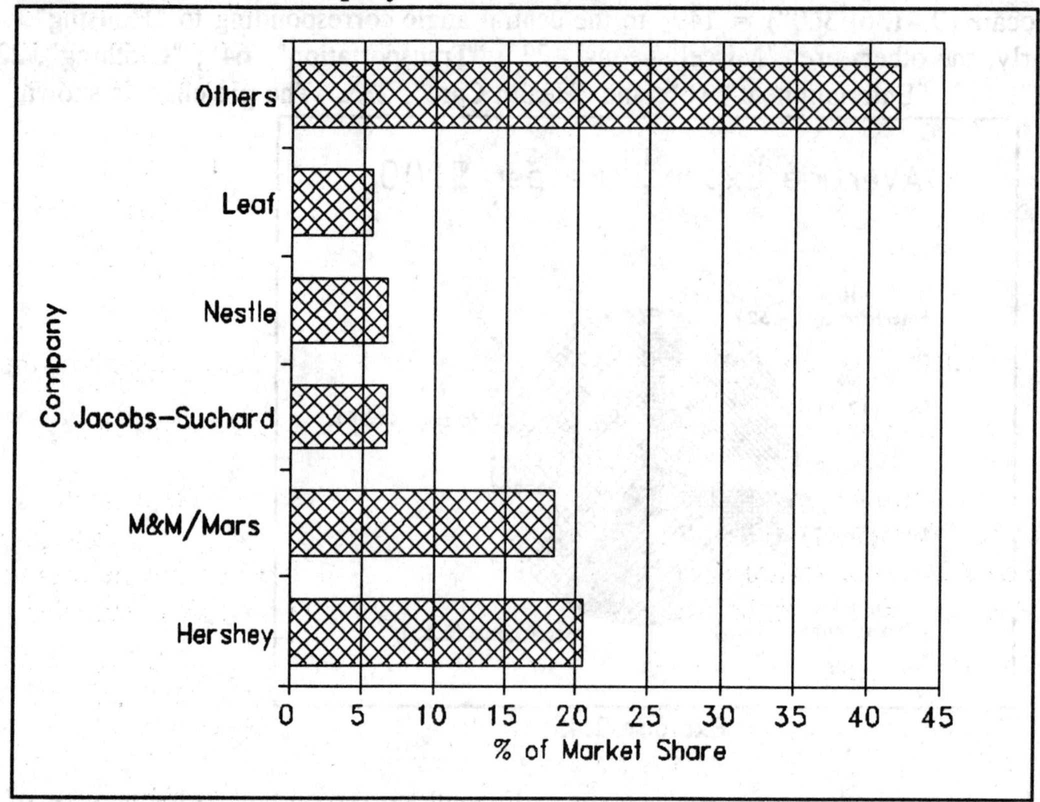

Exercise 2.49

2.51 Vertical bars with heights corresponding to the monthly basic rates for cable TV and pay TV are plotted for each year. The result follows:

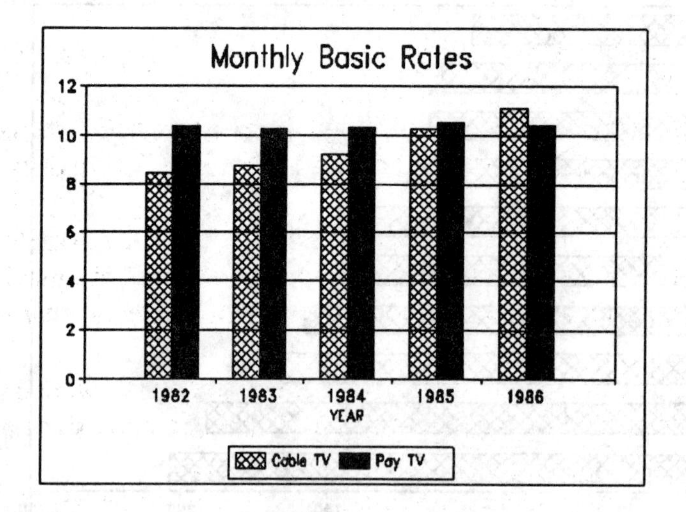

Exercise 2.51

EXERCISES for Section 2.4
Numerical Measures of Central Location

2.53 Since $\Sigma x = n\bar{x}$, the total amount saved is $\Sigma x = (10)(0.75) = \$7.50$.

2.55 **a.** We calculate $\Sigma x = 8 + 3 + 6 + 1 + 8 = 26$. Therefore the sample mean is:

$$\bar{x} = \frac{\Sigma x}{n} = \frac{26}{5} = 5.2.$$

b. After being arranged in numerical order, the sample is 1, 3, 6, 8, 8. The median is the middle number which is 6.
c. The mode is 8 since it appears most frequently in the sample.

2.57 **a.** We calculate $\Sigma x = 36 + 49 + 58 + 27 + 63 + 32 + 41 = 306$. Therefore, the mean age of the 7 mathematics department members is $\bar{x} = \frac{\Sigma x}{n} = \frac{306}{7} = 43.7.$

b. The data would be regarded as a sample if it were used to estimate the mean of all faculty at the school. It would be denoted by $\bar{x}$ and would be a statistic.
c. The data would be regarded as a population if it were used to determine the mean age of the mathematics faculty at the particular school. It would be denoted by μ and would be a parameter.

2.59 **a.** We calculate $\Sigma x = 55{,}000 + 52{,}000 + 49{,}000 + 54{,}000 + 198{,}000 + 45{,}000 = \$453{,}000$. Therefore, the mean salary of the six attorneys is

$$\bar{x} = \frac{\Sigma x}{n} = \frac{\$453{,}000}{6} = \$75{,}500.$$

b. After being arranged in numerical order, the median is the middle number. Since there is an even number of sample points, the two center points are averaged. Therefore, the median is $(52{,}000 + 54{,}000)/2 = \$53{,}000$.
c. The median more accurately describes the "typical" salary at this firm because most of the salaries are relatively "close" to the median.
d. The firm's president would most likely use the mean because the salary of $198,000 causes the mean to be quite high, thus creating an image of prosperity concerning his staff.

2.61 Since the mean weight of the 197 dancers was 156 pounds, the total weight of the dancers was $\Sigma x = n\bar{x} = 197 \cdot 156 = 30{,}732$ pounds which exceeds the maximum weight of 30,000 pounds. Therefore, the patrons' safety was being jeopardized at that time.

2.63 The mean exam grade for the 74 students is a weighted mean, where we use the mean grades of the sections on the final examination as the x's and the class sizes as weights. Therefore,

$$\bar{x} = \frac{\Sigma xw}{\Sigma w} = \frac{(78.0)\cdot(28) + (84.6)\cdot(25) + (70)\cdot(21)}{28 + 25 + 21} = \frac{5769}{74} = 77.96.$$

2.65 The student's grade point average for the semester is a weighted mean, where we use the numerical coding of the grades earned as the x's and the credit hours as weights. Therefore,

$$\bar{x} = \frac{\Sigma xw}{\Sigma w} = \frac{2 \cdot 3 + 4 \cdot 4 + 3 \cdot 4 + 2 \cdot 3 + 1 \cdot 1}{3 + 4 + 4 + 3 + 1} = \frac{41}{15} = 2.73.$$

2.67 The mean age of this group of drivers may be computed as a weighted mean, where we use the class marks 27, 32, 37, 42, 47 as the x's and the frequency of arrests as weights. Therefore,

$$\bar{x} = \frac{\Sigma xw}{\Sigma w} = \frac{27 \cdot 220 + 32 \cdot 158 + 37 \cdot 111 + 42 \cdot 72 + 47 \cdot 49}{220 + 158 + 111 + 72 + 49} = \frac{20,430}{610} = 33.5.$$

MINITAB LAB ASSIGNMENTS

2.69(M)a.

```
MTB > # Exercise 2.69(M)
MTB > # The data for the percentages of motorists who use
MTB > # their seat belts were entered into C1 using SET C1.
MTB > MEAN C1
   MEAN     =       44.800
```

b. If the number of motorists in each state were known, then the percentage of motorists in the 50 states who use their seat belts could be found as a weighted mean, where we would use the number of motorists in each of the states as the weights.

2.71(M)

```
MTB > # Exercise 2.71(M)
MTB > # The data for the label weights of the candy bars were
MTB > # entered into C2
MTB > MEAN C2
   MEAN     =       74.043
MTB > MEDIAN C2
   MEDIAN =        74.100
```

EXERCISES for Section 2.5

Numerical Measures of Variability

2.73 **a.** The range is the difference between the largest and smallest values in the sample so *range* = largest - smallest = 9 - 0 = 9.

b. We construct a table:

x	$(x - \bar{x})$	$(x - \bar{x})^2$
9	5	25
1	-3	9
0	-4	16
4	0	0
6	2	4
$\Sigma x = 20$	$SS(x) = \Sigma(x-\bar{x})^2 = 54$	

The sample mean is $\overline{x} = \dfrac{\Sigma x}{n} = \dfrac{20}{5} = 4$.

The sample variance is $s^2 = \dfrac{SS(x)}{n-1} = \dfrac{54}{5-1} = 13.5$.

c. The sample standard deviation is $s = \sqrt{s^2} = \sqrt{13.5} = 3.7$.

2.75 **a.** We construct a table:

x	$(x - \overline{x})$	$(x - \overline{x})^2$
5	1.57143	2.46939
8	4.57143	20.89797
2	-1.42857	2.04081
1	-2.42857	5.89795
7	3.57143	12.75511
1	-2.42857	5.89795
0	-3.42857	11.75509
$\Sigma x = 24$	$SS(x) = \Sigma(x-\overline{x})^2 = 61.71427$	

Note: The sample mean is $\overline{x} = \dfrac{\Sigma x}{n} = \dfrac{24}{7} = 3.42857$

b. We construct a table:

x	x^2
5	25
8	64
2	4
1	1
7	49
1	1
0	0
$\Sigma x = 24$	$\Sigma x^2 = 144$

Therefore, using the shortcut formula:

$$SS(x) = \Sigma x^2 - \frac{(\Sigma x)^2}{n} = 144 - \frac{(24)^2}{7} = 144 - 82.29 = 61.71.$$

c. We now find the sample standard deviation: $s = \sqrt{\dfrac{SS(x)}{n-1}} = \sqrt{\dfrac{61.71}{7-1}} = 3.2$.

2.77 The range is the difference between the largest and smallest values in the sample. Therefore: *range* = largest - smallest = 13.14 - 2.43 = 10.71. We construct a table:

x	x^2
13.14	172.6596
2.71	7.3441
2.67	7.1289
2.46	6.0516
2.43	5.9049
$\Sigma x = 23.41$	$\Sigma x^2 = 199.0891$

We may now calculate the standard deviation as follows:

$$SS(x) = \Sigma x^2 - \frac{(\Sigma x)^2}{n} = 199.0891 - \frac{(23.41)^2}{5} = 199.0891 - 109.60562 =$$

89.48348. Thus, the sample standard deviation is

$$s = \sqrt{\frac{SS(x)}{n-1}} = \sqrt{\frac{89.48348}{5-1}} = 4.730.$$

2.79 **a.** We note that for sample A: $\Sigma x = (5 + 8 + 9 + 1 + 0 + 10 + 2) = 35$, so

$\overline{x} = \frac{\Sigma x}{n} = \frac{35}{7} = 5$ and, likewise, for sample B,

$\Sigma x = (1 + 4 + 4 + 6 + 11 + 4 + 5) = 35$, so $\overline{x} = \frac{\Sigma x}{n} = \frac{35}{7} = 5$,

and the sample means are the same. In addition, we see that *range* A = 10 - 0 = 10 and that *range* B = 11 - 1 = 10, so the ranges are also the same.

b. Sample A seems to be more variable.

c. In order to calculate the standard deviation, we construct a table for sample A:

x	$(x - \overline{x})$	$(x - \overline{x})^2$
5	0	0
8	3	9
9	4	16
1	−4	16
0	−5	25
10	5	25
2	−3	9
$\Sigma x = 35$	$SS(x) = \Sigma(x-\overline{x})^2 = 100$	

Therefore, the sample variance is: $s^2 = \frac{SS(x)}{n-1} = \frac{100}{7-1} = 16.6667$ and the

standard deviation for sample A is: $s = \sqrt{s^2} = \sqrt{16.6667} = 4.08$. Similarly, for

sample B, the sample variance is $s^2 = \frac{SS(x)}{n-1} = \frac{56}{7-1} = 9.3333$ and the standard deviation for sample B is: $s = \sqrt{s^2} = \sqrt{9.3333} = 3.06$. This confirms that sample A is more variable.

d. The standard deviation provides a better indication of variability. All data values are involved in the calculation of standard deviation while only two, the high and the low, are used to calculate the range.

2.81 **a.** For the data set {0.013, 0.009, 0.011, 0.013, 0.009}, we calculate $\Sigma x = 0.055$ and $\Sigma x^2 = 0.000621$. Therefore, SS(x) =

$$\Sigma x^2 - \frac{(\Sigma x)^2}{n} = 0.000621 - \frac{(0.055)^2}{5} = 0.000621 - 0.000605 = 0.000016.$$

The sample standard deviation is: $s = \sqrt{\frac{SS(x)}{n-1}} = \sqrt{\frac{0.000016}{5-1}} = 0.002.$

b. When we multiply each value by 1000, we get the data set {13, 9, 11, 13, 9} for which $\Sigma x = 55$ and $\Sigma x^2 = 621$. We now calculate the standard deviation as follows:

$SS(x) = \Sigma x^2 - \frac{(\Sigma x)^2}{n} = 621 - \frac{(55)^2}{5} = 621 - 605 = 16.$ The sample standard

deviation is: $s = \sqrt{\frac{SS(x)}{n-1}} = \sqrt{\frac{16}{5-1}} = 2.$ If we divide this result by 1000 we again get 0.002, the standard deviation of the original sample.

2.83 In order to calculate the standard deviation, we group the data and construct a table:

x	f	xf	x^2f
0	11	0	0
1	5	5	5
2	9	18	36
3	6	18	54
4	3	12	48
5	3	15	75
6	3	18	108
7	0	0	0
8	2	16	128
Sum	n = 42	102	454

We may now calculate the standard deviation as follows:

$$SS(x) = \Sigma x^2 f - \frac{(\Sigma xf)^2}{n} = 454 - \frac{(102)^2}{42} = 454 - 247.71 = 206.29.$$

The sample standard deviation is: $s = \sqrt{\frac{SS(x)}{n-1}} = \sqrt{\frac{206.29}{42-1}} = 2.2.$

MINITAB LAB ASSIGNMENTS

2.85(M)

```
MTB > # Exercise 2.85(M)
MTB > # The data were entered into C1 using SET C1
MTB > MEAN C1
   MEAN     =       3.2734
MTB > MEDIAN C1
   MEDIAN =        3.2400
MTB > STDEV C1
   ST.DEV. =       0.29066
```

2.87(M)

```
MTB > # Exercise 2.87(M)
MTB > # The data were read into C1 and C2 using READ C1 C2
MTB > NAME C1 '1980' C2 '1987'

MTB > MEAN C1
   MEAN     =       78.791
MTB > MEAN C2
   MEAN     =       122.73

MTB > MEDIAN C1
   MEDIAN =        75.400
MTB > MEDIAN C2
   MEDIAN =        115.30

MTB > STDEV C1
   ST.DEV. =        17.943
MTB > STDEV C2
   ST.DEV. =        31.700
```

EXERCISES for Section 2.6

Understanding the Significance of Standard Deviation

2.89 Mound-shaped data sets can be described with the Empirical Rule.

2.91 If the Empirical Rule were applicable, then:
a. approximately 68% of the measurements would lie within the interval from $\bar{x} - s$ to $\bar{x} + s$.
b. approximately 95% of the measurements would lie within the interval from $\bar{x} - 2s$ to $\bar{x} + 2s$.
c. approximately 100% of the measurements would lie within the interval from $\bar{x} - 3s$ to $\bar{x} + 3s$.

2.93 We note that $\bar{x} - 4s$ = \$343 - (4)(\$35) = \$203 and that $\bar{x} + 4s$ = \$343 + (4)(\$35) = \$483, so $k = 4$ and we calculate $1 - 1/k^2 = 1 - 1/(4)^2 = 0.9375$. Therefore, by Chebyshev's Theorem, at least 93.75% of the policies have premiums between \$203 and \$483.

2.95 **a.** If $k = 2$, then $\bar{x} - 2s = 28 - (2)(0.8) = 26.4$ and $\bar{x} + 2s = 28 + (2)(0.8) = 29.6$. Therefore, by the Empirical Rule, approximately 95% of the loaves will have weights between 26.4 oz. and 29.6 oz.
b. If $k = 3$, then $\bar{x} - 3s = 28 - (3)(0.8) = 25.6$ and $\bar{x} + 3s = 28 + (3)(0.8) = 30.4$. Therefore, by the Empirical Rule, nearly all of the loaves will have weights between 25.6 oz. and 30.4 oz.
c. First note that $28.8 = \bar{x} + s$. Therefore, by the Empirical Rule, approximately $(1 - 0.68)/2 = 0.16 = 16\%$ of the loaves will have weights more than 28.8 oz.

2.97 **a.** First note that $\bar{x} - s = 6.32 - 2.17 = 4.15$ and $\bar{x} + s = 6.32 + 2.17 = 8.49$. We count 35 states with unemployment rates between 4.15 and 8.49; i.e. 70% of the data are within 1 standard deviation of the mean.
b. $\bar{x} - 2s = 6.32 - 2(2.17) = 1.98$ and $\bar{x} + 2s = 6.32 + 2(2.17) = 10.66$. We count 47 states with unemployment rates between 1.98 and 10.66; i.e. 94% of the data are within 2 standard deviations of the mean.
c. $\bar{x} - 3s = 6.32 - 3(2.17) = -0.19$ and $\bar{x} + 3s = 6.32 + 3(2.17) = 12.83$. All 50 states have unemployment rates between -0.19 and 12.83; i.e. 100% of the data are within 3 standard deviations of the mean.
d. We note that the percentages for this data are nearly the same as the percentages given by the Empirical Rule (68%, 95%, and $\approx$100%, resp.).
e. The estimates given by Chebyshev's Theorem of "at least 75%" (compared to the actual value of 94%) and "at least 89%" (compared to the actual value of 100%) are very conservative for this data.

2.99 For the ages of 40 blood donors, the range is $R = 64 - 21 = 43$; so $s \approx 43/4 = 10.75$, which is close to the actual value of 10.96.

EXERCISES for Section 2.7
Measures of Relative Position

2.101 The z-score is: $z = \dfrac{x - \bar{x}}{s} = \dfrac{price - 58}{8}$.
The z-scores for these restaurants are:

Restaurant	x = Price	z-score
1	50	(50 - 58)/8 = -1.000
2	45	(45 - 58)/8 = -1.625
3	60	(60 - 58)/8 = 0.250

2.103 If we solve $z = \dfrac{x - \bar{x}}{s}$ for x we get $x = \bar{x} + z \cdot s = 58 + (-2.25) \cdot 8 = 40$, so the cost is 40 cents.

2.105 **a.** The z-score for the resident tuition at the University of Pennsylvania is:

$$z = \frac{x - \bar{x}}{s} = \frac{16650 - 14747}{2006} = 0.95.$$

b. The z-score for the non-resident tuition at Temple University is:

$$z = \frac{x - \bar{x}}{s} = \frac{18084 - 17048}{1430} = 0.72.$$

c. Compared to the other schools, the non-resident tuition at Temple is relatively cheaper.

2.107 We consider the following diagram:

25%	25%	25%	25%

104 150 195

a. From the diagram we see that the percentage of scores on this test between 104 and 195 is approximately 50%.
b. The percentage of scores on this test above 104 is approximately 75%.
c. The percentage of scores on this test above 195 is approximately 25%.

2.109 **a.** The 5-number summary is made up of the minimum value, the first quartile value, the median, the third quartile value, and the maximum value. In MINITAB these values correspond to MIN, Q1, MEDIAN, Q3, and MAX, respectively. Therefore, the 5-number summary would be 21.00, 31.00, 37.50, 44.50, 64.00.
b. *Range* = MAX - MIN = 64.00 - 21.00 = 43.00.
c. The interquartile range is Q3 - Q2 = 44.50 - 31.00 = 13.50.

2.111 The 5-number summary for the data consists of:
MIN = 0, Q_1 = 200, Q_2 = 400, Q_3 = 600, and MAX = 1000.
IQR = Q_3 - Q_1 = 600 - 200 = 400.
Range = MAX - MIN = 1000 - 0 = 1000.

MINITAB LAB ASSIGNMENTS

2.113(M)

```
MTB > # Exercise 2.113(M)
MTB > # The data were read into C1 using SET C1
MTB > LET C2 = (C1 - MEAN(C1))/STDEV(C1)
MTB > NAME C1 'GPA' C2 'z-score'
MTB > print C1 C2
 ROW     GPA    z-score

   1    3.29    0.05710
   2    3.67    1.36446
   3    3.28    0.02269
   4    2.89   -1.31908
   5    3.01   -0.90623
   6    2.94   -1.14706
   7    3.31    0.12590
   8    2.80   -1.62872
   9    3.54    0.91720
  10    3.15   -0.42456
```

```
11   3.08  -0.66540
12   3.01  -0.90623
13   3.17  -0.35576
14   3.36   0.29793
15   3.57   1.02042
16   3.39   0.40114
17   3.32   0.16031
18   3.19  -0.28695
19   2.95  -1.11265
20   3.19  -0.28695
21   2.84  -1.49110
22   3.16  -0.39016
23   3.21  -0.21814
24   3.26  -0.04612
25   3.36   0.29793
26   3.37   0.33233
27   3.39   0.40114
28   3.75   1.63970
29   3.41   0.46995
30   3.97   2.39659
31   2.87  -1.38789
32   2.99  -0.97503
33   3.02  -0.87182
34   3.09  -0.63099
35   3.87   2.05255
36   3.14  -0.45897
37   3.18  -0.32135
38   3.24  -0.11493
39   3.41   0.46995
40   3.61   1.15803
41   2.94  -1.14706
42   3.06  -0.73420
43   2.94  -1.14706
44   3.66   1.33006
45   3.77   1.70850
46   3.72   1.53648
47   3.51   0.81399
```

2.115(M)

```
MTB > # Exercise 2.115(M)
MTB > # The data are in C1 from Exercise 2.113
MTB > BOXPLOT C1
```

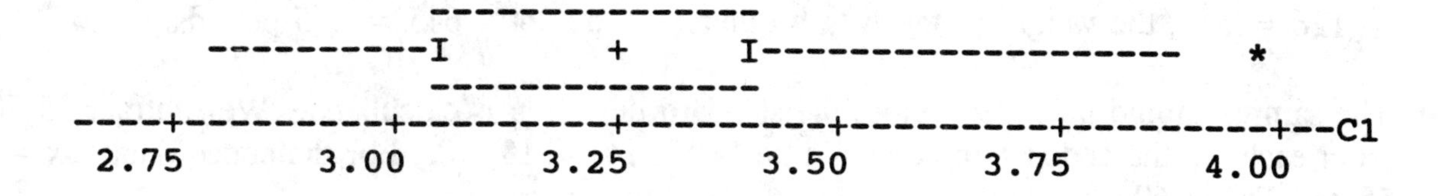

```
                          ---------------
               ----------I       +      I-------------------    *
                          ---------------

    ----+---------+---------+---------+---------+---------+--C1
       2.75      3.00      3.25      3.50      3.75      4.00
```

REVIEW EXERCISES

Chapter 2

2.117 **a.** We calculate: $\Sigma x = 13.0 + 14.3 + 13.6 + 13.8 + 12.8 + 12.9 + 15.2 + 14.0 + 12.0 + 11.9 = 133.5$. Therefore, the mean speed is $\bar{x} = \frac{\Sigma x}{n} = \frac{133.5}{10} = 13.35.$

b. After being arranged in numerical order, the sample is 11.9, 12.0, 12.8, 12.9, 13.0, 13.6, 13.8, 14.0, 14.3, 15.2. Since there is an even number of sample points, the two center numbers must be averaged. Therefore, the median is (13.0 + 13.6)/2 = 13.3.

c. *Range* = maximum value - minimum value = 15.2 - 11.9 = 3.3

d. We calculate: $\Sigma x = 133.5$, $\Sigma x^2 = 1791.79$. Therefore:

$$SS(x) = \Sigma x^2 - \frac{(\Sigma x)^2}{n} = 1791.79 - \frac{(133.5)^2}{10} = 1791.79 - 1782.225 = 9.565.$$

The sample standard deviation is: $s = \sqrt{\frac{SS(x)}{n-1}} = \sqrt{\frac{9.565}{9}} = 1.03.$

e.

Stem	Leaf
11	9
12	8 9 0
13	0 6 8
14	3 0
15	2

2.118 We are given $\Sigma x = 780$ and $\Sigma x^2 = 17,526$. Therefore, the sample mean is:

$\bar{x} = \frac{\Sigma x}{n} = \frac{780}{50} = 15.6.$ In addition, we calculate:

$$SS(x) = \Sigma x^2 - \frac{(\Sigma x)^2}{n} = 17526 - \frac{(780)^2}{50} = 17526 - 12168 = 5358.$$

The sample standard deviation is: $s = \sqrt{\frac{SS(x)}{n-1}} = \sqrt{\frac{5358}{49}} = 10.46.$

2.119 Since the mean weight of the five brothers is 172 pounds, the total weight of the brothers is 172 • 5 = 860 pounds. Since the total weight of four of the brothers is 173 + 149 + 197 + 126 = 645, the weight of the fifth brother must be 860 - 645 = 215 pounds.

2.120 The sample should be coded before the standard deviation is calculated. We subtract 35,500 from each, so the coded sample would be 14, 7, 11, 1, 15, 10. For the coded data: $\Sigma x = 58$ and $\Sigma x^2 = 692$, so that:

$$SS(x) = \Sigma x^2 - \frac{(\Sigma x)^2}{n} = 692 - \frac{(58)^2}{6} = 692 - 560.67 = 131.33.$$

Thus the standard deviation is $s = \sqrt{\frac{SS(x)}{n-1}} = \sqrt{\frac{131.33}{5}} = 5.1$

which is the same as for the original data.

2.121 For milk: $z = \frac{x - \bar{x}}{s} = \frac{89 - 63}{8} = 3.25.$

For hamburg: $z = \frac{x - \bar{x}}{s} = \frac{2.12 - 1.75}{0.12} = 3.08.$

Therefore, milk is relatively more overpriced than hamburg at this store.

2.122 The mean age of all workers employed by this company is a weighted mean, where we use the mean ages of the workers at these sites as the x's and the numbers of employees at the three locations as weights. Therefore,

$$\bar{x} = \frac{\Sigma xw}{\Sigma w} = \frac{42 \cdot 78 + 37 \cdot 103 + 35 \cdot 86}{78 + 103 + 86} = \frac{10097}{267} = 37.8.$$

2.123 The mean number of hours worked by all employees may be computed as a weighted mean, where we use the class marks 28, 33, 38, 43, 48, 53 as the x's and the number of workers as weights. Therefore,

$$\bar{x} = \frac{\Sigma xf}{\Sigma f} = \frac{28 \cdot 7 + 33 \cdot 18 + 38 \cdot 65 + 43 \cdot 90 + 48 \cdot 48 + 53 \cdot 12}{7 + 18 + 65 + 90 + 48 + 12} = \frac{10,070}{240} = 42.0.$$

2.124 We use the tens digit for the stem and the units digit for the leaf. The stem-and-leaf display for the number of assault cases is as follows:

Stem	Leaf
2	5 7
3	9 8 0 2 4
4	3 2 1 8 0 1 9 8 2 6 6 2 0 0 0 2 7 8 2 3
5	2 1 5 2 1 9 1 2
6	1

2.125 We construct a number line which includes the range 25 to 61 and place dots above the points which correspond to the data values. The result follows:

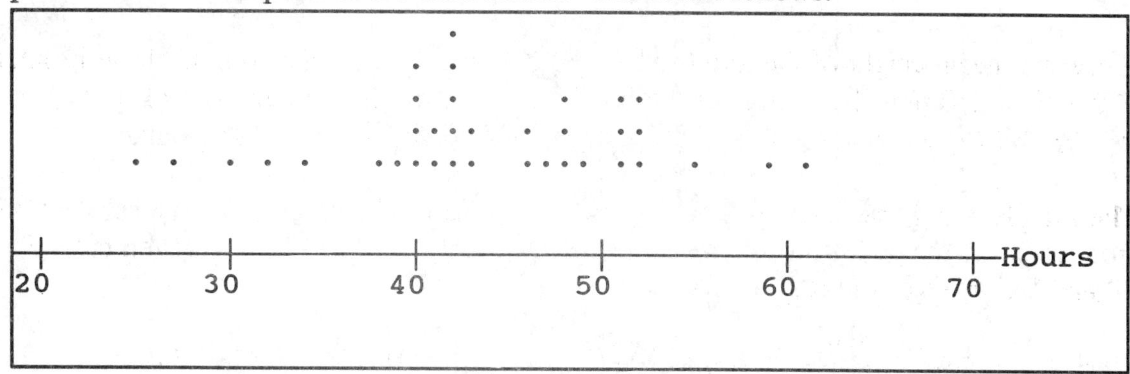

Exercise 2.125

2.126 We group the data into a frequency distribution using classes 0-999, 1000-1999, etc. We obtain:

Miles Since Last Oil Change	Count of Police Cars
0 - 999	3
1000 - 1999	4
2000 - 2999	3
3000 - 3999	5
4000 - 4999	6
5000 - 5999	3
6000 - 6999	0
7000 - 7999	4
8000 - 8999	1

We construct a histogram by plotting class boundaries along the horizontal axis and bars with heights corresponding to the class frequencies in the vertical direction. The result follows: (other answers are possible)

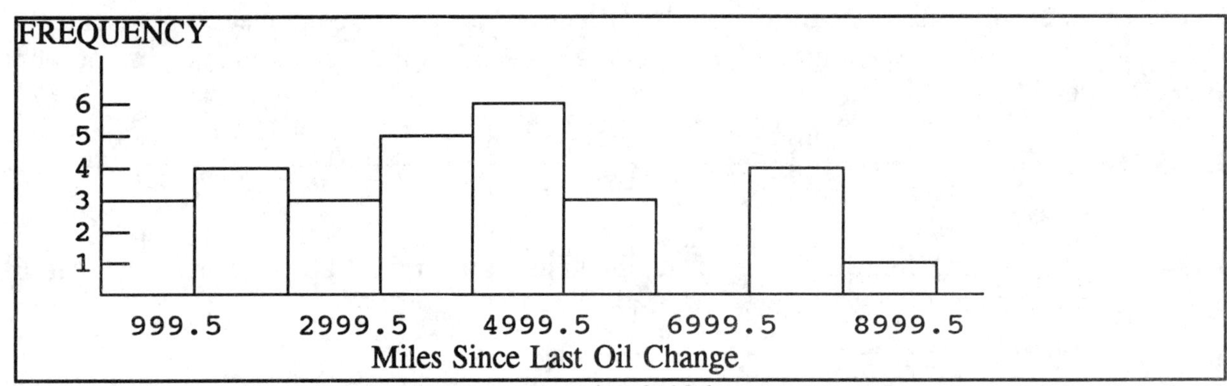

Exercise 2.126

2.127 **a.** We group the data into a frequency distribution using classes 0, 1, 2, 3, 4, 5. We obtain:

Number of Hits	Frequency
0	3
1	9
2	18
3	9
4	8
5	1

b. In order to find the mean and standard deviation, we construct a table as follows:

x	f	xf	x^2f
0	3	0	0
1	9	9	9
2	18	36	72
3	9	27	81
4	8	32	128
5	1	5	25
Sum	48	109	315

The sample mean is: $\bar{x} = \frac{\Sigma xf}{\Sigma f} = \frac{109}{48} = 2.3.$ We finally calculate the standard deviation as follows:

$$SS(x) = \Sigma x^2 f - \frac{(\Sigma xf)^2}{n} = 315 - \frac{(109)^2}{48} = 315 - 247.521 = 67.479.$$

The sample standard deviation is: $s = \sqrt{\frac{SS(x)}{n-1}} = \sqrt{\frac{67.479}{48-1}} = 1.2.$

2.128 The modal rating is Meritorious (M) because it appears most frequently (14 times) in the sample.

2.129 A bar is drawn for each company with heights corresponding to their percentage of total sales. The result is shown:

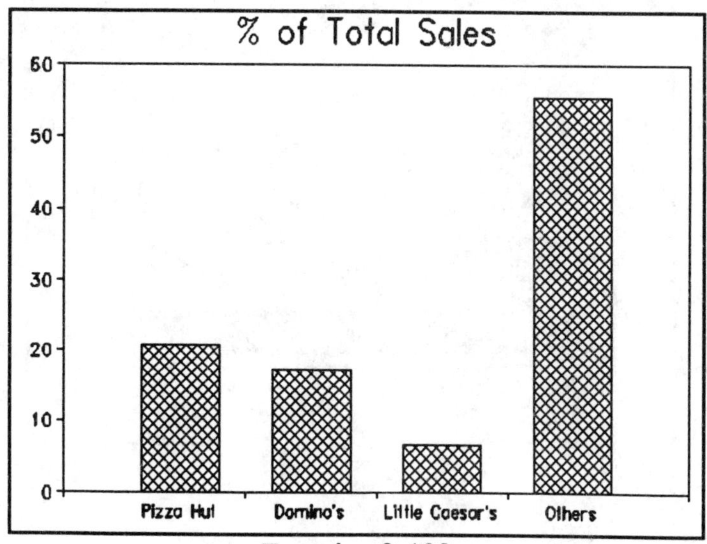

Exercise 2.129

2.130 A slice of the circle, with central angle corresponding to expenditures, is allocated for each of the various categories. The relative frequency for "Pizza Hut" is 20.7/100 = 0.207, so we allocate (0.207)(360°) = 74.5° to the central angle corresponding to "Pizza Hut". Similarly, the others are: "Domino's", 61.2°; "Little Caesar's", 24.1°; and "Others", 200.1°. The pie chart is shown below.

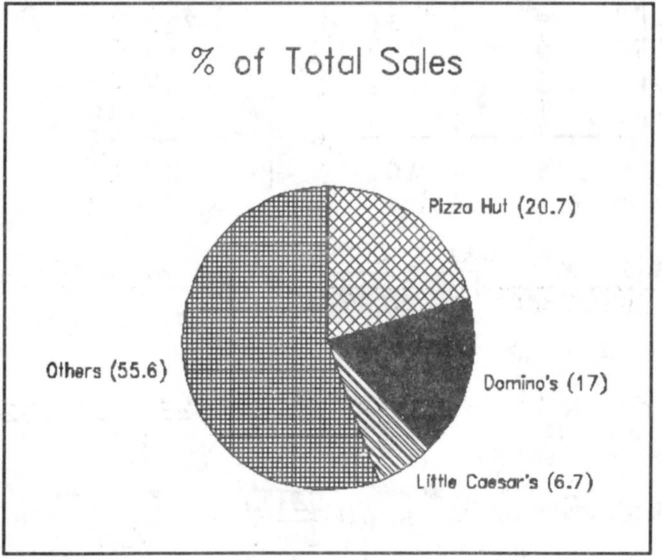

Exercise 2.130

2.131 A horizontal bar is drawn for each source of electricity with lengths corresponding to their percentage of use. The result is shown below.

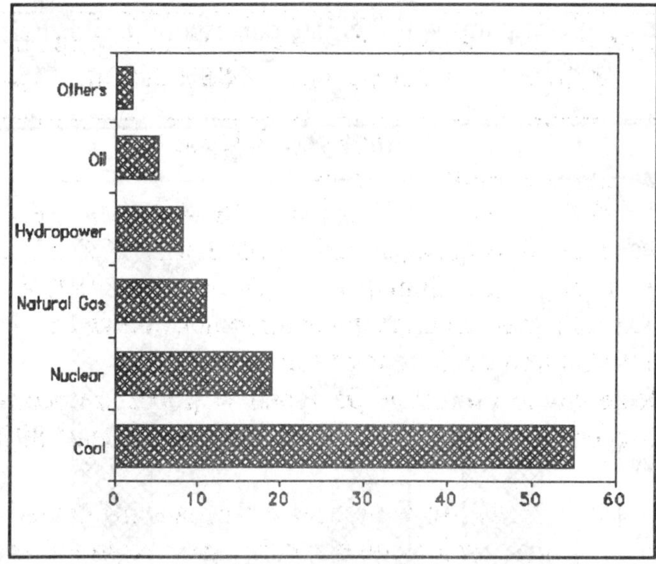

Exercise 2.131

2.132a. We first construct a relative frequency distribution as follows:

Class Limits	Frequency	Relative Frequency
225 - 274	2	0.04
275 - 324	5	0.10
325 - 374	12	0.24
375 - 424	10	0.20
425 - 474	9	0.18
475 - 524	6	0.12
525 - 574	4	0.08
575 - 624	2	0.04

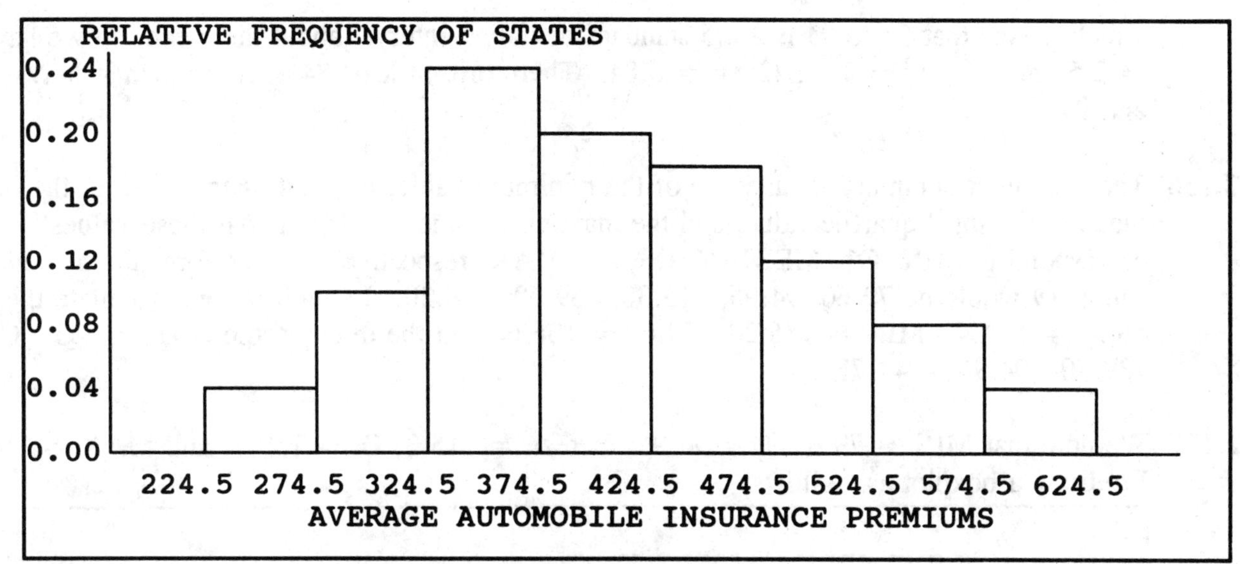

Exercise 2.132 a

This distribution is slightly skewed to the right.

b. We note $\bar{x}$ - 2s = 413 - 2(89) = 235 and $\bar{x}$ + 2s = 413 + 2(89) = 591. We count 48 states with insurance premiums between 235 and 591; i.e. 96% of the data are within 2 standard deviations of the mean. Similarly, $\bar{x}$ - 3s = 413 - 3(89) = 146 and $\bar{x}$ + 3s = 413 + 3(89) = 680. All 50 states have insurance premiums between 146 and 680; i.e. 100% of the data are within 3 standard deviations of the mean.

c. The Empirical Rule would predict ≈ 95% and ≈ 100%, respectively.

d. Chebyshev's Theorem would predict at least 75% and at least 89%, respectively.

2.133 The range = 604 - 244 = \$360, so we estimate the standard deviation as:
s ≈ range/4 = 360/4 = \$90, which compares favorably with the actual value of \$89.

2.134 **a.** We are given: mound-shaped distribution with a mean of 74 and a standard deviation of 8.

We note that for x = 66, $z = \dfrac{x - \bar{x}}{s} = \dfrac{66 - 74}{8} = -1$

and for x = 82, $z = \dfrac{x - \bar{x}}{s} = \dfrac{82 - 74}{8} = 1.$

Therefore, we can conclude that ≈ 68% of the scores are between 66 and 82.

b. For x = 58, $z = \frac{x - \bar{x}}{s} = \frac{58 - 74}{8} = -2.$

Therefore, x = 58 corresponds to 2 standard deviations below the mean; so we can conclude that ≈ 2.5% of the scores are below 58.

c. For x = 90, $z = \frac{x - \bar{x}}{s} = \frac{90 - 74}{8} = 2.$

Therefore, x = 90 corresponds to 2 standard deviations above the mean; so we can also conclude that ≈ 2.5% of the scores are above 90.

2.135 We note that for x = 54, $z = \frac{x - \bar{x}}{s} = \frac{54 - 74}{8} = -2.5,$

and for x = 94, $z = \frac{x - \bar{x}}{s} = \frac{94 - 74}{8} = 2.5,$

which means that 54 to 94 is a 2.5 standard deviation interval about the mean, so we have k = 2.5 and $1 - 1/k^2 = 1 - 1/(2.5)^2 = 0.84$. Therefore, at least 84% of the data is between 54 and 94.

2.136 The 5-number summary is made up of the minimum value, the first quartile value, the median, the third quartile value, and the maximum value. In MINITAB these values correspond to MIN, Q1, MEDIAN, Q3, and MAX, respectively. Therefore, the 5-number summary would be 79.60, 94.95, 115.30, 139.70, 186.20. In addition, we calculate the range = MAX - MIN = 186.20 - 79.60 = 106.60 and the interquartile range = Q3 - Q1 = 139.70 - 94.95 = 44.75.

2.137 We note that MIN = 79.6, Q_1 = 94.95, median = 115.3, Q_3 = 139.7, and MAX = 186.2. We draw a boxplot as follows:

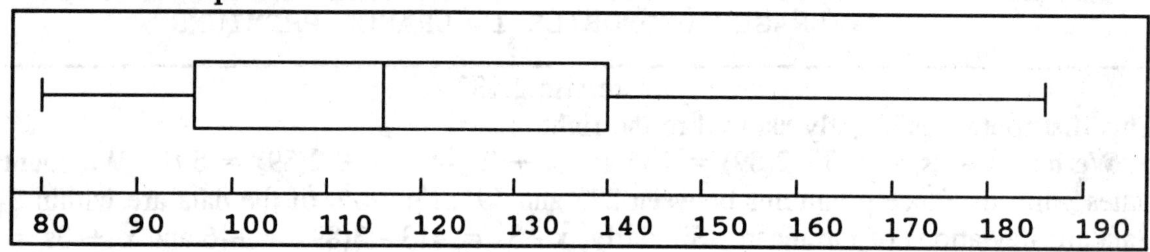

MINITAB LAB ASSIGNMENTS

2.138(M)

```
MTB > # Exercise 2.138(M)
MTB > # The data were read into C1 using SET C1
MTB > # a. dotplot:
MTB > NAME C1 'Age'
MTB > DOTPLOT C1
```

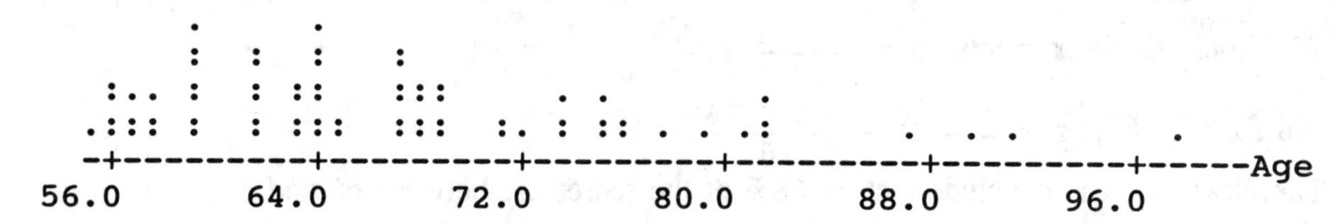

b.

```
MTB > # b. stem-and-leaf display:
MTB > STEM C1

Stem-and-leaf of Age        N  = 72
Leaf Unit = 1.0


   18    5 566667778889999999
   35    6 22222233334444444
  (16)   6 5577777788889999
   21    7 112444
   15    7 5556689
    8    8 1222
    4    8 7
    3    9 01
    1    9 8
```

c.

```
MTB > # c. 5-number summary:
MTB > DESCRIBE C1

                 N      MEAN    MEDIAN    TRMEAN     STDEV    SEMEAN
Age             72     67.39     65.00     66.61      9.25      1.09

               MIN       MAX        Q1        Q3
Age          55.00     98.00     59.75     73.50

MTB > # 5-number summary is: 55.00, 59.75, 65.00, 73.50, 98.00
```

2.139(M)

```
MTB > # Exercise 2.139(M)
MTB > # The data are in C1 from Exercise 2.138.
MTB > # Boxplot:
MTB > BOXPLOT C1


              ----------------
       -------I    +         I-----------------------        *
              ----------------
    --+---------+---------+---------+---------+---------+----Age
     56.0      64.0      72.0      80.0      88.0      96.0
```

2.140(M)

```
MTB > # Exercise 1.140(M)
MTB > # The data were entered into C1 using SET C1
MTB > HISTOGRAM C1;
SUBC> INCREMENT 1;
SUBC> START 35.
```

```
Histogram of C1   N = 35

Midpoint   Count
   35.00       1  *
   36.00       6  ******
   37.00       3  ***
   38.00       7  *******
   39.00       4  ****
   40.00       0
   41.00       2  **
   42.00       1  *
   43.00       3  ***
   44.00       2  **
   45.00       2  **
   46.00       3  ***
   47.00       1  *
```

2.141(M)

```
MTB > # Exercise 2.141(M)
MTB > # The z-scores were entered into C1 using SET C1
MTB > LET C2 = 34576 + C1*4593
MTB > NAME C1 'z-score' C2 'Salary'
MTB > PRINT C1 C2

 ROW  z-score    Salary

   1     1.25   40317.2
   2     0.69   37745.2
   3    -1.06   29707.4
   4     2.97   48217.2
   5     0.89   38663.8
   6     1.27   40409.1
   7    -1.87   25987.1
   8     0.02   34667.9
   9     0.98   39077.1
  10    -1.52   27594.6
  11    -2.65   22404.6
  12    -0.92   30350.4
  13     1.06   39444.6
  14    -1.22   28972.5
  15     0.52   36964.4
  16     3.02   48446.9
```

CHAPTER 3
MODELING BIVARIATE DATA

Section 3.1
Straight Lines and the SS( ) Notation

3.1 The slope is the coefficient of x which is $b_1 = 8$; the y-intercept is the constant term in the equation which is $b_0 = 3$.

3.3 The slope is the coefficient of x which is $b_1 = 4$; the y-intercept is the constant term in the equation which is $b_0 = 0$.

3.5 The slope is the coefficient of x which is $b_1 = 0$; the y-intercept is the constant term in the equation which is $b_0 = 10$.

3.7 We first solve for y and get: $y = -8 + \frac{4}{3}x$. The slope is the coefficient of x which is $b_1 = \frac{4}{3}$; the y-intercept is the constant term in the equation which is $b_0 = -8$.

3.9

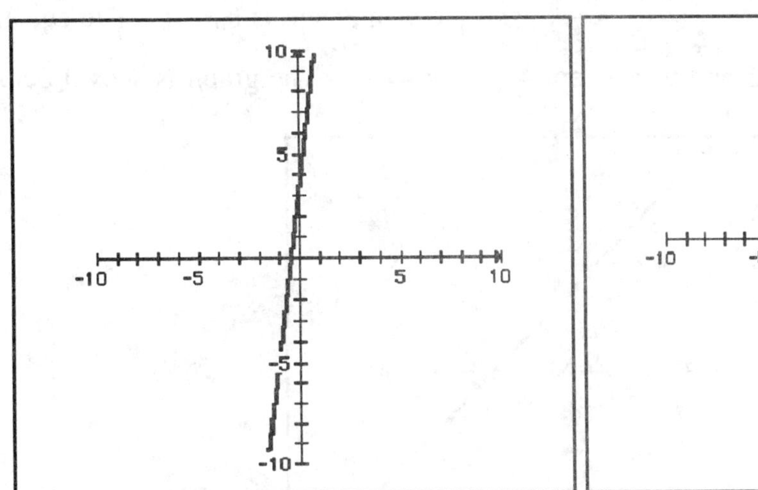

Exercise 3.1: $y = 3 + 8x$

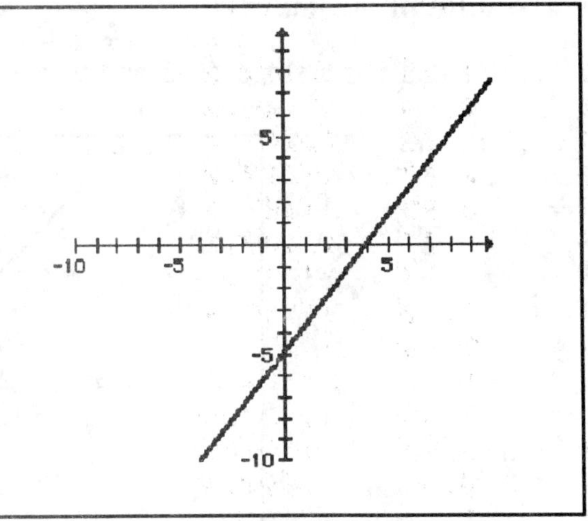

Exercise 3.2: $5x - 4y = 20$

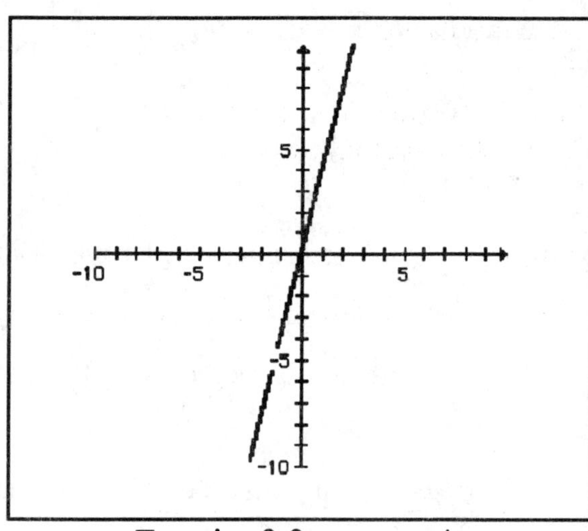

Exercise 3.3: $y = 4x$

Exercise 3.4: $2x + 3y = 10$

3.11 The slope of the line is $\dfrac{\Delta y}{\Delta x} = \dfrac{3 - 0}{0 - 7} = -\dfrac{3}{7}$.

3.13 The equation of the line that has a slope of 10 and y-intercept 7 is $y = 7 + 10x$.

3.15 The slope of the line is $\dfrac{\Delta y}{\Delta x} = \dfrac{0 - 4}{4 - 0} = -1$. and the y-intercept is $b_0 = 4$. The equation of the line that has a slope of -1 and y-intercept 4 is $y = 4 - x$. The graph is shown below.

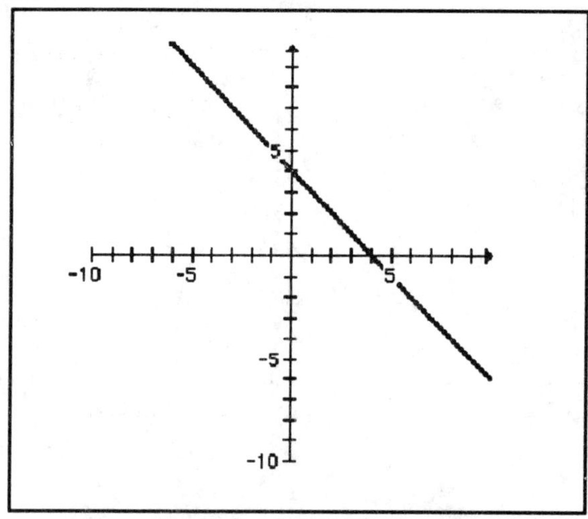

Graph for Exercise 3.15

3.17 **a.** The y-intercept is the fixed fee of \$35.00 and the slope is the cost per mile of 30 cents per mile = \$0.30. The equation of the line that has a slope of 0.3 and y-intercept 35.00 is: $y = 35.00 + 0.30x$.

b. See part a. The y-intercept is the fixed fee of \$35.00 and the slope is the cost per mile of 30 cents per mile = \$0.30. Therefore, $b_0 = 35.00$, $b_1 = 0.30$.

c. The additional cost for the second day is the slope of the line multiplied by the number of miles driven = (0.30)(100) = \$30.00.

3.19 **a.** The sales increase for each increase of \$1 in advertising expenditures is the slope of the line; that is the coefficient of x which is $b_1 = 1.5$.
b. The increase in sales if advertising is increased by \$1,000 is the slope of the line multiplied by the increase in advertising = (1.5)(1000) =\$1500.00.

3.21 We construct a table:

x	y	x^2	y^2	xy
5	10	25	100	50
0	4	0	16	0
4	7	16	49	28
-1	2	1	4	-2
3	6	9	36	18
$\Sigma x = 11$	$\Sigma y = 29$	$\Sigma x^2 = 51$	$\Sigma y^2 = 205$	$\Sigma xy = 94$

Therefore:

$$SS(x) = \Sigma x^2 - \frac{(\Sigma x)^2}{n} = 51 - \frac{(11)^2}{5} = 51 - 24.2 = 26.8$$

$$SS(y) = \Sigma y^2 - \frac{(\Sigma y)^2}{n} = 205 - \frac{(29)^2}{5} = 205 - 168.2 = 36.8$$

$$SS(xy) = \Sigma xy - \frac{(\Sigma x)(\Sigma y)}{n} = 94 - \frac{(11)(29)}{5} = 94 - 63.8 = 30.2.$$

3.23 We construct a table:

x	y	x^2	y^2	xy
39	10	1521	100	390
37	15	1369	225	555
35	16	1225	256	560
15	30	225	900	450
25	21	625	441	525
38	14	1444	196	532
18	34	324	1156	612
60	3	3600	9	180
36	10	1296	100	360
40	14	1600	196	560
$\Sigma x = 343$	$\Sigma y = 167$	$\Sigma x^2 = 13229$	$\Sigma y^2 = 3579$	$\Sigma xy = 4724$

Therefore:

$$SS(x) = \Sigma x^2 - \frac{(\Sigma x)^2}{n} = 13229 - \frac{(343)^2}{10} = 13229 - 11764.9 = 1464.1$$

$$SS(y) = \Sigma y^2 - \frac{(\Sigma y)^2}{n} = 3579 - \frac{(167)^2}{10} = 3579 - 2788.9 = 790.1$$

$$SS(xy) = \Sigma xy - \frac{(\Sigma x)(\Sigma y)}{n} = 4724 - \frac{(343)(167)}{10} = 4724 - 5728.1 = -1004.1$$

EXERCISES for Sections 3.2 and 3.3

Scatter Diagrams & Linear Regression

3.25 **a.**

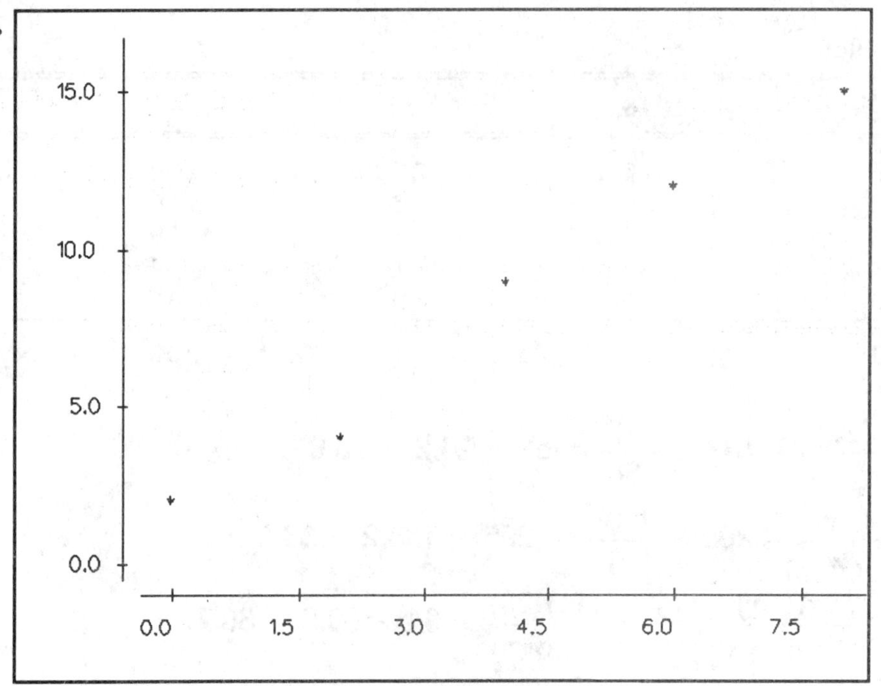

Exercise 3.25

b. From the scatter diagram, it appears that a straight line does provide a good fit to the data points.

3.27 From the data we calculate $\Sigma x = 20$, $\Sigma y = 42$, $\Sigma x^2 = 120$, $\Sigma y^2 = 470$, $\Sigma xy = 236$. We calculate:

$$SS(x) = \Sigma x^2 - \frac{(\Sigma x)^2}{n} = 120 - \frac{(20)^2}{5} = 120 - 80 = 40 \text{ and}$$

$$SS(xy) = \Sigma xy - \frac{(\Sigma x)(\Sigma y)}{n} = 236 - \frac{(20)(42)}{5} = 236 - 168 = 68.$$

Therefore, the slope of the least squares line is: $b_1 = \frac{SS(xy)}{SS(x)} = \frac{68}{40} = 1.7$, and the y-intercept is $b_0 = \bar{y} - b_1\bar{x} = \frac{42}{5} - 1.7\frac{20}{5} = 1.6$.

The least squares line is: $y = 1.6 + 1.7x$.

3.29 From Exercise 3.22, $SS(x) = 112.8333$; $SS(xy) = 254.3333$. The slope of the least squares line is: $b_1 = \frac{SS(xy)}{SS(x)} = \frac{254.3333}{112.8333} = 2.25406$, and the y-intercept is

$$b_0 = \bar{y} - b_1\bar{x} = \frac{34}{6} - 2.25406\frac{13}{6} = 0.7829.$$

The least squares line is $y = 0.7829 + 2.2541x$.

3.31 **a.** From the data we calculate $\Sigma x=8$; $\Sigma y=10$; $\Sigma x^2=26$; $\Sigma y^2=42$; $\Sigma xy=32$. We calculate:

$$SS(x) = \Sigma x^2 - \frac{(\Sigma x)^2}{n} = 26 - \frac{(8)^2}{4} = 26 - 16 = 10 \text{ and}$$

$$SS(xy) = \Sigma xy - \frac{(\Sigma x)(\Sigma y)}{n} = 32 - \frac{(8)(10)}{4} = 32 - 20 = 12.$$

The slope of the least squares line is: $b_1 = \frac{SS(xy)}{SS(x)} = \frac{12}{10} = 1.2$, and the y-intercept is

$b_0 = \bar{y} - b_1\bar{x} = \frac{10}{4} - 1.2\frac{8}{4} = 0.1$. Therefore, the least squares line is $y = 0.1 + 1.2x$.

b.

x	y	Pred y	error	$error^2$
0	1	0.1	0.9	0.81
1	0	1.3	-1.3	1.69
3	4	3.7	0.3	0.09
4	5	4.9	0.1	0.01

c. SSE $= \Sigma(\text{error})^2 = 0.81 + 1.69 + 0.09 + 0.01 = 2.60$.
Also we note that Σerror $= 0.9 + (-1.3) + 0.3 + 0.1 = 0$.

d.

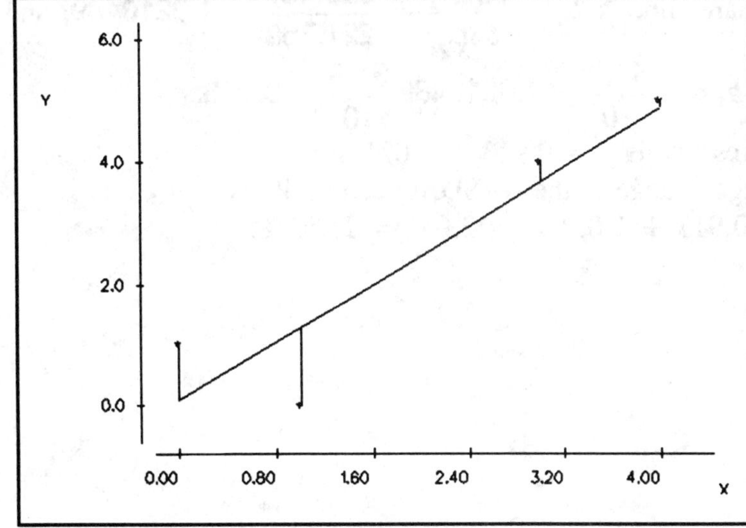

Exercise 3.31 d.

3.33 **a.**

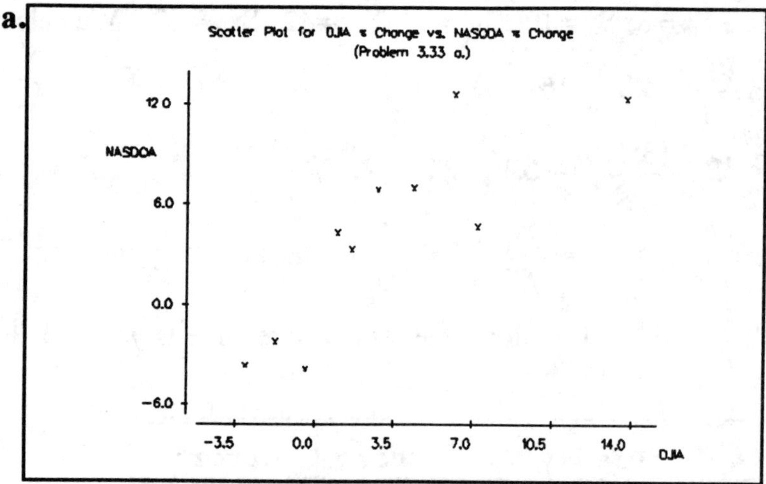

Exercise 3.33 a.

b. From the data we find $\Sigma x=31.9$; $\Sigma y=41.7$; $\Sigma x^2=323.53$; $\Sigma y^2=495.37$; $\Sigma xy=359.46$. We calculate:

$$SS(x) = \Sigma x^2 - \frac{(\Sigma x)^2}{n} = 323.53 - \frac{(31.9)^2}{10} = 323.53 - 101.761 = 221.769$$

$$SS(xy) = \Sigma xy - \frac{(\Sigma x)(\Sigma y)}{n} = 359.46 - \frac{(31.9)(41.7)}{10} = 359.46 - 133.023 = 226.437.$$

The slope of the least squares line is $b_1 = \frac{SS(xy)}{SS(x)} = \frac{226.437}{221.769} = 1.0210489$, and the y-intercept is $b_0 = \bar{y} - b_1\bar{x} = \frac{41.7}{10} - 1.0210489\frac{31.9}{10} = 0.913.$

Therefore, the least squares line is y = 0.913 + 1.021x.

c. The predicted percentage change in the NASDAQ composite is:

y = 0.913 + 1.021x = 0.913 + 1.021(2) = 2.955 = 2.955%.

MINITAB LAB ASSIGNMENTS

3.35(M)

```
MTB > # Exercise 3.35(M)
MTB > # The data were read into C1 and C2
MTB > NAME C1 'x' C2 'y'
MTB > PLOT C2 C1

 y       -   *
         -         *
         -
    2.10+                 *      *
         -                            *
         -                                  *
         -
         -
    1.40+                                         *
         -
         -
         -
         -
    0.70+                                               *
         -
         -                                                    *
         -
          ------+---------+---------+---------+---------+---------+x
              40.0      48.0      56.0      64.0      72.0      80.0
```

3.37(M)

```
MTB > # Exercise 3.37(M)
MTB > # The data were read into columns C1 and C2
MTB > NAME C1 'x' C2 'y'
MTB > PLOT C2 C1
```

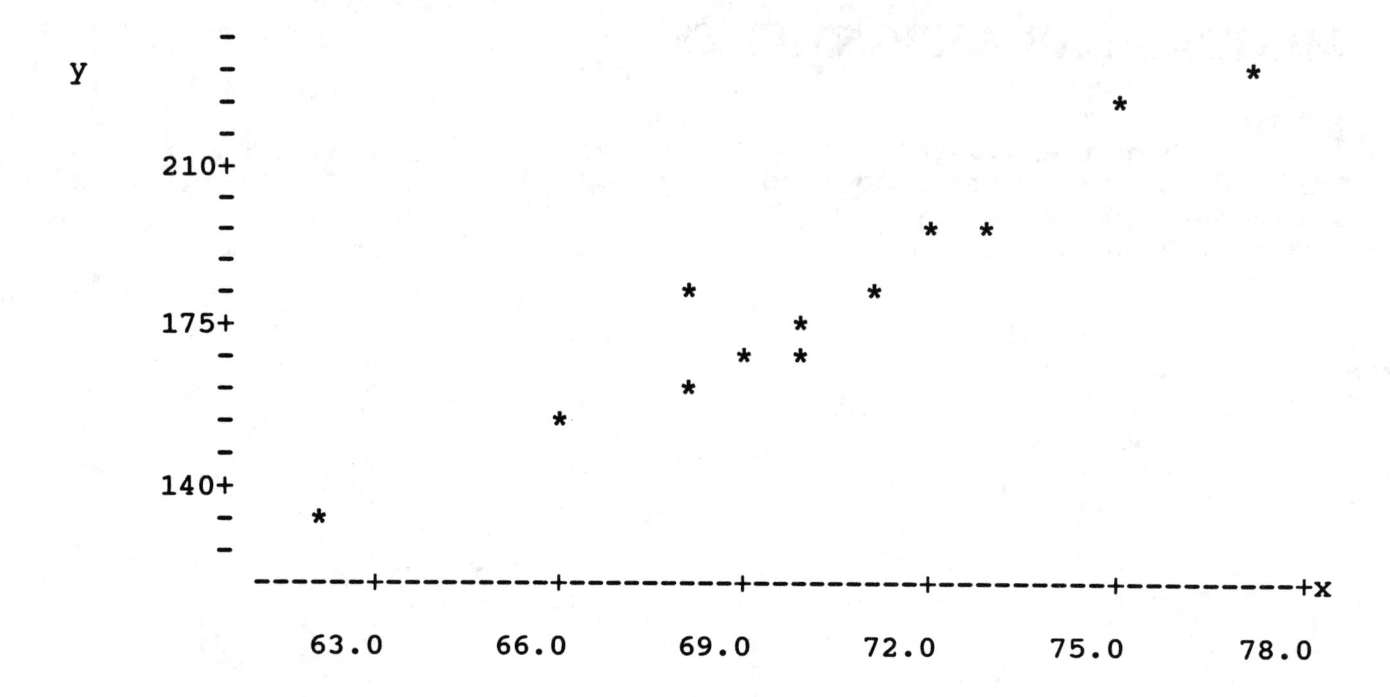

3.39(M)

```
MTB > # Exercise 3.39(M)
MTB > # The data were read into columns C1 and C2 in problem 3.37(M)
MTB > REGRESSION C2 1 C1

The regression equation is
y = - 289 + 6.70 x

Predictor        Coef        Stdev     t-ratio        p
Constant      -288.59        43.36       -6.66    0.000
x              6.7028       0.6178       10.85    0.000

s = 8.218        R-sq = 92.2%      R-sq(adj) = 91.4%

Analysis of Variance

SOURCE        DF          SS          MS          F        p
Regression     1      7948.4      7948.4     117.70    0.000
Error         10       675.3        67.5
Total         11      8623.7

MTB > # The regression equation is y = -289 + 6.70 x.
```

EXERCISES for Section 3.4

Coefficient of Correlation

3.41 X is an adult's age, and y is the time required to run a mile: Positively correlated.

3.43 X is a company's advertising expenditures, and y is its sales revenue: Positively correlated.

3.45 X is the horsepower of a truck's engine, and y is the load capacity of the truck: Positively correlated.

3.47 X is an adult's age, and y is his/her systolic blood pressure: Positively correlated.

3.49

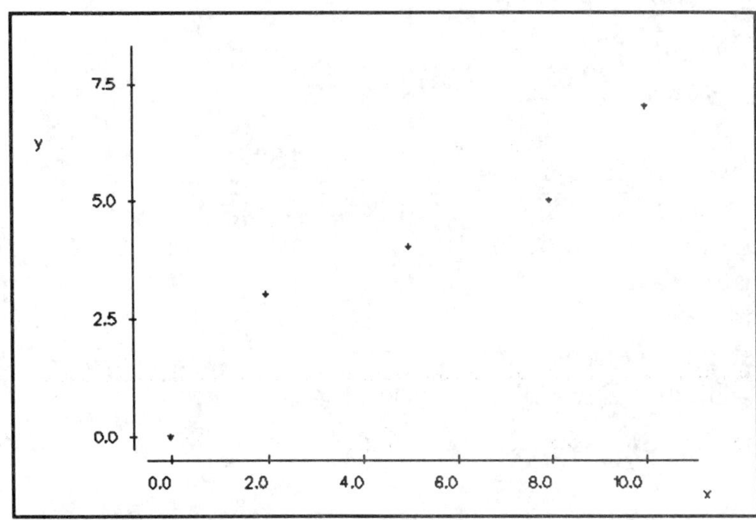

Exercise 3.49

In order to calculate the correlation coefficient for the data points we calculate: $\Sigma x=25$; $\Sigma y=19$; $\Sigma x^2=193$; $\Sigma y^2=99$; $\Sigma xy=136$. Therefore:

$$SS(x) = \Sigma x^2 - \frac{(\Sigma x)^2}{n} = 193 - \frac{(25)^2}{5} = 68$$

$$SS(y) = \Sigma y^2 - \frac{(\Sigma y)^2}{n} = 99 - \frac{(19)^2}{5} = 26.8$$

$$SS(xy) = \Sigma xy - \frac{(\Sigma x)(\Sigma y)}{n} = 136 - \frac{(25)(19)}{5} = 41.$$

The correlation coefficient is $r = \dfrac{SS(xy)}{\sqrt{SS(x)\,SS(y)}} = \dfrac{41}{\sqrt{(68)(26.8)}} = 0.96.$

3.51 In order to find the correlation coefficient for the data points we calculate: $\Sigma x=343$; $\Sigma y=167$; $\Sigma x^2=13229$; $\Sigma y^2=3579$; $\Sigma xy=4724$. Therefore:

$$SS(x) = \Sigma x^2 - \frac{(\Sigma x)^2}{n} = 13229 - \frac{(343)^2}{10} = 1464.1$$

$$SS(y) = \Sigma y^2 - \frac{(\Sigma y)^2}{n} = 3579 - \frac{(167)^2}{10} = 790.1$$

$$SS(xy) = \Sigma xy - \frac{(\Sigma x)(\Sigma y)}{n} = 4724 - \frac{(343)(167)}{10} = -1004.1.$$

The correlation coefficient is $r = \dfrac{SS(xy)}{\sqrt{SS(x)\,SS(y)}} = \dfrac{-1004.1}{\sqrt{(1464.1)(790.1)}} = -0.93.$

Therefore, as the miles increase, the tread depth tends to decrease.

3.53 We are given: $\Sigma x=3.627$; $\Sigma y=1133$; $\Sigma x^2=0.940577$; $\Sigma y^2=92815$; $\Sigma xy=293.746$. Therefore:

$$SS(x) = \Sigma x^2 - \frac{(\Sigma x)^2}{n} = 0.940577 - \frac{(3.627)^2}{14} = 0.000924928$$

$$SS(y) = \Sigma y^2 - \frac{(\Sigma y)^2}{n} = 92815 - \frac{(1133)^2}{14} = 1122.93$$

$$SS(xy) = \Sigma xy - \frac{(\Sigma x)(\Sigma y)}{n} = 293.746 - \frac{(3.627)(1133)}{14} = 0.218071.$$

The correlation coefficient is: $r = \frac{SS(xy)}{\sqrt{SS(x)SS(y)}} = \frac{0.218071}{\sqrt{(0.000924928)(1122.93)}} = 0.21.$

We would expect this to be positive; however, the linear relationship is very weak between the number of wins and the batting average.

3.55 In order to calculate the correlation coefficient for the data points we calculate: $\Sigma x=513$; $\Sigma y=15.32$; $\Sigma x^2=30741$; $\Sigma y^2=30.298$; $\Sigma xy=797.84$. Therefore:

$$SS(x) = \Sigma x^2 - \frac{(\Sigma x)^2}{n} = 30741 - \frac{(513)^2}{9} = 1500$$

$$SS(y) = \Sigma y^2 - \frac{(\Sigma y)^2}{n} = 30.298 - \frac{(15.32)^2}{9} = 4.219956$$

$$SS(xy) = \Sigma xy - \frac{(\Sigma x)(\Sigma y)}{n} = 797.84 - \frac{(513)(15.32)}{9} = -75.4.$$

The correlation coefficient is $r = \frac{SS(xy)}{\sqrt{SS(x)SS(y)}} = \frac{-75.4}{\sqrt{(1500)(4.219956)}} = -0.95.$

This suggests that as one's age increases, the estimated years of added life decrease almost linearly.

MINITAB LAB ASSIGNMENTS

3.57(M)

```
MTB > # Exercise 3.57 (M)
MTB > # The data were read into C1 and C2
MTB > NAME C1 'Avg.' C2 'Wins'
MTB > CORRELATION C1 C2

Correlation of Avg. and Wins = 0.214
```

3.59(M)a.

```
MTB > # Exercise 3.59 (M)
MTB > # The data were read into C1 and C2
MTB > NAME C1 'DJIA' C2 'SALES'
MTB > # a. The scatter plot:
```

```
MTB > PLOT C2 C1

      75+
        -
 SALES  -                                                          *
        -                                                              *
        -                                                       *
      60+                                    *         *
        -                                    *   *
        -                          *
        -
        -                        *
      45+               *   *
        -             *
        -         *       *
        -     *
        -         *
      30+
        -
          ------+---------+---------+---------+---------+---------+DJIA

              1500      1750      2000      2250      2500      2750

MTB > # b. Least Squares line:
MTB > REGRESSION C2 1 C1

The regression equation is
Sales = - 2.39 + 0.0270 DJIA

Predictor        Coef       Stdev    t-ratio        p
Constant       -2.393       4.270      -0.56    0.584
DJIA         0.027027    0.002119      12.75    0.000

s = 3.362       R-sq = 92.1%     R-sq(adj) = 91.5%

Analysis of Variance

SOURCE       DF          SS          MS         F        p
Regression    1      1839.1      1839.1    162.69    0.000
Error        14       158.3        11.3
Total        15      1997.3

Unusual Observations
Obs.     DJIA      Sales        Fit Stdev.Fit  Residual    St.Resid
  8      1532     32.300     39.013     1.261    -6.713      -2.15R

R denotes an obs. with a large st. resid.

MTB > # The least squares line is y = -2.39 + 0.0270 x

MTB > # c. The Coefficient of Correlation:
MTB > CORR C1 C2

Correlation of DJIA and Sales = 0.960
```

REVIEW EXERCISES

CHAPTER 3

3.60 We first solve for y and get: $y = 5 - \frac{5}{2}x$.
The slope is the coefficient of x which is b_1 = -2.5; the y-intercept is the constant term in the equation which is b_0 = 5. The graph appears to the right.

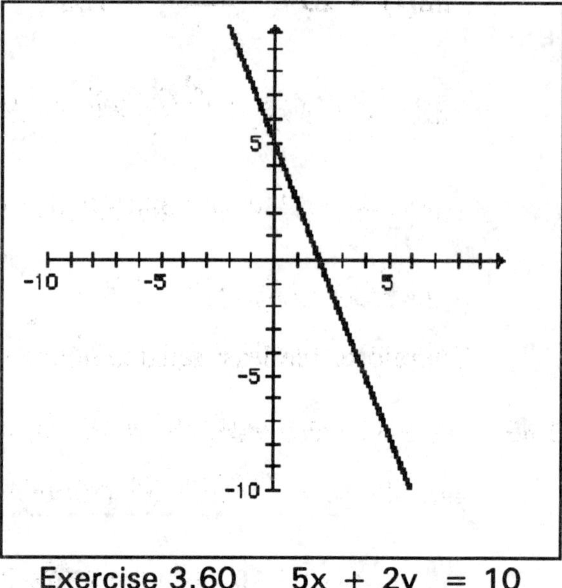

Exercise 3.60   5x + 2y = 10

3.61 The equation of the line with slope -7/3 and y-intercept 1/6 is: $y = \frac{1}{6} - \frac{7}{3}x$.

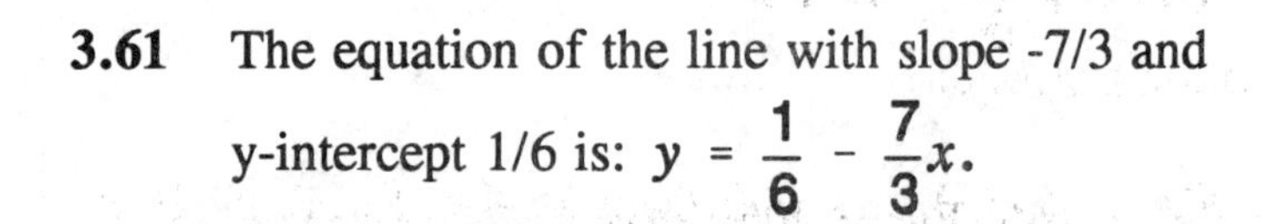

3.62 **a.** The equation that gives the relation between y and x is y = 135 + 75x. (See part b).
b. The slope is 75, since he/she receives $75 for each car that he/she sells. The y-intercept is 135 since each salesperson receives a weekly salary of $135 even if no sales are made.

3.63 If two brothers differ in age by two years, then, since the slope of the line y = 13.2 + 5.0x is 5.0, their scores are expected to differ by (5.0)(2) = 10.

3.64

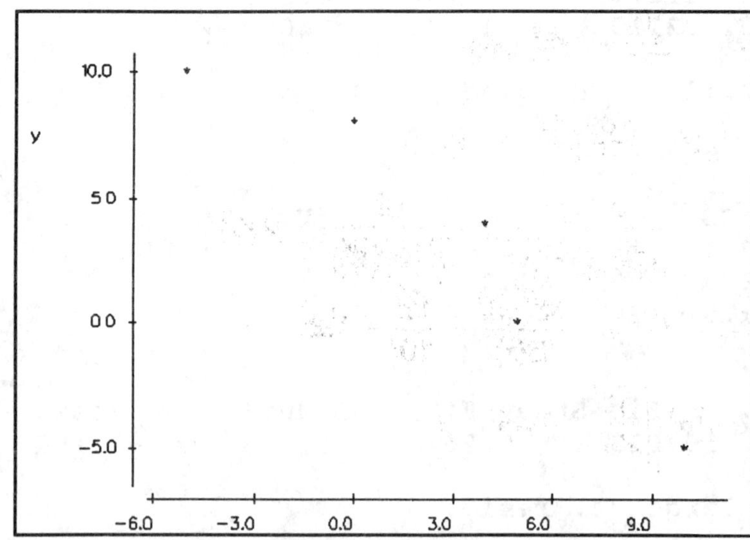

Scatter Diagram for Exercise 3.64

3.65 From the data we calculate $\Sigma x=14$; $\Sigma y=17$; $\Sigma x^2=166$; $\Sigma y^2=205$; $\Sigma xy=-84$. We calculate:

$$SS(x) = \Sigma x^2 - \frac{(\Sigma x)^2}{n} = 166 - \frac{(14)^2}{5} = 126.8 \text{ and}$$

$$SS(xy) = \Sigma xy - \frac{(\Sigma x)(\Sigma y)}{n} = -84 - \frac{(14)(17)}{5} = -131.6.$$

The slope of the least squares line is: $b_1 = \frac{SS(xy)}{SS(x)} = \frac{-131.6}{126.8} = -1.03785$, and the y-intercept is $b_0 = \bar{y} - b_1\bar{x} = \frac{17}{5} - (-1.03785)\frac{14}{5} = 6.306$.
Therefore, the least squares line is y = 6.306 - 1.038x.

3.66 We have calculated: $\Sigma x = 14$; $\Sigma y = 17$; $\Sigma x^2 = 166$; $\Sigma y^2 = 205$; $\Sigma xy = -84$, SS(x) = 126.8 and SS(xy) = -131.6. In addition: $SS(y) = \Sigma y^2 - \frac{(\Sigma y)^2}{n} = 205 - \frac{(17)^2}{5} = 147.2$.

Therefore, the correlation coefficient is $r = \frac{SS(xy)}{\sqrt{SS(x)SS(y)}} = \frac{-131.6}{\sqrt{(126.8)(147.2)}} = -0.96$.
The data points lie in nearly a straight line with negative slope.

3.67 We calculate: $\Sigma x = 8$; $\Sigma y = 10$; $\Sigma x^2 = 26$; $\Sigma y^2 = 42$; $\Sigma xy = 32$. Therefore:

$$SS(x) = \Sigma x^2 - \frac{(\Sigma x)^2}{n} = 26 - \frac{(8)^2}{4} = 10,$$

$$SS(y) = \Sigma y^2 - \frac{(\Sigma y)^2}{n} = 42 - \frac{(10)^2}{4} = 17, \text{ and}$$

$$SS(xy) = \Sigma xy - \frac{(\Sigma x)(\Sigma y)}{n} = 32 - \frac{(8)(10)}{4} = 12.$$

The correlation coefficient is: $r = \frac{SS(xy)}{\sqrt{SS(x)SS(y)}} = \frac{12}{\sqrt{(10)(17)}} = 0.92$.

The slope of the least squares line is: $b_1 = \frac{SS(xy)}{SS(x)} = \frac{12}{10} = 1.2$.

We confirm that $b_1\sqrt{\frac{SS(x)}{SS(y)}} = 1.2\sqrt{\frac{10}{17}} = 0.92 = r$.

3.68 **a.**

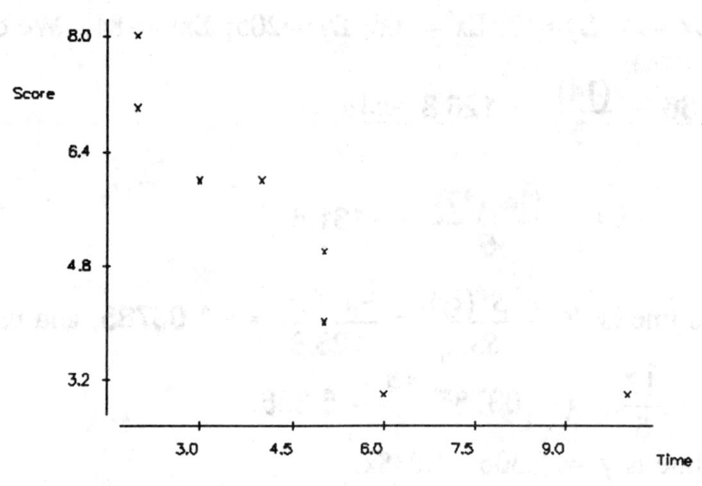

b. We calculate: $\Sigma x=37$; $\Sigma y=42$; $\Sigma x^2=219$; $\Sigma y^2=244$; $\Sigma xy=165$. Therefore:

$$SS(x) = \Sigma x^2 - \frac{(\Sigma x)^2}{n} = 219 - \frac{(37)^2}{8} = 47.875,$$

$$SS(y) = \Sigma y^2 - \frac{(\Sigma y)^2}{n} = 244 - \frac{(42)^2}{8} = 23.5, \text{ and}$$

$$SS(xy) = \Sigma xy - \frac{(\Sigma x)(\Sigma y)}{n} = 165 - \frac{(37)(42)}{8} = -29.25.$$

The slope of the least squares line is: $b_1 = \frac{SS(xy)}{SS(x)} = \frac{-29.25}{47.875} = -0.610966,$

and the y-intercept is $b_0 = \bar{y} - b_1\bar{x} = \frac{42}{8} - (-0.610966)\frac{37}{8} = 8.076.$

The equation of the least squares line is: y = 8.076 - 0.611x.

c. The correlation coefficient is $r = \frac{SS(xy)}{\sqrt{SS(x)SS(y)}} = \frac{-29.25}{\sqrt{(47.875)(23.5)}} = -0.87.$

This indicates that the subtest scores tend to decrease as the time after injury increases.

d. The predicted score of a patient who was injured seven weeks ago is

$\hat{y} = 8.076 - (0.611)(7) = 3.80 \approx 4.$

3.69 **a.** We are given that for the data: $\Sigma x = 435.54$; $\Sigma y = 38.5$; $\Sigma x^2 = 13836$; $\Sigma y^2 = 224.45$; $\Sigma xy = 1194.7$. Therefore:

$$SS(x) = \Sigma x^2 - \frac{(\Sigma x)^2}{n} = 13836 - \frac{(435.54)^2}{15} = 1189.66,$$

$$SS(y) = \Sigma y^2 - \frac{(\Sigma y)^2}{n} = 224.45 - \frac{(38.5)^2}{15} = 125.633, \text{ and}$$

$$SS(xy) = \Sigma xy - \frac{(\Sigma x)(\Sigma y)}{n} = 1194.7 - \frac{(435.54)(38.5)}{15} = 76.814.$$

The correlation coefficient is: $r = \frac{SS(xy)}{\sqrt{SS(x)SS(y)}} = \frac{76.814}{\sqrt{(1189.66)(125.633)}} = 0.199.$

b. No, the relationship does not appear to be significant, so it does not appear that higher priced stocks tend to have the larger percentage increases.

3.70 **a.**

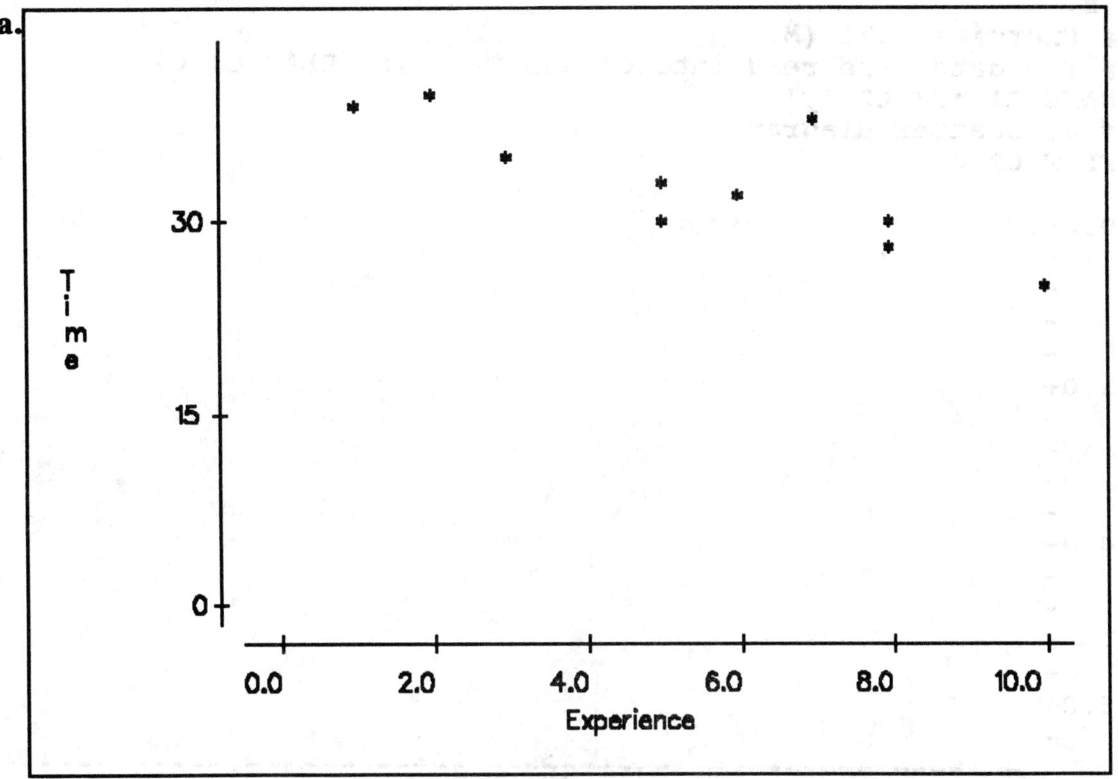

Exercise 3.70

b. For this data: $\Sigma x=55$; $\Sigma y=330$; $\Sigma x^2=377$; $\Sigma y^2=11112$; $\Sigma xy=1711$. Therefore:

$$SS(x) = \Sigma x^2 - \frac{(\Sigma x)^2}{n} = 377 - \frac{(55)^2}{10} = 74.5,$$

$$SS(y) = \Sigma y^2 - \frac{(\Sigma y)^2}{n} = 11112 - \frac{(330)^2}{10} = 222, \text{ and}$$

$$SS(xy) = \Sigma xy - \frac{(\Sigma x)(\Sigma y)}{n} = 1711 - \frac{(55)(330)}{10} = -104.$$

The slope of the least squares line is: $b_1 = \frac{SS(xy)}{SS(x)} = \frac{-104}{74.5} = -1.395973,$

and the y-intercept is $b_0 = \bar{y} - b_1\bar{x} = \frac{330}{10} - (-1.395973)\frac{55}{10} = 40.678.$

The equation of the least squares line is y = 40.678 - 1.396x.

c. The correlation coefficient is: $r = \frac{SS(xy)}{\sqrt{SS(x)SS(y)}} = \frac{-104}{\sqrt{(74.5)(222)}} = -0.81.$

MINITAB LAB ASSIGNMENTS

3.71(M)a.

```
MTB > # Exercise 3.71 (M)
MTB > # The data were read into C1 and C2 using READ C1 C2
MTB > NAME C1 'x' C2 'y'
MTB > # a. Scatter diagram:
MTB > PLOT C2 C1

     40.0+                 *
         -            *
  y      -                                *
         -
         -
     35.0+                     *
         -
         -                               *
         -                                   *
         -
     30.0+                               *              *
         -
         -                                              *
         -
         -
     25.0+                                                        *
         -
          --+---------+---------+---------+---------+---------+----x

          0.0       2.0       4.0       6.0       8.0      10.0
```

b.

```
MTB > # b.The least squares line:
MTB > REGRESSION C2 1 C1

The regression equation is
y = 40.7 - 1.40 x

Predictor        Coef       Stdev    t-ratio        p
Constant       40.678       2.204      18.45    0.000
x             -1.3960      0.3590      -3.89    0.005

s = 3.099       R-sq = 65.4%     R-sq(adj) = 61.1%
```

```
Analysis of Variance

SOURCE         DF          SS          MS        F        p
Regression      1      145.18      145.18    15.12    0.005
Error           8       76.82        9.60
Total           9      222.00

Unusual Observations
Obs.        x          y        Fit  Stdev.Fit  Residual   St.Resid
 10       7.0     38.000     30.906      1.118     7.094      2.45R

R denotes an obs. with a large st. resid.

MTB > # The equation of the least squares line is y = 40.7 - 1.40 x.
```

c.

```
MTB > # c. The coefficient of correlation:
MTB > CORRELATION of C1 vs. C2

Correlation of x and y = -0.809
```

3.72(M)

```
MTB > # Exercise 3.72 (M)
MTB > # The data were read into C1 and C2 using READ C1 C2
MTB > NAME C1 'x' C2 'y'
MTB > PLOT C2 C1
```

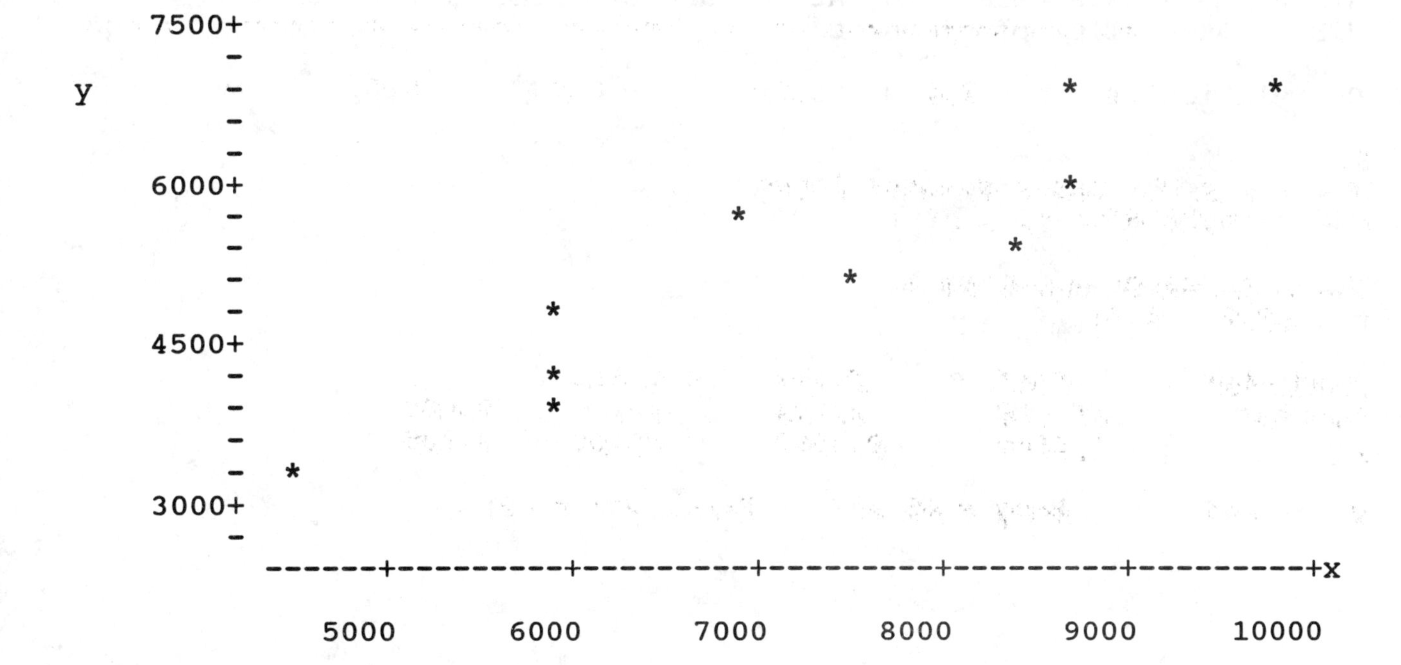

3.73(M)

```
MTB > # Exercise 3.73 (M)
MTB > # The data were read into C1 and C2 using READ C1 C2
MTB > REGRESS C2 1 C1

The regression equation is
y = 179 + 0.694 x

Predictor        Coef       Stdev    t-ratio        p
Constant        179.3       672.5       0.27    0.797
x             0.69367     0.09102       7.62    0.000

s = 458.5        R-sq = 87.9%     R-sq(adj) = 86.4%

Analysis of Variance

SOURCE       DF          SS          MS         F        p
Regression    1    12211388    12211388     58.08    0.000
Error         8     1682120      210265
Total         9    13893508

MTB > # The equation of the least squares line to predict hot dog
MTB > # sales is: y = 179 + 0.694 x.
```

3.74(M)

```
MTB > # Exercise 3.74 (M)
MTB > # The data were read into C1 and C2 using READ C1 C2
MTB > CORRELATION of C1 vs. C2

Correlation of x and y = 0.938
```

CHAPTER 4
PROBABILITY: MEASURING UNCERTAINTY

EXERCISES for Section 4.1
Probability, Sample Spaces, and Events

4.1 The following is a sample space that shows the different ways a test with two true-false questions can be answered: TT, TF, FT, FF .

4.3 The following is a sample space showing the different ways that three true-false questions can be answered: TTT, TTF, TFT, TFF, FTT, FTF, FFT, FFF.

4.5 Let E be the event that a student who has a complete lack of knowledge of the two multiple choice questions referred to in Exercise 4.4 answers each question correctly by guessing. Since the outcomes in the sample space are equally likely, we may apply the theoretical method of assigning a probability to E. Therefore $P(E) = \frac{e}{s} = \frac{1}{9}$.

4.7 **a.** The following is a sample space for the experiment of selecting one card from a standard deck of 52 and, if the card is red, tossing a coin; if it is black, tossing a die: RH, RT, B1, B2, B3, B4, B5, B6.
b. The points in the sample space are not equally likely.

4.9 We define the event E: a birth in the United States will be by Caesarean section. Using the relative frequency definition for the given data, we calculate:

$$P(E) \approx \frac{906,000}{3,757,000} = 0.241.$$

4.11 The relative frequency definition was used since the estimate is based upon a proportion of occurrences for empirical results.

4.13 The sample space used does not have equally likely points. Therefore, the theoretical method of assigning probabilities is not valid.

4.15 We are given that P(A) = P(B) = P(C) = 2P(D). Since P(A) + P(B) + P(C) + P(D) = 1, we have 2P(D) + 2P(D) + 2P(D) + P(D) = 7P(D) = 1. Therefore, the subjective probabilities being assigned by the investor are P(D) = $\frac{1}{7}$ and P(A) = P(B) = P(C) = $\frac{2}{7}$.

4.17 **a.-f.** The results will vary from experiment to experiment.

MINITAB LAB ASSIGNMENTS

Note: Answers to the following Minitab problems will vary.

4.19(M)

```
MTB > # Exercise 4.19(M)
MTB > RAND 200 C1;
SUBC> INTEGERS 0 1.
MTB > LET K1 = SUM(C1)
```

```
MTB > PRINT K1
K1        104.000
MTB > # Therefore there are 104 males in 200 births.
```

4.21(M)

```
MTB > # Exercise 4.21(M)
MTB > RANDOM 400 C1;
SUBC> INTEGERS 1 to 4.
MTB > TALLY C1

        C1   COUNT
         1      94
         2     115
         3      99
         4      92
        N=     400
```

We would expect each of the counts to be 100 on the average.

EXERCISES for Section 4.2

Basic Techniques for Counting Sample Points

4.23 $7! = 7 \cdot 6 \cdot 5 \cdot 4 \cdot 3 \cdot 2 \cdot 1 = 5040.$

4.25 $\dfrac{100!}{99!} = \dfrac{100 \cdot 99!}{99!} = 100.$

4.27 $C(10,3) = \dfrac{10!}{3! \cdot 7!} = \dfrac{10 \cdot 9 \cdot 8 \cdot 7!}{3 \cdot 2 \cdot 1 \cdot 7!} = 120.$

4.29 $C(12,9) = \dfrac{12!}{9! \cdot 3!} = \dfrac{12 \cdot 11 \cdot 10 \cdot 9!}{9! \cdot 3 \cdot 2 \cdot 1} = 220.$

4.31 $C(88,87) = \dfrac{88!}{87! \cdot 1!} = \dfrac{88 \cdot 87!}{87! \cdot 1} = 88.$

4.33 $C(88,0) = \dfrac{88!}{0! \cdot 88!} = \dfrac{88!}{1 \cdot 88!} = 1.$

4.35 Each of the five multiple-choice questions has four alternatives. Therefore, by the multiplication rule, the quiz may be answered in $4 \cdot 4 \cdot 4 \cdot 4 \cdot 4 = 1024$ different ways.

4.37 Each of the five true-false questions has two alternatives, and each of the five multiple-choice questions has four alternatives. Therefore, by the multiplication rule, the quiz may be answered in $2 \cdot 2 \cdot 2 \cdot 2 \cdot 2 \cdot 4 \cdot 4 \cdot 4 \cdot 4 \cdot 4 = 32{,}768$ different ways.

4.39 Suppose the World Series is tied after four games. The following tree diagram shows the different ways that the two teams can complete the remaining games:
(A = American League champion; N = National League champion.)

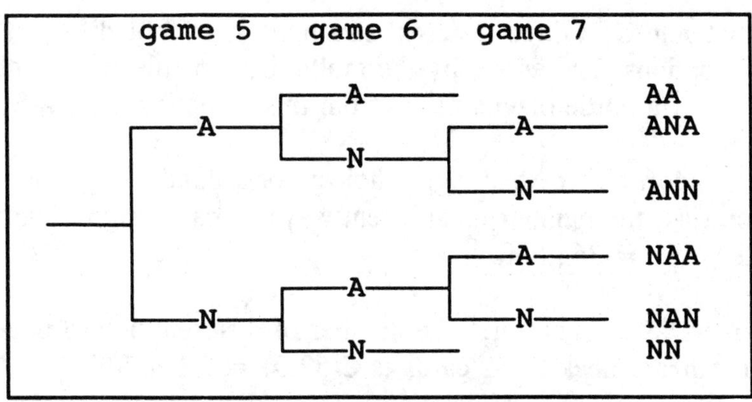

Tree Diagram for Exercise 4.39

4.41 There are four rooms available for three courses that are scheduled at the same time, so there are four rooms available for the first course, three for the second and two for the third. Therefore, by the multiplication rule, the number of possible room assignments that can be made to the three courses is $4 \cdot 3 \cdot 2 = 24$.

4.43 The following is a tree diagram which displays the different possible room assignments to the three courses, where the rooms are denoted by R1, R2, R3, and R4.

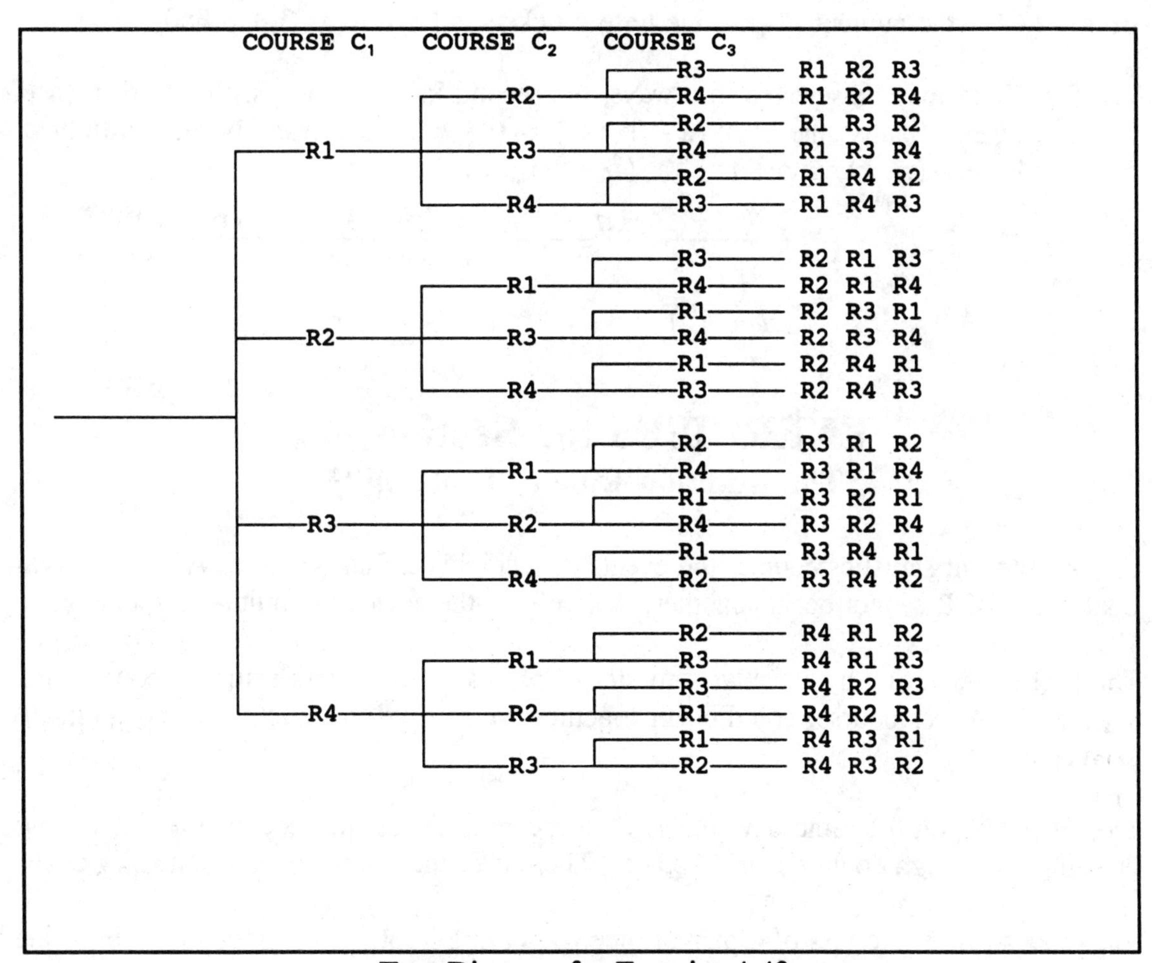

Tree Diagram for Exercise 4.43

4.45 The mail order company sells vitamin C in three different size bottles, in four different potencies, and with two coating options. Therefore, by the multiplication rule, the number of different ways a customer can order a bottle of vitamin C from this company is $3 \cdot 4 \cdot 2 = 24$.

4.47 The manager has 9 choices for leadoff batter, then eight choices for second batter, etc. Therefore, by the multiplication rule, the number of different ways he can assign a batting order is $9 \cdot 8 \cdot 7 \cdot 6 \cdot 5 \cdot 4 \cdot 3 \cdot 2 \cdot 1 = 9! = 362{,}880$.

4.49 The order of the cards is not important. Therefore, the number of different 5-card poker hands that can be dealt from a standard deck of 52 cards is $C(52,5) = 2{,}598{,}960$.

4.51 Let E be the event of being dealt a 5-card hand consisting of three aces and a pair of kings. From problem 4.50, the number of outcomes in E is $e = 24$. The number of outcomes in the sample space is $s = C(52,5) = 2{,}598{,}960$, which is the number of 5-card poker hands that can be dealt from a standard deck of 52 cards. Since the outcomes in the sample space are equally likely, we may apply the theoretical method of assigning a probability to E.

Therefore $P(E) = \frac{e}{s} = \frac{24}{2{,}598{,}960} = 0.000009234.$

4.53 The order is not important. Therefore, if in a lottery, a player selects six different numbers from 1 to 40, the number of possible lottery tickets is $C(40,6) = 3{,}838{,}380$.

4.55 The first thing may be selected in n ways, the second in (n - 1) ways, the third in (n - 2) ways,......,and, finally, the kth in $[n - (k - 1)] = (n - k + 1)$ ways. By the multiplication rule, we have:

$$P(n,k) = n \cdot (n-1) \cdot (n-2) \cdot \ldots \cdot (n-k+1)$$

$$= \frac{n \cdot (n-1) \cdot (n-2) \cdot \ldots \cdot (n-k+1) \cdot (n-k)!}{(n-k)!}$$

$$= \frac{n!}{(n-k)!}$$

EXERCISES for Section 4.3
The Addition Rule of Probability

4.57 In your present statistics course, the events of receiving a final grade of A and receiving a final grade of B cannot occur together. Therefore, the events are mutually exclusive.

4.59 The events of, during a particular semester, a professor teaching a statistics course and teaching a physics course, could occur together. Therefore, the events are not mutually exclusive.

4.61 The events of, on the same day, receiving a speeding ticket and receiving a ticket for going through a stop sign could occur together. Therefore, the events are not mutually exclusive.

4.63 The events of, in one toss of a pair of dice, observing a total of 10 spots and observing a total of at least 8 spots could occur together. Therefore, the events are not mutually exclusive.

4.65 The probability that a randomly selected respondent has at least a college degree is, by the addition rule for mutually exclusive events, 0.243 + 0.138 + 0.181 = 0.562.

4.67 Consider the experiment of randomly selecting one card from a standard 52-card deck.
a. Let E denote the event that the card is both red and a club. Then P(E) = 0.
b. Let E denote the event that the card is either red or black. Then P(E) = 1. (Note: other answers are possible.)

4.69 Let A denote the event that an engineer for this corporation is a U.S. citizen, and let B denote the event that an engineer for this corporation is located in the United States. From the table, we note that 400 of the 500 engineers are U.S. citizens, that 350 of the 500 engineers are located in the United States, and that 280 of the 500 engineers are both. We apply the addition role of probability to find :

$$P(A \text{ or } B) = P(A) + P(B) - P(AB) = \frac{400}{500} + \frac{350}{500} - \frac{280}{500} = \frac{470}{500} = 0.94.$$

4.71 Let L denote the event that a member of the sales force at a large insurance company has a laptop computer, and let D denote the event that he or she has a desktop computer. We are given P(L) = 0.40, P(D) = 0.65, and P (LD) = 0.24. We use the addition rule:
P(L or D) = P(L) + P(D) - P(LD) = 0.40 + 0.65 - 0.24 = 0.81 = 81%.

4.73 Let D denote the event that a new car sale is dealer-financed, and C the event that the new car sale is a cash sale. We are given P(D) = 0.37 and P(C) = 0.18. The events D and C are mutually exclusive, so we apply the addition rule to find the probability that the next purchase of a new car at this dealership will be either a cash sale or dealer-financed:
P(D or C) = P(D) + P(C) = 0.37 + 0.18 = 0.55.

4.75 Let A denote the event that a student completed the introductory statistics course at a certain college with a grade of A, and similarly define the events B and C. We are given P(A) = 0.15, P(B) = 0.24, and P(C) = 0.32. By the addition rule, the probability that a randomly selected student from this group received a grade of C or better is P(A or B or C) = P(A) + P(B) + P(C) = 0.15 + 0.24 + 0.32 = 0.71.

4.77 Let A denote the event that a randomly selected card from a 52-card deck will be an ace; B, the event that the card is black; and H, the event that the card is a heart. Then by this generalized addition formula:
P(A or B or H) = [P(A) + P(B) + P(H)] - [P(AB) + P(AH) + P(BH)] + P(ABH)

$$= \left[\frac{1}{13} + \frac{1}{2} + \frac{1}{4}\right] - \left[\frac{1}{26} + \frac{1}{52} + 0\right] + 0 = \frac{10}{13} = 0.769.$$

EXERCISES for Section 4.4

The Multiplication & Conditional Rules of Probability

4.79 The event of a student receiving an A in Calculus I does affect her chances of obtaining an A in Calculus II. Therefore, the events are dependent.

4.81 We note $\frac{3}{10} = P(E \mid F) = P(E) = 0.3$. Therefore, the events are independent.

4.83 The event of tossing a tail on the first flip of a coin is not related to tossing a tail on the second toss. Therefore, the events are independent.

4.85 The event of buying a lottery ticket is prerequisite to winning the lottery. Therefore, the events are dependent.

4.87 We note that $0.24 = P(T \mid S) \neq P(T) = 0.21$. Therefore, the events are dependent.

4.89 **a.** $P(Y \mid D)$ = the probability that the company will show a profit for the year if December sales are good.
b. $P(D \mid Y)$ = the probability that, if the company shows a profit for the year, December sales are good.
c. $P(DY)$ = the probability that December sales are good and the company will show a profit for the year.

4.91 We are given that $P(A) = 0.5$ and $P(B) = 0.4$, where A and B are independent events. Therefore: $P(A \mid B) = P(A) = 0.5$, and $P(B \mid A) = P(B) = 0.4$.

4.93 We are given independent events A and B with $P(A) = 0.8$ and $P(B) = 0.3$. Therefore, by the multiplication rule for independent events: $P(AB) = P(A)P(B) = (0.8)\cdot(0.3) = 0.24$.
By the addition rule of probability we determine:
$P(A \text{ or } B) = P(A) + P(B) - P(AB) = 0.8 + 0.3 - 0.24 = 0.86$.

4.95 Six of ten packages are underweight. If four are selected,

a. P(all will be underweight) $= \frac{6}{10}\cdot\frac{5}{9}\cdot\frac{4}{8}\cdot\frac{3}{7} = 0.0714$.

b. P(none will be underweight) $= \frac{4}{10}\cdot\frac{3}{9}\cdot\frac{2}{8}\cdot\frac{1}{7} = 0.00476$.

4.97 Let F denote the event that a woman is chosen, and let E denote the event of selecting a woman who is 75 or older. We note from the table that 12882 of the 21129 people are female and that 3486 are both female and 75 or older.

a. The probability that a woman is chosen is $P(F) = \frac{12,882}{21,129} = 0.6097$.

b. The probability of selecting a woman who is 75 or older is

$P(FE) = \frac{3486}{21129} = 0.1650$.

c. If a woman is chosen, the probability that she is 75 or older is:

$$P(E|F) = \frac{P(EF)}{P(F)} = \frac{0.1650}{0.6097} = 0.271.$$

d. Let M denote the event that a man is chosen, and let G denote the event of selecting a man who is 75 or older. We note from the table that 8247 of the 21129 people are male and that 865 are both male and 75 or older. Therefore, $P(M) = \frac{8247}{21129} = 0.3903$, and

$P(MG) = \frac{865}{21129} = 0.0409$. By the conditional rule for probability, if a man is chosen, the probability that he is 75 or older is $P(G|M) = \frac{P(GM)}{P(M)} = \frac{0.0409}{0.3903} = 0.105$.

4.99 Let F denote the event that a student will pass the final examination and C the event that a student will pass the course. We are given P(F) = 0.80 and P(C | F) = 0.95. The probability that a student will pass both the final exam and the course is: P(FC) = P(F)·P(C | F) = (0.80)·(0.95) = 0.76.

4.101 Let F denote the event that a student is a freshman, and T denote the event that the student utilized the services of the college's tutoring center. We are given P(F) = 0.31, P(T) = 0.16 and P(F | T) = 0.58. Therefore, P(TF) = P(T)· P(F | T) = (0.16)·(0.58) = 0.0928. For last semester's freshmen, the percentage that used the tutoring center was $P(T|F) = \frac{P(TF)}{P(F)} = \frac{0.0928}{0.31} = 0.299$ = 29.9%.

EXERCISES for Sections 4.5 and 4.6

The Complement Rule of Probability
Combining the Rules of Probability

4.103 The birth of a wire-haired fox terrier is a female.

4.105 At least one component in an optical scanner does not function properly.

4.107 All seven department faculty are married.

4.109 All 10 in a class of 10 took the exam.

4.111 Fewer than eight cards in a 13-card bridge hand are red.

4.113 A fair die is tossed twice. P(at least one ace) = 1 - P(none) = $1 - (\frac{5}{6})^2 = 1 - \frac{25}{36} = \frac{11}{36}$.

4.115 Ninety percent of 1988 domestic cars were equipped with air conditioning. If four are randomly selected, P(at least one has air conditioning) = 1 - P(none) = 1 - P(0) = $1 - (0.10)^4$ = 1 - (0.0001) = 0.9999.

4.117 If two cards are randomly selected from a 52-card deck, the probability of obtaining an ace (A) and a king (K) in any order is: p = P(AK) + P(KA) = P(A)·P(K | A) + P(K)·P(A | K) = $\frac{4}{52}\cdot\frac{4}{51} + \frac{4}{52}\cdot\frac{4}{51} = 0.012$.

4.119 **a.** P(at least one will make a hole-in-one) = 1 - P(0) =

$1 - \left(\frac{3749}{3750}\right)^4 = 1 - 0.99893 = 0.00107.$

b. P(all four will make a hole-in-one) = $\left(\frac{1}{3750}\right)^4 = 5.06 \times 10^{-15}$.

4.121 We are given: P(A) = 0.55, P(B) = 0.60, P(AB) = 0.25.

a. $P(\bar{A}) = 1 - P(A) = 1 - 0.55 = 0.45.$

b. P(A or B) = P(A) + P(B) - P(AB) = 0.55 + 0.60 - 0.25 = 0.90.

c. We note from the diagram $P(A\bar{B}) = 0.30.$

d. $P(A \text{ or } \bar{B}) = P(A) + P(\bar{B}) - P(A\bar{B}) = 0.55 + 0.40 - 0.30 = 0.65.$

e. $P(A|B) = \frac{P(AB)}{P(B)} = \frac{0.25}{0.60} = 0.417.$

f. $P(A|\bar{B}) = \frac{P(A\bar{B})}{P(\bar{B})} = \frac{0.30}{0.40} = 0.75.$

4.123 Let S denote the event that a college senior achieves a satisfactory score on the Medical College Admission Test and let E be the event that he will prepare for the test. We are given $P(S \mid E) = 0.70$, $P(S \mid \bar{E}) = 0.20$ and $P(E) = 0.75$. We wish to find P(S) where P(S) $= P(ES \text{ or } \bar{E}S) = P(ES) + P(\bar{E}S) = P(E) \cdot P(S \mid E) + P(\bar{E}) \cdot P(S \mid \bar{E}) = (0.75) \cdot (0.70) + (0.25) \cdot (0.20) = 0.575$. See the tree diagram below.

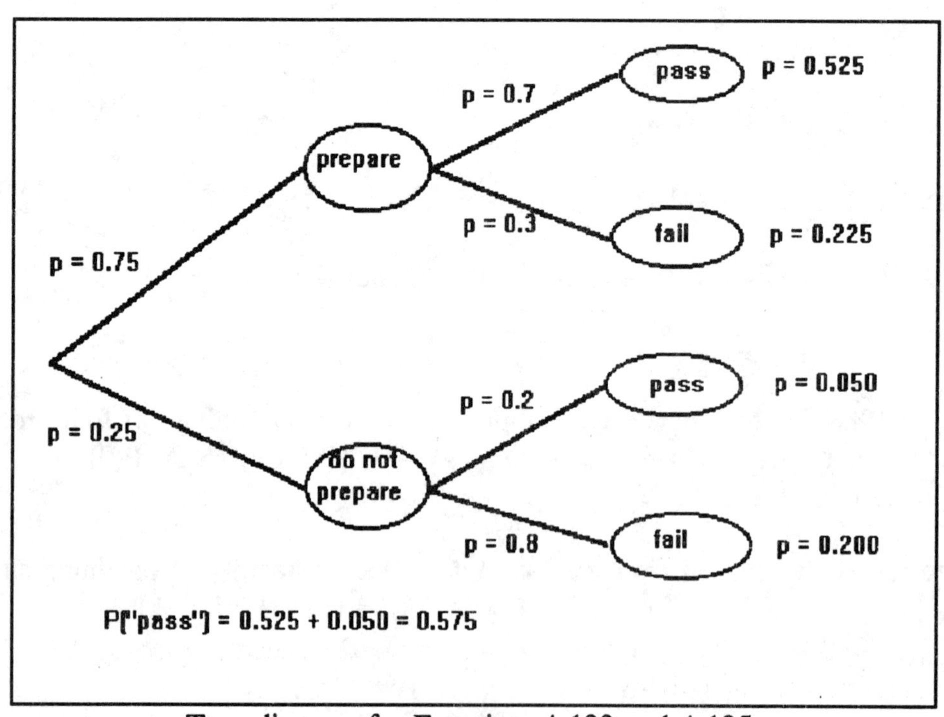

Tree diagram for Exercises 4.123 and 4.125.

4.125 Let S denote the event that a college senior achieves a satisfactory score on the Medical College Admission Test and let E be the event that he will prepare for the test. We are given $P(S \mid E) = 0.70$, $P(S \mid \bar{E}) = 0.20$ and $P(E) = 0.75$. In problem 4.123, we found that $P(S) = 0.575$. We wish to find $P(E \mid S)$. Using the conditional rule for probability

$$P(E|S) = \frac{P(ES)}{P(S)} = \frac{P(E)\cdot P(S|E)}{P(S)} = \frac{0.75\cdot 0.70}{0.575} = 0.913.$$

REVIEW EXERCISES
CHAPTER 4

4.127 A and B are mutually exclusive events for which $P(A) = 0.2$ and $P(B) = 0.4$. We apply the addition rule for mutually exclusive events to find:
$P(A \text{ or } B) = P(A) + P(B) = 0.2 + 0.4 = 0.6$.

4.128 A and B are independent events for which $P(A) = 0.2$ and $P(B) = 0.4$. By the multiplication rule for independent events $P(AB) = P(A)\cdot P(B) = (0.2)\cdot(0.4) = 0.08$. Using the addition rule for probability, we find: $P(A \text{ or } B) = P(A) + P(B) - P(AB) = 0.2 + 0.4 - 0.08 = 0.52$.

4.129 **a.** $C(5,3) = \frac{5!}{3!\cdot 2!} = \frac{5\cdot 4\cdot 3!}{3!\cdot 2\cdot 1} = 10.$

b. $C(15,13) = \frac{15!}{13!\cdot 2!} = \frac{15\cdot 14\cdot 13!}{13!\cdot 2\cdot 1} = 105.$

c. $C(15,2) = \frac{15!}{2!\cdot 13!} = \frac{15\cdot 14\cdot 13!}{2\cdot 1\cdot 13!} = 105.$

d. $C(1000,999) = \frac{1000!}{999!\cdot 1!} = \frac{1000\cdot 999!}{999!\cdot 1} = 1000.$

4.130 A super combo consists of any choice of five from twelve ingredients. Since the ingredient order is not important, one can order a super combo in $C(12,5)$ ways, where

$$C(12,5) = \frac{12!}{5!\cdot 7!} = 792.$$

4.131 We define the event L to mean that an adult likes liver. Using the relative frequency definition, we calculate $P(L) \approx \frac{204}{586} = 0.348.$

4.132 Events A and B are such that $P(A) = 0.83$, $P(B) = 0.56$, and $P(AB) = 0.49$. Therefore,

$$P(B|A) = \frac{P(BA)}{P(A)} = \frac{0.49}{0.83} = 0.590 \text{ and } P(A|B) = \frac{P(AB)}{P(B)} = \frac{0.49}{0.56} = 0.875.$$

4.133 $P(A \mid B) = 0.875 \neq 0.83 = P(A)$, so the events are not independent. Also, $P(AB) = 0.49 \neq 0$, so the events are not mutually exclusive.

4.134 Since identical twins constitute one-third of all twin births, the probability that a set of twins is identical is ⅓. If the maternity ward of a hospital has two sets of twins, then

a. P(both sets are identical twins) $= (\frac{1}{3})^2 = \frac{1}{9}$

b. P(at least one set is identical twins) = 1 - P(none) $= 1 - (\frac{2}{3})^2 = 1 - \frac{4}{9} = \frac{5}{9}$.

4.135 The probability that a space shuttle will receive a potentially dangerous impact is $\frac{1}{30}$. For the next four space shuttles, P(at least one will receive a potentially dangerous impact) = 1 - P(none) $= 1 - (\frac{29}{30})^4 = 1 - 0.873 = 0.127$.

4.136 Let S denote shorefront and $\bar{S}$ denote off-shore lot; possible sizes are 1, 2, or 5 acres. The following is a sample space to show the possible selections: S1, S2, S5, $\bar{S}1$, $\bar{S}2$, $\bar{S}5$.

4.137 **a.** The events of a company experiencing a decrease in sales for 1992 and the company enjoying an increase in profits for 1992 could occur together. Therefore, the events are not mutually exclusive.
b. In 10 tosses of a coin, the events of obtaining 6 heads and obtaining more than 4 heads could occur together. Therefore, the events are not mutually exclusive.
c. In 10 tosses of a coin, the events of obtaining 6 heads and obtaining fewer than 4 heads could not occur together. Therefore, the events are mutually exclusive.
d. The events of a family purchasing an RCA television and purchasing a Toshiba video cassette recorder could occur together. Therefore, the events are not mutually exclusive.

4.138 By the multiplication rule for counting, the number of different arrangements in which the four trees can be planted is $4 \cdot 3 \cdot 2 \cdot 1 = 4! = 24$.

4.139 We are given that P(A) = 0.35, P(B) = 0.45 and P(AB) = 0.20.

a. P(A or B) = P(A) + P(B) - P(AB) = 0.35 + 0.45 - 0.20 = 0.60.

b. We see from the Venn diagram that $P(\bar{A}B) = 0.25$.

c. $P(\bar{A}$ or $B) = P(\bar{A}) + P(B) - P(\bar{A}B) = 0.65 + 0.45 - 0.25 = 0.85$.

d. $P(\bar{A}|B) = \frac{P(\bar{A}B)}{P(B)} = \frac{0.25}{0.45} = 0.556$.

4.140 By the multiplication rule for counting, the number of different ways the three couples can be seated if there are no restrictions on their placement is $6 \cdot 5 \cdot 4 \cdot 3 \cdot 2 \cdot 1 = 6! = 720$.

4.141 By the multiplication rule for counting, the number of different ways the three couples can be seated if each couple must sit together is $6 \cdot 1 \cdot 4 \cdot 1 \cdot 2 \cdot 1 = 48$.

Review Exercises

4.142 Wade Boggs reached base in 43.0% of his plate appearances. For a game in which he was at bat five times: P(he reached base at least once) = 1 - P(none) = 1 - $(0.57)^5$ = 1 - 0.06 = 0.94.

4.143 A buyer can choose from four ranges, six refrigerators, two dishwashers, and five microwaves. By the multiplication rule, the number of different ways the buyer can select these appliances is $4 \cdot 6 \cdot 2 \cdot 5 = 240$.

4.144 **a.** The event of a Florida State student receiving an A in statistics is probably not related to his girlfriend at the University of Alaska receiving an A in statistics. Therefore, the events are independent.
b. The events of two close friends in the same statistics class each receiving an A for the course are probably related, so the events are not independent.
c. The probability of the event of obtaining a black card is ½, which is also the probability of the event of obtaining a black card if the card is known to be an ace. Therefore, the events are independent.

4.145 Since the order of selection does not matter, the number of ways two pups can be selected from a litter of eight is given by C(8,2) = 28.

4.146 We choose two of the 8 retrievers in C(8,2) = 28 ways, and two of the 10 shepherds in C(10,2) = 45 ways. By the multiplication rule the total number of ways the four pups can be chosen is C(8,2)·C(10,2) = 28·45 = 1260.

4.147 The investment club consists of 11 men and 9 women and five members will be selected at random from the 20.
a. The total number of ways the sample can be selected is C(20,5) = 15504.
b. In order to find the number of ways the sample can be selected so that it consists of three men and two women, we note that the three men can be selected in C(11,3) = 165 ways, and the two women in C(9,2) = 36 ways. By the multiplication rule, the total number of such samples is 165·36 = 5940.
c. Let E denote the event that the sample of five will consist of three men and two women. Then by the theoretical method of assigning probabilities:

$$P(E) = \frac{e}{s} = \frac{5940}{15504} = 0.383.$$

4.148 The component will be activated unless both switches fail, where each switch has a probability of 0.002 of failing. By the complement rule of probability, P(component will be activated) = 1 - P(both switches fail) = 1 - $(0.002)^2$ = 0.999996.

4.149 We wish to find the probability that exactly one kid of three will request a Nintendo game pak, where the probability of each requesting a pak is 0.50.
P(exactly one) = P(first only) + P(second only) + P(third only)
= (0.5)·(0.5)·(0.5) + (0.5)·(0.5)·(0.5) + (0.5)·(0.5)·(0.5) = 0.375.

4.150 **a.** There are a total of 48 + 42 = 90 defective boats among the 400. If D is the event that a boat is defective, then by the theoretical method of assigning probabilities:

$$P(D) = \frac{e}{s} = \frac{90}{400} = 0.225.$$

b. Let F denote the event that the boat is a four person boat. We note that

$P(DF) = \frac{42}{400} = 0.105$, and that $P(F) = \frac{150}{400} = 0.375$.

Therefore, $P(D|F) = \frac{P(DF)}{P(F)} = \frac{0.105}{0.375} = 0.28$.

4.151 Let T denote the event that the boat in Exercise 4.150 is a 2-person boat and D the event that it is defective. Then by the theoretical method of assigning probabilities:

$$P(D) = \frac{e}{s} = \frac{90}{400} = 0.225, \quad P(T) = \frac{e}{s} = \frac{250}{400} = 0.625,$$

$$P(DT) = \frac{48}{400} = 0.12.$$

Using the addition rule, we find the probability that it will either be a 2-person boat or defective: P(T or D) = P(T) + P(D) - P(DT) = 0.625 + 0.225 - 0.12 = 0.73.

4.152 P(at least one) = 1 - P(none) = 1 - $(0.40)^3$ = 1 - (0.064) = 0.936.

4.153 Thirty percent of all residential students at a university have a refrigerator in their room. Four students are randomly chosen.

a. By the multiplication rule for independent events, P(all have a refrigerator in their room) = $(0.30)^4$ = 0.0081.

b. P(at least one has a refrigerator) = 1 - P(none) = 1 - $(0.70)^4$ = 1 - (0.2401) = 0.7599.

c. P(fewer than four have a refrigerator) = 1 - P(all four have a refrigerator) = 1 - $(0.30)^4$ = 1 - (0.0081) = 0.9919.

4.154 Define M to be the event that an employee is in management and S the event that he/she is a participant in the company's stock purchase plan. We are given P(M) = 0.20, P(S | M) = 0.88. We wish to find the probability that an employee at this company is in management and participates in the stock purchase plan, which is:

P(MS) = P(M) $\cdot$P(S | M) = (0.20)$\cdot$(0.88) = 0.176.

4.155 Define S to be the event that a car entering an intersection continues straight and R the event the car turns right. Further, define C to be the event the car will have a collision. We are given P(S) = 0.80, P(R) = 0.20, P(C | S) = 0.0004, P(C | R) = 0.0036. Note that C = SC or RC, with SC and RC mutually exclusive. Applying the addition and multiplication rules we have: P(C) = P(SC or RC) = P(SC) + P(RC) = P(S)$\cdot$P(C | S) +P(R)$\cdot$P(C | R) = (0.80)$\cdot$(0.0004) + (0.20)$\cdot$(0.0036) = 0.00104.

4.156 Let A denote the event the instructor is given promotion and T the event that she receives tenure. We are given P(T) = 0.80, P(AT) = 0.60 and wish to find the probability she assigns to being promoted if she is granted tenure. By the conditional rule of probability,

$$P(A|T) = \frac{P(AT)}{P(T)} = \frac{0.60}{0.80} =0.75.$$

4.157 Let A denote the event that the bottle was filled by machine A, and similarly define events B and C. Also let U denote the event that the bottle is underfilled. We note that U = AU or BU or CU where AU, BU, and CU are mutually exclusive.
Applying the addition and multiplication rules we have: (see the diagram)
$P(U) = P(A) \cdot P(U \mid A) + P(B) \cdot P(U \mid B) + P(C) \cdot P(U \mid C) =$
$(0.60) \cdot (0.01) + (0.30) \cdot (0.02) + (0.10) \cdot (0.05) = 0.006 + 0.006 + 0.005 = 0.017.$

4.158 We use the conditional rule for probability to calculate the probability that a bottle of soda, as described in exercise 4.157, was filled by machine B, if purchased and found to be underfilled. We found P(U) = 0.017, and note that $P(BU) = P(B) \cdot P(U \mid B) =$ $(0.30) \cdot (0.02) = 0.006$. Therefore, $P(B|U) = \frac{P(BU)}{P(U)} = \frac{0.006}{0.017} = 0.353.$

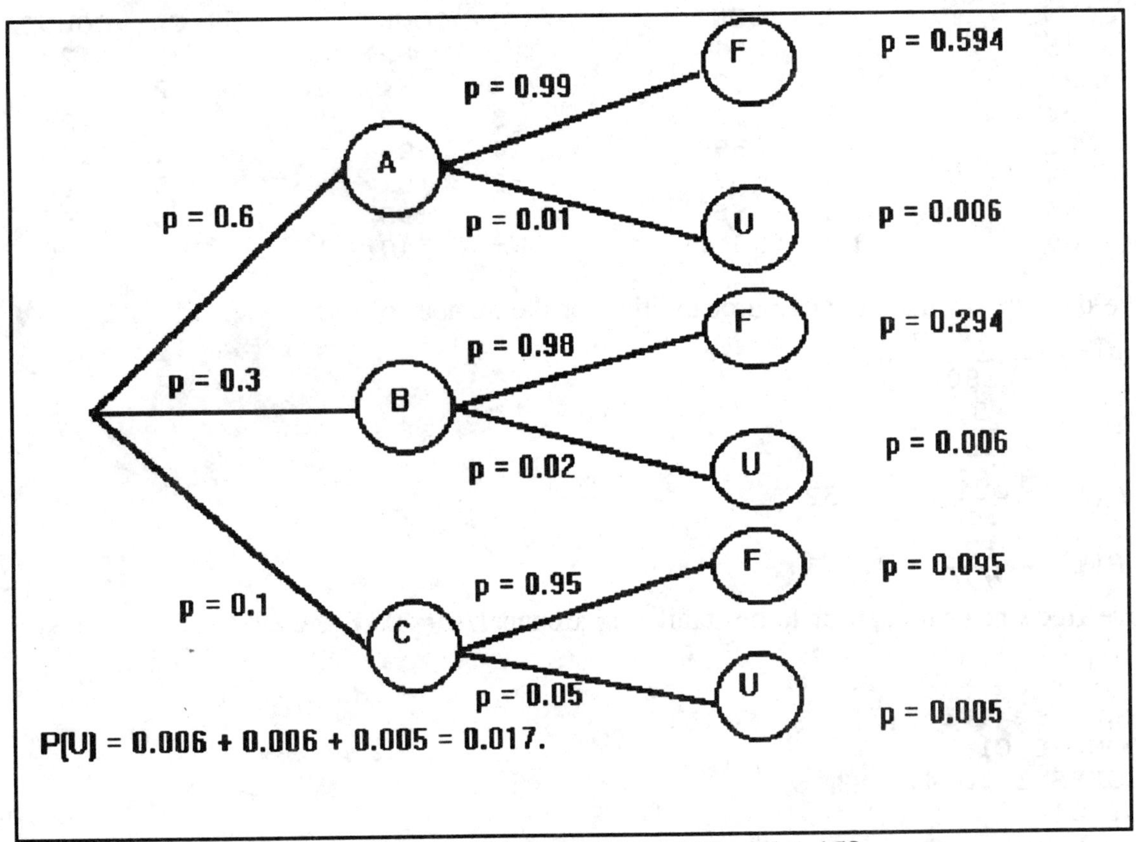

Tree diagram for Exercises 4.157 and 4.158

4.159 P(at least two people in a group of 23 have the same birthday) = 1 - P(all people in a group of 23 have different birthdays) $= 1 - \frac{365}{365} \cdot \frac{364}{365} \cdot \frac{363}{365} \cdots \frac{344}{365} \cdot \frac{343}{365} = 0.507.$

4.160 P(at least two people in a group of 15 have the same birthday) = 1 - P(all people in a group of 15 have different birthdays) $= 1 - \frac{365}{365} \cdot \frac{364}{365} \cdot \frac{363}{365} \cdots \frac{352}{365} \cdot \frac{351}{365} = 0.253.$

MINITAB LAB ASSIGNMENTS

(Results of the following problems will vary)

4.161(M)

```
MTB > # Exercise 4.161(M)
MTB > RANDOM 180 C1; # part a
SUBC> INTEGERS 1 to 6.
MTB > RANDOM 360 C2; # part b
SUBC> INTEGERS 1 to 6.
MTB > RANDOM 540 C3; # part c
SUBC> INTEGERS 1 to 6.
MTB > RANDOM 720 C4; # part d
SUBC> INTEGERS 1 to 6.
MTB > TALLY C1-C4 # part e
```

C1	COUNT	C2	COUNT	C3	COUNT	C4	COUNT
1	35	1	51	1	82	1	123
2	26	2	57	2	84	2	135
3	24	3	61	3	82	3	125
4	21	4	59	4	96	4	120
5	28	5	67	5	101	5	120
6	46	6	65	6	95	6	97
N=	180	N=	360	N=	540	N=	720

From the above data we calculate relative probabilities for the number of ones:

From part a: $P(1) = \frac{35}{180} = 0.194.$

From part b: $P(1) = \frac{51}{360} = 0.142.$

From part c: $P(1) = \frac{82}{540} = 0.152.$

From part d: $P(1) = \frac{123}{720} = 0.171.$

Yes, the relative frequencies do appear to be stabilizing around 1/6 = 0.167.

4.162(M)

```
MTB > # Exercise 4.162(M)
MTB > RANDOM 30 C1;
SUBC> INTEGERS 1 to 4.
MTB > TALLY C1
```

C1	COUNT
1	14
2	5
3	8
4	3
N=	30

Note : In the above 1 = "club"; 2 = "diamond"; 3 = "heart"; 4 = "spade".

4.163(M)

```
MTB > # Exercise 4.163(M)
MTB > RANDOM 40 C1;
SUBC> INTEGERS 1 to 3.
MTB > TALLY C1

      C1  COUNT
       1     13
       2     15
       3     12
      N=     40
```

4.164(M)

```
MTB > # Exercise 4.164(M)
MTB > SET C1
DATA> 1:10
DATA> END
MTB > SAMPLE 3 C1 C2
MTB > NAME C2 'Sample'
MTB > PRINT C2

Sample
   6    4    9
```

4.165(M)

```
MTB > # Exercise 4.165(M)
MTB > SET C1
DATA> 1:64
DATA> END
MTB > SAMPLE 32 C1 C2
MTB > NAME C2 'Sample'
MTB > PRINT C2

Sample
     8    41    23    53    40    18    35    54    19    63    29    32
    22    36    12    42     7    31    57    50    30     4    20    21
    43    10    51    58    25    39    28     6
```

CHAPTER 5
RANDOM VARIABLES AND THEIR DISTRIBUTIONS

EXERCISES for Section 5.1
Random Variables

5.1 Since weight is a measurement, the random variable is continuous.

5.3 Since length is a measurement, the random variable is continuous.

5.5 Since a college student's grade is a count of the number of points earned, the random variable is discrete.

5.7 Since the depth of snow is a measurement, the random variable is continuous.

5.9 Since temperature is a measurement, the random variable is continuous.

5.11 Since the volume of ice is a measurement, the random variable is continuous.

5.13 Since travel time is a measurement, the random variable is continuous.

5.15 Since the number of stories in a building is a count, the random variable is discrete.

5.17 Since the number of hairs is a count, the random variable is discrete.

5.19 Since the width of a laptop computer is a measurement, the random variable is continuous.

5.21 Since the distance traveled by the hiker is a measurement, the random variable is continuous.

5.23 Since weight is a measurement, the random variable is continuous.

5.25 Since the retail price of a wedge of cheddar cheese is a count of the number of cents required to buy it, the random variable is discrete.

5.27 Since the number of display days is a count, the random variable is discrete.

5.29 The random variable x is discrete and is the number of copies sold; x is a number from the set $\{0,1,\ldots, 725{,}000\}$.

5.31 The random variable x is continuous and is the length in miles of the coastline damaged by the oil spill; $0 \leq x \leq 13$.

5.33 The random variable x is discrete and is the number of Spanish I students who enroll in Spanish II; x is a number from the set $\{0,1,\ldots,97\}$.

5.35 The random variable x is discrete and is the number of switch activations until it fails; x is a number from the set $\{1,2,3,\ldots\}$.

EXERCISES for Section 5.2
Discrete Probability Distributions

5.37 Since $\Sigma P(x) = 1$, we calculate: $P(4) = 1 - \{P(3) + P(5) + P(6) + P(7)\} = 1 - (0.04 + 0.16 + 0.38 + 0.15) = 1 - 0.73 = 0.27$.

5.39 Probability spike graph:

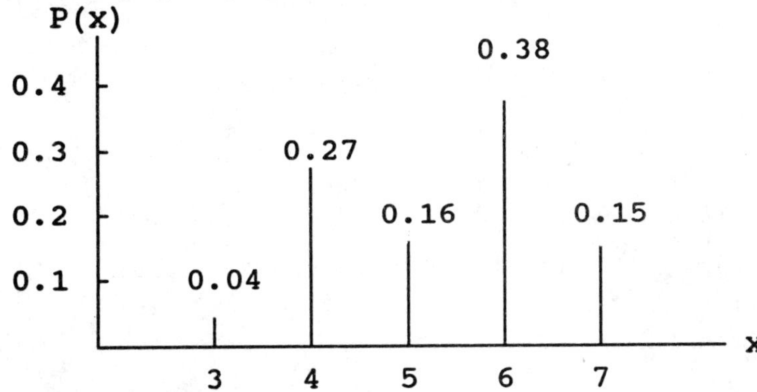

5.41 Probability Histogram:

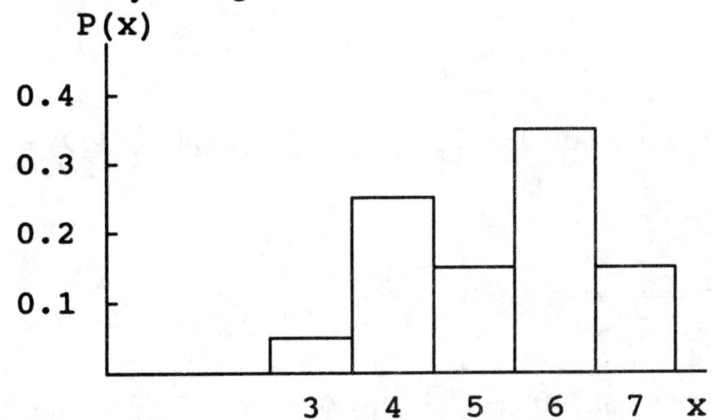

5.43 Yes, this is a valid probability distribution; we note that $\Sigma P(x) = 1$ and $0 \leq P(x) \leq 1$ for all x.

5.45 Converted to a table we have:

x	-1	0	1	2
P(x)	16/30	9/30	4/30	1/30

Yes, this is a valid probability distribution; we note that $\Sigma P(x) = 1$ and $0 \le P(x) \le 1$ for all x. The spike graph display:

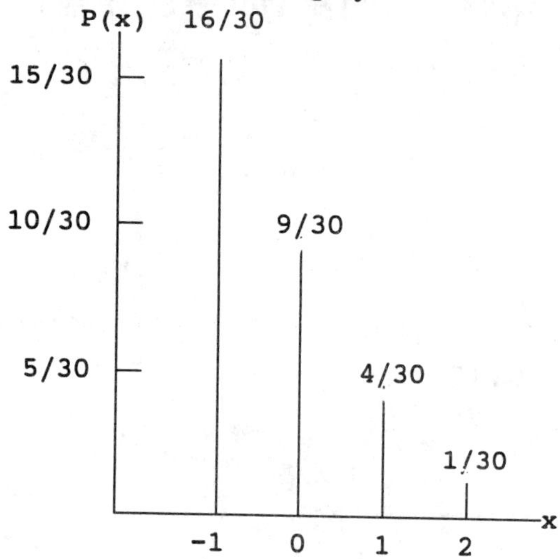

5.47 $P(0) = C(2,0)(0.7)^0(0.3)^2 = 0.09$
$P(1) = C(2,1)(0.7)^1(0.3)^1 = 0.42$
$P(2) = C(2,2)(0.7)^2(0.3)^0 = 0.49$
This is a valid probability distribution; we note that $\Sigma P(x) = P(0) + P(1) + P(2) = 1$, and $0 \le P(x) \le 1$ for all x.
Converted to a table we have:

x	0	1	2
P(x)	0.09	0.42	0.49

5.49 **a.** $P(x \le 1) = P(0) + P(1)$. We note that $P(0) = C(3,0)(0.4)^0(0.6)^3 = 0.216$ and that $P(1) = C(3,1)(0.4)^1(0.6)^2 = 0.432$. Therefore, $P(x \le 1) = 0.216 + 0.432 = 0.648$.
b. $P(x \ge 2) = P(2) + P(3)$. We note that $P(2) = C(3,2)(0.4)^2(0.6)^1 = 0.288$ and that $P(3) = C(3,3)(0.4)^3(0.6)^0 = 0.064$. Therefore, $P(x \ge 2) = 0.288 + 0.064 = 0.352$.

5.51 **a.** The following sample space shows the different ways that a die and a quarter can fall:
{1H, 2H, 3H, 4H, 5H, 6H, 1T, 2T, 3T, 4T, 5T, 6T}
b. Since there are twelve equally likely sample points, we see that the probability of each sample point is 1/12.
c. To each sample point we assign the random variable x, which equals the sum of the number of spots and the number of heads that show:

	1H	2H	3H	4H	5H	6H	1T	2T	3T	4T	5T	6T
	↓	↓	↓	↓	↓	↓	↓	↓	↓	↓	↓	↓
x:	2	3	4	5	6	7	1	2	3	4	5	6

d. We multiply by 1/12 the number of times each value of the random variable occurs in order to obtain the following probability distribution:

x	1	2	3	4	5	6	7
P(x)	1/12	2/12	2/12	2/12	2/12	2/12	1/12

5.53 We apply the special multiplication rule and the rule of complements to obtain the following:
P(0) = (0.18)(0.18) = 0.0324
P(2) = (0.82)(0.82) = 0.6724
P(1) = 1 - [P(0) + P(2)] = 1 - [0.0324 + 0.6724] = 1 - 0.7048 = 0.2952.
Therefore, the probability distribution for x is:

x	0	1	2
P(x)	0.0324	0.2952	0.6724

Note: This problem may also be done by using a tree diagram.

5.55 We apply the general multiplication rule and the rule of complements to obtain the following:
P(0) = (2/5)(1/4) = 0.1
P(2) = (3/5)(2/4) = 0.3.
P(1) = 1 - P(0) - P(2) = 1 - 0.1 - 0.3 = 0.6
Therefore, the probability distribution for x is:

x	0	1	2
P(x)	0.1	0.6	0.3

Note: This problem may also be done by using a tree diagram.

MINITAB LAB ASSIGNMENTS

NOTE: The answers to the following MINITAB problems will vary.

5.57(M)

```
MTB > # Exercise 5.57(M)
MTB > READ C1 C2
DATA> 0 0.34
DATA> 1 0.31
DATA> 2 0.21
DATA> 3 0.11
DATA> 4 0.03
DATA> END
      5 ROWS READ
MTB > RANDOM 162 C3;
SUBC> DISCRETE C1 C2.
MTB > HISTOGRAM C3

Histogram of C3   N = 162
Each * represents 2 obs.

Midpoint   Count
       0      49   *************************
       1      54   ***************************
       2      33   *****************
       3      22   ***********
       4       4   **

MTB > SUM C3
   SUM      =       202.00
```

5.59(M)

```
MTB > # Exercise 5.59(M)
MTB > READ C1 C2
DATA> 1 0.67
DATA> 2 0.33
DATA> END
      2 ROWS READ
MTB > RANDOM 600 C3;
SUBC> DISCRETE C1 C2.
MTB > HISTOGRAM C3

Histogram of C3   N = 600
Each * represents 10 obs.

Midpoint   Count
       1     410  *****************************************
       2     190  *******************

MTB > # Therefore, we have 410 heads and 190 tails.
```

EXERCISES for Sections 5.3 and 5.4

The Mean of a Discrete Random Variable
Measuring the Variability of a Random Variable

5.61 We construct a table:

x	P(x)	xP(x)	$x^2P(x)$
0	0.2	0.0	0.0
1	0.1	0.1	0.1
2	0.1	0.2	0.4
3	0.4	1.2	3.6
4	0.2	0.8	3.2
		μ = 2.3	$\Sigma x^2P(x)$ = 7.3

$\sigma^2 = \Sigma x^2P(x) - \mu^2 = 7.3 - (2.3)^2 = 2.01.$

The standard deviation of x is $\sigma = \sqrt{2.01} = 1.42.$

5.63 We construct a table:

x	P(x)	xP(x)	$x^2P(x)$
-4	0.15	-0.60	2.40
-2	0.21	-0.42	0.84
0	0.34	0.00	0.00
4	0.18	0.72	2.88
8	0.12	0.96	7.68
		μ = 0.66	$\Sigma x^2P(x)$ = 13.80

$\sigma^2 = \Sigma x^2P(x) - \mu^2 = 13.8 - (0.66)^2 = 13.3644.$

The standard deviation of x is $\sigma = \sqrt{13.3644} = 3.66.$

5.65 We construct a table:

x	P(x)	xP(x)		$x^2P(x)$
2	5/32	10/32		20/32
3	10/32	30/32		90/32
4	17/32	68/32		272/32
		μ = 108/32	$\Sigma x^2 P(x)$	= 382/32
		= 27/8		= 191/16

$\sigma^2 = \Sigma x^2P(x) - \mu^2 = 191/16 - (27/8)^2 = 35/64.$

The standard deviation of x is $\sigma = \sqrt{\frac{35}{64}}$ = **0.74.**

5.67 We construct a table:

x	P(x)	xP(x)		$x^2P(x)$
0	$C(2,0)(0.7)^0(0.3)^2 = 0.09$	0.00		0.00
1	$C(2,1)(0.7)^1(0.3)^1 = 0.42$	0.42		0.42
2	$C(2,2)(0.7)^2(0.3)^0 = 0.49$	0.98		1.96
		μ = 1.40	$\Sigma x^2 P(x)$	= 2.38

The variance is $\sigma^2 = \Sigma x^2P(x) - \mu^2 = 2.38 - (1.40)^2 = 0.42$, so we see that the standard deviation of x is $\sigma = \sqrt{0.42}$ = **0.65.**

5.69 We read the values of P(x) from the probability spike graph and construct a table:

x	P(x)	xP(x)	x - μ	$(x - \mu)^2$	$(x - \mu)^2P(x)$
2	0.1	0.2	-4	16	1.6
4	0.2	0.8	-2	4	0.8
6	0.4	2.4	0	0	0.0
8	0.2	1.6	2	4	0.8
10	0.1	1.0	4	16	1.6
		μ = 6.0			σ^2 = 4.8

The mean of x is $\mu = \Sigma xP(x) = 6.0$, the variance is $\sigma^2 = \Sigma(x - \mu)^2P(x) = 4.8$, and the standard deviation is $\sigma = \sqrt{4.8}$ = **2.19.**

5.71 Using the results from problem 5.69, we calculate: $\mu - 2\sigma = 6 - 2(2.19) = 1.62$ and $\mu + 2\sigma = 6 + 2(2.19) = 10.38$. Thus, $P(1.62 \leq x \leq 10.38) = P(2) + P(4) + P(6) + P(8) + P(10) = 1$.

5.73 Over the long run, the average number of scoops per cone purchased will be the population mean: $\mu = \Sigma xP(x) = 1(0.50) + 2(0.28) + 3(0.17) + 4(0.05) = 1.77$.

5.75 Using the shortcut formula we first find the variance: $\sigma^2 = \Sigma x^2P(x) - \mu^2 = [1^2(0.50) + 2^2(0.28) + 3^2(0.17) + 4^2(0.05)] - (1.77)^2 = 0.8171$. Therefore, the standard deviation is:

$\sigma = \sqrt{0.8171}$ = **0.90.**

5.77 The expected loss is the mean μ = 15,000(0.001) + 5,000(0.02) = 15 + 100 = \$115. In order to make a \$30 profit, the insurance company must charge \$115 + \$30 = \$145.

MINITAB LAB ASSIGNMENTS

5.79(M)

```
MTB > # Exercise 5.79(M)
MTB > SET C1
DATA> 2:78/2
DATA> END
MTB > LET C2 = C1/1560
MTB > LET C3= C1*C2
MTB > SUM C3                    # THIS IS THE MEAN OF X
    SUM     =      52.667
```

5.81(M)

```
MTB > # Exercise 5.81(M)
MTB > READ C1 C2
DATA> 54 0.03
DATA> 56 0.11
DATA> 65 0.15
DATA> 66 0.08
DATA> 69 0.18
DATA> 74 0.21
DATA> 76 0.02
DATA> 79 0.09
DATA> 81 0.01
DATA> 87 0.05
DATA> 98 0.07
DATA> END
      11 ROWS READ
MTB > LET C3 = C1*C2
MTB > LET C4 = (C1**2)*C2
MTB > LET K1 = SUM (C3)
MTB > PRINT K1 #MEAN
K1         71.4200
MTB > LET K2 = SUM(C4) - K1**2
MTB > PRINT K2 #VARIANCE
K2       114.344
```

REVIEW EXERCISES
CHAPTER 5

5.83 Over the long run, the mean number of deaths per ten lobsters is the population mean: μ = ΣxP(x) = 0(0.48) + 1(0.28) + 2(0.14) + 3(0.06) + 4(0.03) + 5(0.01) = 0.91.

5.84 **a.** The random variable of likely interest is the number of monitors in the lot of 100 which operate satisfactorily.
b. Since the random variable is a count, it is discrete.

c. The range of possible values for the random variable is the set $\{0,1,\ldots,100\}$.

5.85 **a.** The random variable of likely interest is x, the amount of beverage in the cup.
b. Since the amount of beverage in the cup is a measurement, the random variable is continuous.
c. The range of possible values for the random variable is the interval: $0 \leq x \leq 8$.

5.86 Since the velocity of a falling comet is a measurement, the random variable is continuous.

5.87 Since the number of outpatient visits during 1995 to a Veterans Administration clinic is a count, the random variable is discrete.

5.88 Since the number of tosses of a silver dollar until the first head is a count, the random variable is discrete.

5.89 Since the weight of pepperoni on a pizza is a measurement, the random variable is continuous.

5.90 Since the number of pieces of pepperoni on a pizza is a count, the random variable is discrete.

5.91 Since the down time of a mainframe computer is a measurement, the random variable is continuous.

5.92 Since the amount of fructose in a can of soda is a measurement, the random variable is continuous.

5.93 Since the systolic blood pressure of an athlete is a measurement, the random variable is continuous.

5.94 Since the volume of air inhaled is a measurement, the random variable is continuous.

5.95 Since the daily cost of stay at a hospital is a count (of the number of dollars), the random variable is discrete.

5.96 Since $\Sigma P(x) = 1$, we calculate: $P(18) = 1 - \{P(15) + P(16) + P(17) + P(19)\} = 1 - (0.11 + 0.23 + 0.09 + 0.13) = 1 - 0.56 = 0.44$.

5.97 This is a valid probability distribution since $\Sigma P(x) = 0.36 + 0.30 + 0.18 + 0.10 + 0.03 + 0.02 + 0.01 = 1$, and $0 \leq P(x) \leq 1$ for all x.

5.98 This is not a valid probability distribution since $\Sigma P(x) = 0.2 + 0.3 + 0.4 + 0.2 = 1.1$, which does not equal 1.

5.99 This is a valid probability distribution since $\Sigma P(x) = 1/15 + 2/15 + 3/15 + 4/15 + 5/15 = 1$ and $0 \leq P(x) \leq 1$ for all x.

5.100 Since each numerator is the square of the value of x, we conclude that $P(x) = \frac{x^2}{55}$ for $x=1,2,3,4,5$ is a formula which specifies the probability distribution.

5.101

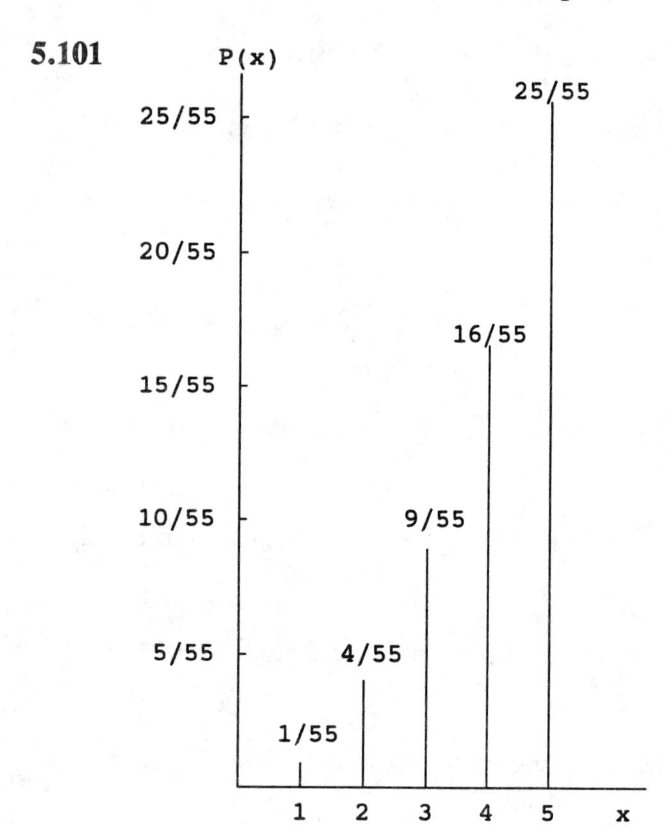

5.102

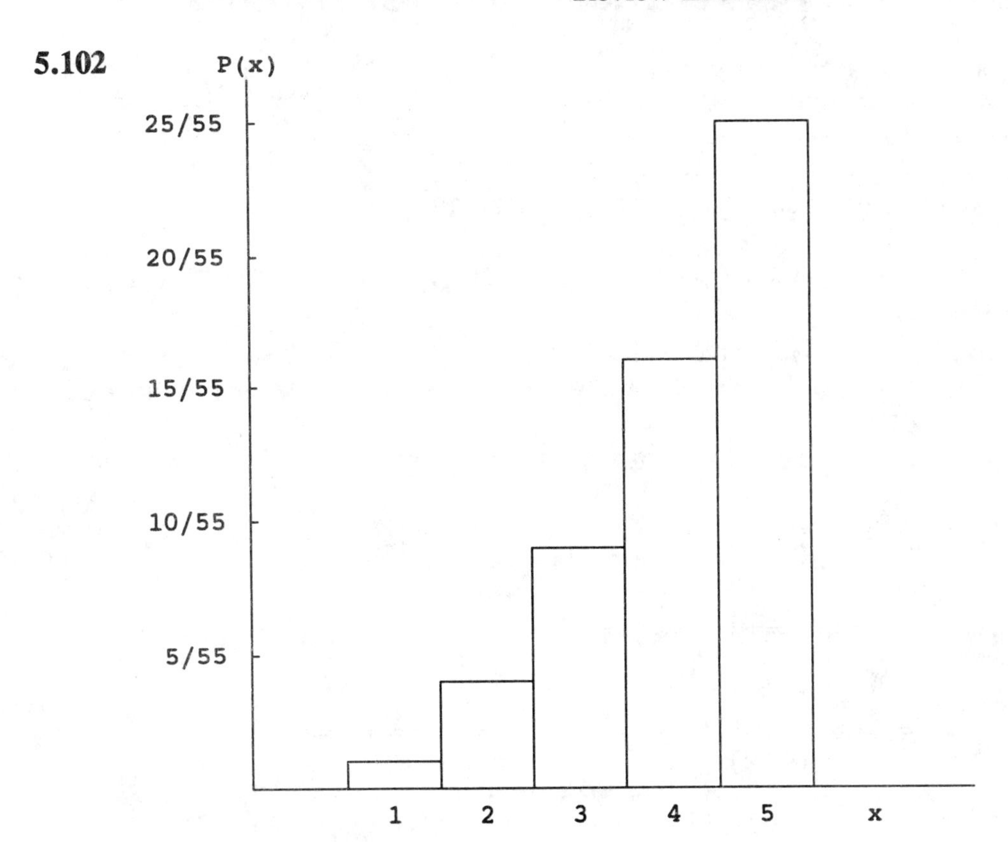

5.103 We calculate: $P(0) = C(4,0)(0.2)^0(0.8)^4 = 1 \cdot 1 \cdot (0.4096) = 0.4096$
$P(1) = C(4,1)(0.2)^1(0.8)^3 = 4 \cdot (0.2) \cdot (0.512) = 0.4096$
$P(2) = C(4,2)(0.2)^2(0.8)^2 = 6 \cdot (0.04) \cdot (0.64) = 0.1536$
$P(3) = C(4,3)(0.2)^3(0.8)^1 = 4 \cdot (0.008) \cdot (0.8) = 0.0256$
$P(4) = C(4,4)(0.2)^4(0.8)^0 = 1 \cdot (0.0016) \cdot 1 = 0.0016$
a. Therefore $P(x < 2) = P(0) + P(1) = 0.4096 + 0.4096 = 0.8192$.
b. $P(x \geq 3) = P(3) + P(4) = 0.0256 + 0.0016 = 0.0272$.

5.104 We construct a table:

x	P(x)	xP(x)	$x^2P(x)$
0	0.4096	0.0	0.0
1	0.4096	0.4096	0.4096
2	0.1536	0.3072	0.6144
3	0.0256	0.0768	0.2304
4	0.0016	0.0064	0.0256
		μ = 0.8000	$\Sigma x^2P(x)$ = 1.2800

$\sigma^2 = \Sigma x^2P(x) - \mu^2 = 1.28 - (0.8)^2 = 0.64$.
Therefore, the standard deviation of x is $\sigma = \sqrt{0.64} = 0.8$.

5.105 We apply the special multiplication rule and the rule of complements to obtain the following:
$P(0) = (¼)(¼) = 1/16$,
$P(2) = (¾)(¾) = 9/16$ and
$P(1) = 1 - [P(0) + P(2)] = 1 - [1/16 + 9/16] = 1 - 10/16 = 6/16$.

Therefore, the probability distribution for x is:

x	0	1	2
P(x)	1/16	6/16	9/16

Note: This problem may also be done by using a tree diagram.

5.106 **a.** The sample space contains the points: AA, AB, AC, AD, BA, BB, BC, BD, CA, CB, CC, CD, DA, DB, DC, DD.

b. Since there are 16 equally likely sample points, the probability of each is 1/16.

c. We assign each sample point a random variable value x, where x denotes the number of questions answered correctly:

AA,	AB,	AC,	AD,	BA,	BB,	BC,	BD,	CA,	CB,	CC,	CD,	DA,	DB,	DC,	DD
↓	↓	↓	↓	↓	↓	↓	↓	↓	↓	↓	↓	↓	↓	↓	↓
0	1	0	0	1	2	1	1	0	1	0	0	0	1	0	0

d. Therefore, the probability distribution of the random variable x is:

x	0	1	2
P(x)	9/16	6/16	1/16

e. We construct a table:

x	P(x)	xP(x)	$x^2P(x)$
0	9/16	0	0
1	6/16	6/16	6/16
2	1/16	2/16	4/16
		μ = 8/16 = 1/2	$\Sigma x^2 P(x)$ = 10/16 = 5/8.

$\sigma^2 = \Sigma x^2P(x) - \mu^2 = 5/8 - (1/2)^2 = 3/8$. Therefore, the mean of x is $\mu = ½$, and the standard deviation of x is $\sigma = \sqrt{\frac{3}{8}} = 0.61$.

5.107 We apply the special multiplication rule and the rule of complements to obtain the following:
P(0) = (0.2)(0.2)(0.2) = 0.008,
P(3) = (0.8)(0.8)(0.8) = 0.512,
P(1) = (0.8)(0.2)(0.2) + (0.2)(0.8)(0.2) + (0.2)(0.2)(0.8) = 0.096, and
P(2) = 1 - [P(0) + P(1) + P(3)] = 1 - [0.008 + 0.096 + 0.512] = 1 - 0.616 = 0.384.
Therefore, the probability distribution for x is:

x	0	1	2	3
P(x)	0.008	0.096	0.384	0.512

Note: This problem may also be done by using a tree diagram.

5.108 We construct a table:

x	P(x)	xP(x)		$x^2P(x)$
1	0.2	0.2		0.2
3	0.1	0.3		0.9
5	0.1	0.5		2.5
7	0.3	2.1		14.7
9	0.3	2.7		24.3
		μ = 5.8	$\Sigma x^2P(x)$	= 42.6

$\sigma^2 = \Sigma x^2P(x) - \mu^2 = 42.6 - (5.8)^2 = 8.96$. Therefore, the mean of x is $\mu = 5.8$, and the standard deviation of x is $\sigma = \sqrt{8.96} = 2.99$.

5.109 We construct a table:

x	P(x)	xP(x)		$x^2P(x)$
0	0.27	0.0		0.0
1	0.21	0.21		0.21
2	0.34	0.68		1.36
3	0.18	0.54		1.62
		μ = 1.43	$\Sigma x^2P(x)$	= 3.19

$\sigma^2 = \Sigma x^2P(x) - \mu^2 = 3.19 - (1.43)^2 = 1.1451$. Therefore, the standard deviation of x is $\sigma = \sqrt{1.1451} = 1.07$.

5.110 We read the probability values from the probability histogram and construct a table:

x	P(x)	xP(x)		$x^2P(x)$
2	0.1	0.2		0.4
3	0.1	0.3		0.9
4	0.3	1.2		4.8
5	0.3	1.5		7.5
6	0.1	0.6		3.6
7	0.1	0.7		4.9
		μ = 4.5	$\Sigma x^2P(x)$	= 22.1

$\sigma^2 = \Sigma x^2P(x) - \mu^2 = 22.1 - (4.5)^2 = 1.85$. Therefore, the mean of x is $\mu = 4.5$, and the standard deviation of x is $\sigma = \sqrt{1.85} = 1.36$.

5.111 $P(\mu - 2\sigma \leq x \leq \mu + 2\sigma) = P(4.5 - 2 \cdot 1.36 \leq x \leq 4.5 + 2 \cdot 1.36) =$
$P(1.78 \leq x \leq 7.22) = P(2) + P(3) + \ldots\ldots + P(7) = 1$.

5.112 We calculate the probability values from the probability distribution and construct a table:

x	P(x)	xP(x)	$x^2P(x)$
0	2/9	0	0
1	3/9	3/9	3/9
2	4/9	8/9	16/9
		μ = 11/9	$\Sigma x^2 P(x)$ = 19/9

$\sigma^2 = \Sigma x^2P(x) - \mu^2 = 19/9 - (11/9)^2 = 50/81$. Therefore, the mean of x is $\mu = 11/9$, and

the standard deviation of x is $\sigma = \sqrt{\dfrac{50}{81}} = 0.79.$

5.113 Over the long run, the average number of pecans per cake is the population mean $\mu = \Sigma xP(x) = (7)(0.03) + (8)(0.42) + (9)(0.46) + (10)(0.09) = 0.21 + 3.36 + 4.14 + 0.9 = 8.61$.

5.114 We first calculate $\Sigma x^2P(x) = (7)^2(0.03) + (8)^2(0.42) + (9)^2(0.46) + (10)^2(0.09) = 1.47 + 26.88 + 37.26 + 9 = 74.61$. From Problem 5.113, the mean is $\mu = 8.61$. Therefore, $\sigma^2 = \Sigma x^2P(x) - \mu^2 = 74.61 - (8.61)^2 = 0.4779$. Therefore, the standard deviation is

$\sigma = \sqrt{0.4779} = 0.69.$

5.115 The expected loss is (0.15)($30,000) + (0.05)($15,000) = $5,250. Since the insurance company will charge $1,000 more than its expected loss, the charge should be $1,000 + $5,250 = $6,250.

MINITAB LAB ASSIGNMENTS

5.116(M)

```
MTB > # Exercise 5.116(M)
MTB > SET C1
DATA> 1:6
DATA> END
MTB > SET C2
DATA> 2 1 2 1 2 1
DATA> END
MTB > LET C2 = C2/9
MTB > RANDOM 360 C3;
SUBC> DISC C1 C2.
MTB > HIST C3

Histogram of C3   N = 360
Each * represents 2 obs.

Midpoint   Count
       1      93  ***********************************************
       2      33  *****************
       3      79  ****************************************
       4      46  ***********************
       5      68  **********************************
       6      41  *********************
```

```
MTB > # Results of this problem will vary.
```

5.117(M)

```
MTB > # Exercise 5.117(M)
MTB > READ C1 C2
DATA> 125 0.10
DATA> 135 0.04
DATA> 145 0.14
DATA> 155 0.07
DATA> 165 0.17
DATA> 175 0.20
DATA> 185 0.01
DATA> 195 0.08
DATA> 205 0.01
DATA> 215 0.04
DATA> 225 0.06
DATA> 235 0.08
DATA> END
      12 ROWS READ
MTB > LET C3 = C1*C2
MTB > SUM C3                       # THIS IS THE MEAN OF X
   SUM      =      172.50
```

5.118(M)

```
MTB > # Exercise 5.118(M)
MTB > #  SEE 5.117(M) FOR DATA INPUT
MTB > LET K1 = SUM((C1**2)*C2) - (SUM(C3))**2
MTB > PRINT K1                    # THIS IS THE VARIANCE OF X
K1        1022.74
MTB > LET K2 = SQRT(K1)
MTB > PRINT K2                    # THIS IS THE STANDARD DEVIATION OF X
K2        31.9804
```

5.119(M)

```
MTB > # Exercise 5.119(M)
MTB > SET C1
DATA> 1:100
DATA> END
MTB > LET C2 = C1/5050
MTB > LET C3 = C1*C2
MTB > SUM C3                      # THIS IS THE MEAN OF X
   SUM      =      67.000
```

5.120(M)

```
MTB > # Exercise 5.120(M)
MTB > # SEE 5.119(M) FOR DATA INPUT
MTB > LET K1 = SUM((C1**2)*C2) - (SUM(C3))**2
MTB > LET K2 = SQRT(K1)
MTB > PRINT K2                    # THIS IS THE STANDARD DEVIATION OF X
K2        23.6854
```

5.121(M)a.

```
MTB > # Exercise 5.121(M)
MTB > READ C1 C2
DATA> 0 1
DATA> 1 4
DATA> 2 6
DATA> 3 4
DATA> 4 1
DATA> END
      5 ROWS READ
MTB > LET C2 = C2/16
MTB > LET C3 = C1*C2
MTB > SUM C3                      # THIS IS THE MEAN OF X
   SUM      =       2.0000
MTB > LET K1 = SUM((C1**2)*C2) - (SUM(C3))**2
MTB > LET K2 = SQRT(K1)
MTB > PRINT K2                    # THIS IS THE STANDARD DEVIATION OF X
K2       1.00000
```

b.

```
MTB > RANDOM 400 C3;
SUBC> DISC C1 C2.
```

c.

```
MTB > HIST C3

Histogram of C3   N = 400
Each * represents 5 obs.

Midpoint   Count
       0      22   *****
       1      92   *******************
       2     167   **********************************
       3      96   ********************
       4      23   *****

MTB > # 0  22/400 = 0.0550 is close to the expected 1/16 = 0.0625
MTB > # 1  92/400 = 0.2300 is close to the expected 4/16 = 0.2500
MTB > # 2 167/400 = 0.4175 is close to the expected 6/16 = 0.3750
MTB > # 3  96/400 = 0.2400 is close to the expected 4/16 = 0.2500
MTB > # 4  23/400 = 0.0575 is close to the expected 1/16 = 0.0625
```

d.

```
MTB > MEAN C3
   MEAN     =       2.0150
MTB > STDEV C3
  ST.DEV. =     0.96025

MTB > # The mean of 2.0150 is close to the expected value of 2
MTB > # The standard deviation of 0.96025 is close to the
MTB > # theoretical value of 1.
MTB > # Results of this problem will vary.
```

CHAPTER 6
DISCRETE PROBABILITY DISTRIBUTIONS

EXERCISES for Section 6.1
Binomial Experiments

6.1 This experiment is binomial where success means "tossing a head on a single throw". The values of n and p are 7 and 0.4, respectively.

6.3 Since the population is extremely large, this experiment is binomial where success means "selecting a household with a personal computer". The values of n and p are 500 and 0.15, respectively.

6.5 This experiment is not binomial since the trials are not independent.

6.7 This experiment is binomial where success means "answering a question correctly". The values of n and p are 10 and 0.5, respectively.

6.9 This experiment is binomial where success means "the use of a charge card by a customer". The values of n and p are 3 and 0.70, respectively.

6.11 Since the population is extremely large, this experiment is binomial where success means "the selection of a home which has a microwave oven". The values of n and p are 20 and 0.75, respectively.

EXERCISES for Section 6.2
The Binomial Probability Distribution

6.13 $C(6,4)\cdot(0.1)^4\cdot(0.9)^2 = 15\cdot(0.0001)\cdot(0.81) = 0.001$.

6.15 We are given n = 5 and p = ⅔ so $P(x) = C(5,x)\cdot(\frac{2}{3})^x\cdot(\frac{1}{3})^{(5-x)}$.

Therefore, $P(2) = C(5,2)\cdot(\frac{2}{3})^2\cdot(\frac{1}{3})^3 = 10\cdot\frac{4}{9}\cdot\frac{1}{27} = 0.165$.

6.17 We are given n = 5 and p = ¼, so $P(x) = C(5,x)\cdot(¼)^x\cdot(¾)^{(5-x)}$.
$P(x \leq 2) = P(0) + P(1) + P(2)$, where
$P(0) = C(5,0)\cdot(¼)^0\cdot(¾)^5 = 1\cdot(1)\cdot(0.2373) = 0.23730$,
$P(1) = C(5,1)\cdot(¼)^1\cdot(¾)^4 = 5\cdot(0.25)\cdot(0.3164) = 0.39550$, and
$P(2) = C(5,2)\cdot(¼)^2\cdot(¾)^3 = 10\cdot(0.0625)\cdot(0.4219) = 0.26369$. Therefore,
$P(x \leq 2) = 0.23730 + 0.39550 + 0.26369 = 0.89649 = 0.896$.

6.19 We are given n = 6 and p = 0.2, so $P(x) = C(6,x)\cdot(0.2)^x\cdot(0.8)^{(6-x)}$. We calculate
$P(4) = C(6,4)(0.2)^4(0.8)^2 = 15(0.0016)(0.64) = 0.01536$ and
$P(5) = C(6,5)(0.2)^5(0.8)^1 = 6(0.00032)(0.8) = 0.001536$. Therefore,
$P(4) + P(5) = 0.01536 + 0.001536 = 0.017$.

6.21 From Table I, with $n = 10$ and $p = 0.6$, $P(x \leq 8) = 0.954$.

6.23 From Table I, with $n = 10$ and $p = 0.6$, $P(x = 8) = P(x \leq 8) - P(x \leq 7) = 0.954 - 0.833 = 0.121$.

6.25 From Table I, with $n = 10$ and $p = 0.6$, $P(x \geq 8) = 1 - P(x \leq 7) = 1 - 0.833 = 0.167$.

6.27 From Table I, with $n = 10$ and $p = 0.6$, $P(4 < x < 8) = P(x \leq 7) - P(x \leq 4) = 0.833 - 0.166 = 0.667$.

6.29 From Table I, with $n = 10$ and $p = 0.6$, $P(4 < x \leq 8) = P(x \leq 8) - P(x \leq 4) = 0.954 - 0.166 = 0.788$.

6.31 We are given $n = 5$ and $p = 0.65$, so $P(x) = C(5,x) \cdot (0.65)^x \cdot (0.35)^{(5-x)}$. The probability that three had a VCR is given by P(3), where $P(3) = C(5,3) \cdot (0.65)^3 \cdot (0.35)^2 = 10 \cdot (0.2746) \cdot (0.1225) = 0.336$.

6.33 We are given $n = 6$ and $p = 0.35$, so $P(x) = C(6,x) \cdot (0.35)^x \cdot (0.65)^{(6-x)}$. The probability that two of its next six orders of 2x4s will be for pine is given by P(2), where $P(2) = C(6,2) \cdot (0.35)^2 \cdot (0.65)^4 = 15 \cdot (0.1225) \cdot (0.1785) = 0.328$.

6.35 **a.** We are given $n = 5$ and $p = 0.5$, so $P(x) = C(5,x) \cdot (0.5)^x \cdot (0.5)^{(5-x)}$. The probability that at least one will have power windows may be found by using the binomial probability distribution formula as $P(x \geq 1) = 1 - P(0)$, where $P(0) = C(5,0) \cdot (0.50)^0 \cdot (0.50)^5 = 1 \cdot (1) \cdot (0.03125) = 0.03125$. Therefore, $P(x \geq 1) = 1 - 0.03125 = 0.96875$.
b. Using Table I in the Appendix, we have: $P(x \geq 1) = 1 - P(0) = 1 - P(x \leq 0) = 1 - 0.031 = 0.969$.

6.37 We use Table I with $n = 20$ and $p = 0.05$:
a. The probability that exactly one is afflicted by full-fledged SAD is:
$P(x = 1) = P(x \leq 1) - P(x \leq 0) = 0.736 - 0.358 = 0.378$.
b. The probability that at least one is afflicted by full-fledged SAD is:
$P(x \geq 1) = 1 - P(x \leq 0) = 1 - 0.358 = 0.642$.

6.39 We use Table I with $n = 15$ and $p = 0.20$:
a. The probability that four are snow tires is given by:
$P(x = 4) = P(x \leq 4) - P(x \leq 3) = 0.836 - 0.648 = 0.188$.
b. The probability that at most four are snow tires is given by: $P(x \leq 4) = 0.836$.
c. The probability that at least four are snow tires is given by:
$P(x \geq 4) = 1 - P(x \leq 3) = 1 - 0.648 = 0.352$.
d. The probability that at least two but not more than four are snow tires is given by:
$P(2 \leq x \leq 4) = P(x \leq 4) - P(x \leq 1) = 0.836 - 0.167 = 0.669$.

6.41 We are given $n = 10$ and $p = \frac{1}{78}$ so $P(x) = C(10, x) \cdot (\frac{1}{78})^x \cdot (\frac{77}{78})^{(10-x)}$.
a. P(no shuttle disasters) =
$P(0) = C(10, 0) \cdot (\frac{1}{78})^0 \cdot (\frac{77}{78})^{10} = 1 \cdot 1 \cdot (0.879) = 0.879$.

b. P(one shuttle disaster) =

$P(1) = C(10,1)\cdot(\frac{1}{78})^1\cdot(\frac{77}{78})^9 = 10\cdot(0.01282)\cdot(0.89036) = 0.114.$

c. P(at least one shuttle disaster) = $P(x \geq 1) = 1 - P(0)$, where from above, P(0) = 0.879. Therefore, $P(x \geq 1) = 1 - 0.879 = 0.121$.

MINITAB LAB ASSIGNMENTS

6.43(M)

```
MTB > # Exercise 6.43(M)
MTB > CDF 30;
SUBC> BINO 50 .63.
      K  P( X LESS OR = K)
     30               0.3805
```

The probability that more than 30 in a sample of 50 voters will favor a reduction is 1 - 0.3805 = 0.6195.

6.45(M)a.

```
MTB > # Exercise 6.45(M)
MTB > PDF 7;
SUBC> BINO 50 .12.
      K             P( X = K)
    7.00               0.1467
```

The probability that seven bolts will contain defects is 0.1467.

b.

```
MTB > CDF 6;
SUBC> BINO 50 .12.
      K  P( X LESS OR = K)
    6.00               0.6065
```

The probability that fewer than seven bolts will contain defects is 0.6065.

EXERCISES for Section 6.3

The Mean & Standard Deviation of a Binomial Random Variable

6.47 We are given n = 100 and p = 0.9, so q = 1 - p = 1 - 0.9 = 0.1. Therefore, $\mu = np = 100\cdot(0.9) = 90$, $\sigma^2 = npq = 100\cdot(0.9)\cdot(0.1) = 9$ and $\sigma = \sqrt{9} = 3$.

6.49 We are given n = 180 and p = 1/6, so q = 1 - p = 1 - (1/6) = 5/6. Therefore, $\mu = np = 180\cdot(1/6) = 30$, $\sigma^2 = npq = 180\cdot(1/6)\cdot(5/6) = 25$ and $\sigma = \sqrt{25} = 5$.

6.51 We are given n = 400 and p = 0.5, so q = 1 - p = 1 - 0.5 = 0.5. Therefore, $\mu = np = 400\cdot(0.5) = 200$, $\sigma^2 = npq = 400\cdot(0.5)\cdot(0.5) = 100$ and $\sigma = \sqrt{100} = 10$.

6.53 We are given n = 48 and p = 0.25, so q = 1 - p = 1 - 0.25 = 0.75. Therefore, $\mu = np = 48(0.25) = 12$, $\sigma^2 = npq = 48 \cdot (0.25) \cdot (0.75) = 9$ and $\sigma = \sqrt{9} = 3$.

6.55 We are given n = 60 and p = 0.5, so q = 1 - p = 1 - 0.5 = 0.5. Therefore, $\mu = np = 60 \cdot (0.5) = 30$, $\sigma^2 = npq = 60 \cdot (0.5) \cdot (0.5) = 15$ and $\sigma = \sqrt{15} = 3.87$.

6.57 We are given n = 80 and p = ¾, so q = 1 - p = 1 - ¾ = ¼. Therefore, $\mu = np = 80 \cdot (3/4) = 60$, $\sigma^2 = npq = 80 \cdot (3/4) \cdot (1/4) = 15$ and $\sigma = \sqrt{15} = 3.87$.

6.59 **a.** We construct a table:

x	P(x)	xP(x)	$x^2P(x)$
0	0.0625	0.00	0.00
1	0.2500	0.25	0.25
2	0.3750	0.75	1.50
3	0.2500	0.75	2.25
4	0.0625	0.25	1.00
		$\mu = 2$	$\Sigma x^2 P(x) = 5$

By Formula 5-7: $\sigma^2 = \Sigma x^2P(x) - \mu^2 = 5 - (2)^2 = 1$.
The standard deviation of x is $\sigma = \sqrt{1} = 1$.
b. We are given n = 4 and p = 0.5, so q = 1 - p = 1 - 0.5 = 0.5. Therefore, $\mu = np = 4 \cdot (0.5) = 2$, $\sigma^2 = npq = 4 \cdot (0.5) \cdot (0.5) = 1$ and $\sigma = \sqrt{1} = 1$.

6.61 We are given n = 128 and p = ⅓. Therefore, the number that can be expected to be identical twins is $\mu = np = 128 \cdot (1/3) = 42.67$. (This is a long range average value and therefore may be a noninteger.)

6.63 **a.** We are given n = 84 and p = 25% = 0.25. Therefore, the number of jobs that we would expect to be awarded to minority firms, if the company's claim is correct, is: $\mu = np = 84 \cdot (0.25) = 21$.
b. We note that q = 1 - p = 1 - 0.25 = 0.75. Therefore, $\sigma^2 = npq = 84 \cdot (0.25) \cdot (0.75) = 15.75$ and $\sigma = \sqrt{15.75} = 3.97$.
c. No; the z-value for 9 is $z = \frac{X - \mu}{\sigma} = \frac{9 - 21}{3.97} = -3.02$, which is more than 3 standard deviations below the mean which, by Chebyshev's Theorem, would have a probability value less than $(1/3)^2 = 1/9$.

6.65 **a.** We are given n = 400 and p = 0.5. Therefore, the mean of x is $\mu = np = 400 \cdot (0.5) = 200$.
b. We note that q = 1 - p = 1 - 0.5 = 0.5. Therefore, $\sigma^2 = npq = 400 \cdot (0.5) \cdot (0.5) = 100$ and $\sigma = \sqrt{100} = 10$.
c. By the Empirical rule, we would expect x to fall within 2 standard deviations of the mean about 95% of the time, i.e. between 180 and 220.

EXERCISES for Section 6.4

The Hypergeometric Probability Distribution

6.67 We are given N = 10, S = 5, n = 4 and wish to find P(3).

$$P(x) = \frac{C(S,x)\cdot C(N-S,n-x)}{C(N,n)} = \frac{C(5,x)\cdot C(5,4-x)}{C(10,4)}.$$

Therefore, $P(3) = \dfrac{C(5,3)\cdot C(5,1)}{C(10,4)} = \dfrac{10\cdot 5}{210} = 0.238.$

6.69 We are given N = 9, S = 4, n = 5 and wish to find P(x ≤ 2) = P(0) + P(1) + P(2).

$$P(x) = \frac{C(S,x)\cdot C(N-S,n-x)}{C(N,n)} = \frac{C(4,x)\cdot C(5,5-x)}{C(9,5)}.$$

so:

$$P(0) = \frac{C(4,0)\cdot C(5,5)}{C(9,5)} = \frac{1\cdot 1}{126} = \frac{1}{126}.$$

$$P(1) = \frac{C(4,1)\cdot C(5,4)}{C(9,5)} = \frac{4\cdot 5}{126} = \frac{20}{126}.$$

$$P(2) = \frac{C(4,2)\cdot C(5,3)}{C(9,5)} = \frac{6\cdot 10}{126} = \frac{60}{126}.$$

$$P(x\le 2) = \frac{1}{126} + \frac{20}{126} + \frac{60}{126} = \frac{81}{126} = 0.643.$$

6.71 We consider "success" to mean that a bat has an internal flaw that will result in its breaking on impact with a baseball. Since the baseball player selects five of the ten bats, we have N = 10, S = 3, n = 5 and wish to find P(3), the probability that he will choose the three defective bats:

$$P(x) = \frac{C(S,x)\cdot C(N-S,n-x)}{C(N,n)} = \frac{C(3,x)\cdot C(7,5-x)}{C(10,5)}.$$

Therefore $P(3) = \dfrac{C(3,3)\cdot C(7,2)}{C(10,5)} = \dfrac{1\cdot 21}{252} = \dfrac{21}{252} = 0.083.$

6.73 We consider "success" to mean that one of the numbers selected is chosen by the player. We have N = 40, S = 6, n = 6 and wish to find P(6), the probability that all six numbers are chosen correctly:

$$P(x) = \frac{C(S,x)\cdot C(N-S,n-x)}{C(N,n)} = \frac{C(6,x)\cdot C(34,6-x)}{C(40,6)}.$$

Therefore $P(6) = \dfrac{C(6,6)\cdot C(34,0)}{C(40,6)} = \dfrac{1\cdot 1}{3838380} = (2.6)\cdot 10^{-7}.$

6.75 **a.** We have N = 20, S = 2, and n = 3 and wish to find P(0), the probability that none of the purchased bags contains ten dollars:

$$P(x) = \frac{C(S,x)\cdot C(N-S,n-x)}{C(N,n)} = \frac{C(2,x)\cdot C(18,3-x)}{C(20,3)}.$$

Therefore $P(0) = \frac{C(2,0)\cdot C(18,3)}{C(20,3)} = \frac{1\cdot 816}{1140} = 0.716.$

b. The probability that exactly one purchased bag contains ten dollars is:

$$P(1) = \frac{C(2,1)\cdot C(18,2)}{C(20,3)} = \frac{2\cdot 153}{1140} = 0.268.$$

6.77 Since there are 52 cards in a deck, four of which are aces, we have N = 52, S = 4, and n = 5. We wish to find P(3), the probability of obtaining exactly three aces:

$$P(x) = \frac{C(S,x)\cdot C(N-S,n-x)}{C(N,n)} = \frac{C(4,x)\cdot C(48,5-x)}{C(52,5)}.$$

Therefore $P(3) = \frac{C(4,3)\cdot C(48,2)}{C(52,5)} = \frac{4\cdot 1128}{2598960} = 0.00174.$

6.79 From Exercise 6.71, N = 10, S = 3 and n = 5. Therefore, $p = \frac{S}{N} = \frac{3}{10} = 0.3$, and q = 1 - p = 1 - 0.3 = 0.7. The mean is μ = np = 5$\cdot$(0.3) = 1.5, the variance is

$$\sigma^2 = npq\frac{(N-n)}{(N-1)} = 5\cdot(0.3)\cdot(0.7)\cdot\frac{(10-5)}{(10-1)} = 0.583 \text{ and}$$

$\sigma = \sqrt{0.583} = 0.764.$

6.81 We have N = 13, S = 5, n = 3 and wish to find P(x ≥ 1), the probability that a participant would draw at least one topic for which he/she had prepared. This is 1- P(0),

where $P(x) = \frac{C(S,x)\cdot C(N-S,n-x)}{C(N,n)} = \frac{C(5,x)\cdot C(8,3-x)}{C(13,3)}.$

Therefore $P(0) = \frac{C(5,0)\cdot C(8,3)}{C(13,3)} = \frac{1\cdot 56}{286} = 0.1958$

so P(x ≥ 1) = 1 - P(0) = 1 - 0.1958 = 0.804.

EXERCISES for Section 6.5

The Poisson Probability Distribution

6.83 We are given that the random variable has a Poisson distribution with mean μ = 4.

Therefore $P(x) = \frac{\mu^x\cdot e^{-\mu}}{x!} = \frac{4^x\cdot e^{-4}}{x!}$ so we have

$$P(3) = \frac{4^3\cdot e^{-4}}{3!} = \frac{64\cdot(0.0183)}{6} = 0.195.$$

The Poisson Probability Distribution

6.85 We are given that the random variable has a Poisson distribution with mean $\mu = 3.4$. Therefore $P(x) = \frac{\mu^x \cdot e^{-\mu}}{x!} = \frac{3.4^x \cdot e^{-3.4}}{x!}$ so we have

$$P(4) = \frac{(3.4)^4 \cdot e^{-3.4}}{4!} = \frac{133.6336 \cdot (e^{-3.4})}{24} = 0.186.$$

6.87 We are given that the random variable has a Poisson distribution with mean $\mu = 1.6$. Therefore $P(x) = \frac{\mu^x \cdot e^{-\mu}}{x!} = \frac{(1.6)^x \cdot e^{-1.6}}{x!}$ and $P(x \leq 1) = P(0) + P(1)$:

$$P(0) = \frac{(1.6)^0 \cdot e^{-1.6}}{0!} = \frac{1 \cdot e^{-1.6}}{1} = 0.2019$$

$$P(1) = \frac{(1.6)^1 \cdot e^{-1.6}}{1!} = \frac{1.6 \cdot e^{-1.6}}{1} = 0.3230$$

and $P(x \leq 1) = 0.2019 + 0.3230 = 0.525$.

6.89 We are given that the random variable has a Poisson distribution with mean $\mu = 6.8$ and wish to find P(0), the probability that none of the kernels fail to pop:

$$P(x) = \frac{\mu^x \cdot e^{-\mu}}{x!} = \frac{6.8^x \cdot e^{-6.8}}{x!}$$ and

therefore, $P(0) = \frac{(6.8)^0 \cdot e^{-6.8}}{0!} = \frac{1 \cdot e^{-6.8}}{1} = 0.001.$

6.91 We are given that the author averages 1.2 typing errors per page, which would suggest an average of 6 errors in 5 pages. Therefore, we have a Poisson distribution with mean $\mu = 6$ so $P(x) = \frac{\mu^x \cdot e^{-\mu}}{x!} = \frac{6^x \cdot e^{-6}}{x!}$. The probability that five pages will contain a total of two typing errors is $P(2) = \frac{6^2 \cdot e^{-6}}{2!} = \frac{36 \cdot (0.002479)}{2} = 0.045.$

6.93 Since only one person in 100,000 has the extra bone, in a city of 50,000 people, we have a Poisson distribution with mean $\mu = 50{,}000/100{,}000 = 0.5$. We wish to find $P(x \geq 1)$, the probability that at least one person will have the extra bone. This is 1 - P(0):

$$P(x) = \frac{\mu^x \cdot e^{-\mu}}{x!} = \frac{0.5^x \cdot e^{-0.5}}{x!}$$ and therefore

$P(0) = \frac{(0.5)^0 \cdot e^{-0.5}}{0!} = \frac{1 \cdot e^{-0.5}}{1} = 0.607$ and $P(x \geq 1) = 1 - 0.607 = 0.393$.

6.95a. Since the switchboard receives an average of 72 calls in one hour, in a period of ten minutes, we use a Poisson distribution with mean $\mu = 72/6 = 12$. The probability of fewer than 5 calls in this time period is $P(x < 5)$, where $P(x) = \frac{\mu^x \cdot e^{-\mu}}{x!} = \frac{12^x \cdot e^{-12}}{x!}$. We find $P(x < 5) = P(0) + P(1) + P(2) + P(3) + P(4) = 0.000006 + 0.000074 + 0.000442 + 0.001770 + 0.005309 = 0.008$.

b. For a period of four minutes, we use a Poisson distribution with mean $\mu = 72/15 = 4.8$. The probability of more than two calls during this time is $P(x > 2) = 1 - [P(0) + P(1) + P(2)]$; where $P(x) = \frac{\mu^x \cdot e^{-\mu}}{x!} = \frac{4.8^x \cdot e^{-4.8}}{x!}$. We find $P(x > 2) = 1 - [0.0082 + 0.0395 + 0.0948] = 1 - 0.143 = 0.857$.

6.97 **a.** In Exercise 6.83, we had a Poisson distribution with $\mu = 4$. Therefore, $\sigma^2 = 4$ and $\sigma = \sqrt{4} = 2$.
b. In Exercise 6.84, we had a Poisson distribution with $\mu = 5$. Therefore, $\sigma^2 = 5$ and $\sigma = \sqrt{5} = 2.236$.
c. In Exercise 6.85, we had a Poisson distribution with $\mu = 3.4$. Therefore, $\sigma^2 = 3.4$ and $\sigma = \sqrt{3.4} = 1.844$.
d. In Exercise 6.86, we had a Poisson distribution with $\mu = 1.2$. Therefore, $\sigma^2 = 1.2$ and $\sigma = \sqrt{1.2} = 1.095$.
e. In Exercise 6.87, we had a Poisson distribution with $\mu = 1.6$. Therefore, $\sigma^2 = 1.6$ and $\sigma = \sqrt{1.6} = 1.265$.

MINITAB LAB ASSIGNMENTS

6.99(M)

```
MTB > # Exercise 6.99(M)
MTB > CDF 0;
SUBC> POISSON 0.05.           # 0.05 IS 1/100000 OF 5000.
       K  P( X LESS OR = K)
    0.00              0.9512
```

The probability that at least one will die is 1 - 0.9512 = 0.0488

6.101(M)a.

```
MTB > # Exercise 6.101(M)
MTB > CDF 13;
SUBC> BINO 125 0.1.
       K  P( X LESS OR = K)
   13.00              0.6309
```

The probability that more than 13 will be members is 1 - 0.6309 = 0.3691.

b.

```
MTB > CDF 13;
SUBC> POISSON 12.5. #mean = np = (125)(0.1) = 12.5.
       K  P( X LESS OR = K)
   13.00              0.6278
```

The probability that more than 13 will be members is 1 - 0.6278 = 0.3722.

REVIEW EXERCISES
Chapter 6

6.103 We are given n = 5 and p = 1/4 and wish to find P(3), where
$P(x) = C(5,x) \cdot (1/4)^x \cdot (3/4)^{(5-x)}$.
$P(3) = C(5,3)(1/4)^3(3/4)^2 = 10(0.015625)(0.5625) = 0.088$.

6.104 We are given n = 9 and p = 1/3 and wish to find P(x > 1) = 1 - [P(0) + P(1)]. We calculate:
$P(0) = C(9,0)(1/3)^0(2/3)^9 = 1(1)(512/19683) = 0.0260$.
$P(1) = C(9,1)(1/3)^1(2/3)^8 = 9(1/3)(256/6561) = 0.1171$.
$P(x > 1) = 1 - [0.0260 + 0.1171] = 1 - 0.143 = 0.857$.

6.105 We are given N = 12, S = 7, n = 6 and wish to find P(4).

$$P(x) = \frac{C(S,x) \cdot C(N-S,n-x)}{C(N,n)} = \frac{C(7,x) \cdot C(5,6-x)}{C(12,6)}.$$

Therefore, $P(4) = \frac{C(7,4) \cdot C(5,2)}{C(12,6)} = \frac{35 \cdot 10}{924} = 0.379$.

6.106 We are given N = 13, S = 7, n = 5 and wish to find P(x ≤ 1). This is the same as P(0) + P(1):

$$P(x) = \frac{C(S,x) \cdot C(N-S,n-x)}{C(N,n)} = \frac{C(7,x) \cdot C(6,5-x)}{C(13,5)}.$$

Therefore, $P(0) = \frac{C(7,0) \cdot C(6,5)}{C(13,5)} = \frac{1 \cdot 6}{1287}$ and

$$P(1) = \frac{C(7,1) \cdot C(6,4)}{C(13,5)} = \frac{7 \cdot 15}{1287} = \frac{105}{1287}.$$

$$P(x \le 1) = \frac{6}{1287} + \frac{105}{1287} = \frac{111}{1287} = 0.086.$$

6.107 We are given that the random variable has a Poisson distribution with mean $\mu = 2$. We wish to find P(1 ≤ x ≤ 3) which is P(1)+P(2)+P(3), where $P(x) = \frac{\mu^x \cdot e^{-\mu}}{x!} = \frac{2^x \cdot e^{-2}}{x!}$.

$$P(1) = \frac{2^1 \cdot e^{-2}}{1!} = \frac{2 \cdot (e^{-2})}{1} = 0.2707,$$

$$P(2) = \frac{2^2 \cdot e^{-2}}{2!} = \frac{4 \cdot (e^{-2})}{2} = 0.2707,$$

$$P(3) = \frac{2^3 \cdot e^{-2}}{3!} = \frac{8 \cdot (e^{-2})}{6} = 0.1804,$$

and $P(1 \le x \le 3) = 0.2707 + 0.2707 + 0.1804 = 0.722$.

6.108 We are given that the random variable has a Poisson distribution with mean $\mu = 4.8$. We wish to find P(5), where $P(x) = \frac{\mu^x \cdot e^{-\mu}}{x!} = \frac{4.8^x \cdot e^{-4.8}}{x!}$. We find

$$P(5) = \frac{(4.8)^5 \cdot e^{-4.8}}{5!} = \frac{2548.04 \cdot (e^{-4.8})}{120} = 0.175.$$

6.109 We are given n = 20, p = 0.7 and wish to find $P(9 \le x \le 13) = P(x \le 13) - P(x \le 8)$. From Table I, $P(x \le 13) = 0.392$, $P(x \le 8) = 0.005$ and $P(9 \le x \le 13) = P(x \le 13) - P(x \le 8) = 0.392 - 0.005 = 0.387$.

6.110 We have N = 8, S = 3, n = 4 and wish to find $P(x \ge 1)$, the probability that at least one blade that will fit. This is the same as 1 - P(0):

$$P(x) = \frac{C(S,x) \cdot C(N-S,n-x)}{C(N,n)} = \frac{C(3,x) \cdot C(5,4-x)}{C(8,4)}.$$ Therefore,

$$P(0) = \frac{C(3,0) \cdot C(5,4)}{C(8,4)} = \frac{1 \cdot 5}{70} = 0.0714,$$ and $P(x \ge 1) = 1 - 0.0714 = 0.929$.

6.111 We have N = 13, S = 6, n = 4, where S is the number of boys in the club, and wish to find P(2), the probability that exactly two boys and two girls are chosen for the club:

$$P(x) = \frac{C(S,x) \cdot C(N-S,n-x)}{C(N,n)} = \frac{C(6,x) \cdot C(7,4-x)}{C(13,4)}.$$

$$P(2) = \frac{C(6,2) \cdot C(7,2)}{C(13,4)} = \frac{15 \cdot 21}{715} = 0.441.$$

Note: Using S = 7, where S is the number of girls in the club, will produce the same result, as follows: $P(2) = \frac{C(7,2) \cdot C(6,2)}{C(13,4)} = \frac{21 \cdot 15}{715} = 0.441.$

6.112 Given n = 3000 and p = ¼, the expected number of poor students in the randomly selected sample of first graders is $\mu = np = 3000(¼) = 750$.

6.113 Given n = 10 and p = 0.65, we find $P(x > 7) = P(8) + P(9) + P(10)$, where $P(x) = C(10,x) \cdot (0.65)^x \cdot (0.35)^{(10-x)}$.
$P(8) = C(10,8)(0.65)^8(0.35)^2 = 0.17565$
$P(9) = C(10,9)(0.65)^9(0.35)^1 = 0.07249$
$P(10) = C(10,10)(0.65)^{10}(0.35)^0 = 0.01346$
$P(x > 7) = 0.17565 + 0.07249 + 0.01346 = 0.262$.

6.114 **a.** Given n = 5 and p = 0.75, we find $P(5) = C(5,5)(0.75)^5(0.25)^0 = 0.237$.
b. The probability that none use this method is $P(0) = C(5,0)(0.75)^0(0.25)^5 = 0.001$.

6.115 The random variable has a Poisson distribution with mean $\mu = 2.3$ and we wish to find

P(x ≥ 1), the probability that a newly purchased car will be delivered with at least one defect. This is the same as 1 - P(0), where $P(x) = \frac{\mu^{x} \cdot e^{-\mu}}{x!} = \frac{2.3^{x} \cdot e^{-2.3}}{x!}$. Therefore,

$$P(0) = \frac{2.3^{0} \cdot e^{-2.3}}{0!} = \frac{1 \cdot (e^{-2.3})}{1} = 0.10026$$ and we have

P(x ≥ 1) = 1 - 0.10026 = 0.900.

6.116 Since 4 per 1000 is equivalent to 0.4 per 100, the random variable has a Poisson distribution with mean μ = 0.4. We wish to find P(0), the probability that no one will die before the age of 65 from indoor radon:

$$P(x) = \frac{\mu^{x} \cdot e^{-\mu}}{x!} = \frac{(0.4)^{x} \cdot e^{-0.4}}{x!}$$ so we have

$$P(0) = \frac{(0.4)^{0} \cdot e^{-0.4}}{0!} = \frac{1 \cdot (e^{-0.4})}{1} = 0.670.$$

6.117 With n = 1000, p = 0.68, and q = 1 - p = 1 - 0.68 = 0.32, we find μ = np = 1000(0.68) = 680, σ^2 = npq = 1000(0.68)(0.32) = 217.6 and $\sigma = \sqrt{217.6} = 14.75$.

6.118 **a.** Using a hypergeometric distribution, we have N = 180, S = 18, n = 10 and wish to find P(x ≥ 2), the probability that two or more vehicles will fail the test. This is the same as 1 - [P(0) + P(1)] where

$$P(x) = \frac{C(S,x) \cdot C(N-S,n-x)}{C(N,n)} = \frac{C(18,x) \cdot C(162,10-x)}{C(180,10)}.$$

$$P(0) = \frac{C(18,0) \cdot C(162,10)}{C(180,10)} = \frac{1 \cdot 2.584 \cdot 10^{15}}{7.628 \cdot 10^{15}} = 0.3388$$

$$P(1) = \frac{C(18,1) \cdot C(162,9)}{C(180,10)} = \frac{18 \cdot 1.6890^{14}}{7.628 \cdot 10^{15}} = 0.3985$$

Therefore, P(x ≥ 2) = 1 - [0.3388 + 0.3985] = 1 - 0.7373 = 0.263.

b. We use a binomial distribution with n = 10 and p = 18/180 = 0.1 to find P(x ≥ 2), where $P(x) = C(10,x) \cdot (0.1)^{x} \cdot (0.9)^{(10-x)}$.

$P(0) = C(10,0)(0.1)^{0}(0.9)^{10} = 1(1)(0.3487) = 0.3487$

$P(1) = C(10,1)(0.1)^{1}(0.9)^{9} = 10(0.1)(0.3874) = 0.3874$

Therefore, P(x ≥ 2) = 1 - [0.3487 + 0.3874] = 1 - 0.7361 = 0.264.

6.119 We have n = 10,000, p = 0.5, and q = 1 - p = 1 - 0.5 = 0.5. Therefore, μ = np = 10,000(0.5) = 5,000, σ^2 = npq = 10,000(0.5)(0.5) = 2500, and $\sigma = \sqrt{2500} = 50$.

Since 4800 is 4 standard deviations below the mean, we can conclude that it is very unlikely that the coin is fair.

6.120 We are given μ = np = 56 and σ = 7. Therefore, $\sigma^2 = npq = 7^2 = 49$. Using the information above, we have the equations: np = 56 and npq = 49. Solving simultaneously, we get 56q = 49. Therefore, q = 49/56 = 7/8. Since q = 1 - p, p = 1 - (7/8) = 1/8, so, using np = 56, we find n = 56/p = 56/(1/8) = 448. We conclude n = 448 and p = 1/8.

6.121 We construct a table as suggested:

p	0	0.1	0.2	0.3	0.4	0.5	0.6	0.7	0.8	0.9	1
q	1	0.9	0.8	0.7	0.6	0.5	0.4	0.3	0.2	0.1	0
npq	0	0.09n	0.16n	0.21n	0.24n	0.25n	0.24n	0.21n	0.16n	0.09n	0

From the above table, we speculate that σ^2 = npq is a maximum when p = 0.5.

MINITAB LAB ASSIGNMENTS

6.122(M)

```
MTB > # Exercise 6.122(M)
MTB > CDF 28;
SUBC> BINO 30 0.88.
       K  P( X LESS OR = K)
   28.00                 0.8900
```

The probability that more than 28 will exceed specifications is 1 - 0.8900 = 0.1100.

6.123(M)

```
MTB > # Exercise 6.123(M)
MTB > CDF 2;
SUBC> BINO 48 0.04.
       K  P( X LESS OR = K)
    2.00                 0.6988
```

The probability that more than two will fail to adhere is 1 - 0.6988 = 0.3012.

6.124(M)

```
MTB > # Exercise 6.124(M)
MTB > CDF 7;
SUBC> POIS 14.
       K  P( X LESS OR = K)
    7.00                 0.0316
```

The probability that fewer than 8 calls will be received is 0.0316.

6.125(M)

```
MTB > # Exercise 6.125(M)
MTB > CDF 7;
SUBC> POIS 10.   #one-tenth of one percent of 10,000 = 10.
       K  P( x LESS OR = K)
    7.00                 0.2202
```

The probability that at most 7 will be defective is 0.2202.

CHAPTER 7

NORMAL PROBABILITY DISTRIBUTIONS

EXERCISES for Section 7.2

Normal Probability Distributions

7.1 From Table III, the area under the standard normal curve to the left of $z = 0$ is 0.5000, and to the left of 1.23 is 0.8907. Therefore, the area under the curve between $z = 0$ and $z = 1.23$ is $0.8907 - 0.5000 = 0.3907$.

7.3 From Table III, the area under the standard normal curve to the left of $z = -2.98$ is 0.0014, and to the left of 0 is 0.5000. Therefore, the area under the curve between $z = -2.98$ and $z = 0$ is $0.5000 - 0.0014 = 0.4986$.

7.5 The area under the standard normal curve to the left of $z = -0.51$ is 0.3050, and to the left of 1.12 is 0.8686. Therefore, the area under the curve between $z = -0.51$ and $z = 1.12$ is $0.8686 - 0.3050 = 0.5636$.

7.7 The area under the standard normal curve to the left of $z = -2.31$ is 0.0104, and to the left of -1.47 is 0.0708. Therefore, the area under the curve between $z = -2.31$ and $z = -1.47$ is $0.0708 - 0.0104 = 0.0604$.

7.9 The area under the standard normal curve to the left of $z = 0.20$ is 0.5793, and to the left of 1.31 is 0.9049. Therefore, the area under the curve between $z = 0.20$ and $z = 1.31$ is $0.9049 - 0.5793 = 0.3256$.

7.11 The area under the standard normal curve to the left of $z = 1.48$ is 0.9306.

7.13 The area under the standard normal curve to the left of $z = -0.93$ is 0.1762.

7.15 The area under the standard normal curve to the left of $z = 2.18$ is 0.9854. Therefore, the area under the curve to the right of $z = 2.18$ is $1 - 0.9854 = 0.0146$.

7.17 The area under the standard normal curve to the left of $z = -1.66$ is 0.0485. Therefore, the area under the curve to the right of $z = -1.66$ is $1 - 0.0485 = 0.9515$.

7.19 $P(z \le 2.37)$ is the area under the standard normal curve to the left of $z = 2.37$ which, by Table III, is 0.9911. Therefore, $P(z \le 2.37) = 0.9911$.

7.21 $P(z \le -1.32) = 0.0934$.

7.23 $P(z \ge 1.17) = 1 - 0.8790 = 0.1210$.

7.25 $P(z \ge -2.85) = 1 - 0.0022 = 0.9978$.

7.27 $P(0 \le z \le 0.87) = P(z \le 0.87) - P(z \le 0) = 0.8078 - 0.5000 = 0.3078$.

7.29 $P(-0.55 \le z \le 0) = P(z \le 0) - P(z \le -0.55) = 0.5000 - 0.2912 = 0.2088$.

7.31 $P(-2.15 \leq z \leq 1.96) = P(z \leq 1.96) - P(z \leq -2.15) = 0.9750 - 0.0158 = 0.9592$.

7.33 $P(-0.64 \leq z \leq -0.12) = P(z \leq -0.12) - P(z \leq -0.64) = 0.4522 - 0.2611 = 0.1911$.

7.35 $P(1.65 \leq z \leq 2.04) = P(z \leq 2.04) - P(z \leq 1.65) = 0.9793 - 0.9505 = 0.0288$.

7.37 $P(-3 \leq z \leq 3) = P(z \leq 3) - P(z \leq -3) = 0.9987 - 0.0013 = 0.9974$. By comparison, the value given for $P(-3 \leq z \leq 3)$ by the Empirical Rule is approximately 100% = 1, and by Chebyshev's Theorem, $P(-3 \leq z \leq 3) \geq 1 - (1/3)^2 = 0.8889$.

7.39 Since the area to the left of z is 0.9881, we look in Table III for the z-value that corresponds to a cumulative area of 0.9881. We find that the associated z-value is z = 2.26.

7.41 Since the area to the right of z is 0.3974, we look in Table III for the z-value that corresponds to a cumulative area of 1 - 0.3974 = 0.6026. We find that the associated z-value is z = 0.26.

7.43 Since the area to the right of z is 0.9967, we look in Table III for the z-value that corresponds to a cumulative area of 1 - 0.9967 = 0.0033. We find that the associated z-value is z = -2.72.

7.45 Since the area between 0 and z is 0.4980, we look in Table III for the z-value that corresponds to a cumulative area of 0.5 + 0.4980 = 0.9980. We find that the associated z-value is z = 2.88.

7.47 Since the area between -z and z is 0.4038, and the standard normal distribution is symmetric, the area to the left of -z is $(1/2) \cdot (1 - 0.4038) = 0.2981$. We find from Table III that -z = -0.53 so z = 0.53.

7.49 We need to find the value of z for which 64% of the distribution lies to its left. We look in Table III for the z-value that corresponds to a cumulative area of 0.64. The nearest area to 0.64 in Table III is 0.6406 which is associated with z = 0.36. Therefore, $P_{64} = 0.36$.

7.51 We need to find the value of z for which 25% of the distribution lies to its left. We look in Table III for the z-value that corresponds to a cumulative area of 0.25. The nearest area to 0.25 in Table III is 0.2514, which is associated with z = -0.67. Therefore, $P_{25} = -0.67$.

7.53 We need to find the value of z for which 50% of the distribution lies to the left of it. We look in Table III for the z-value that corresponds to a cumulative area of 0.50. The area 0.50 in Table III is associated with z = 0. Therefore, $P_{50} = 0$.

7.55 Since the area under the standard normal curve to the right of $z_{0.05}$ is 0.05, we look in Table III for the z-value that corresponds to a cumulative area of 1 - 0.05 = 0.95. We note that z = 1.64 corresponds to a cumulative area of 0.9495 and z = 1.65 corresponds to a cumulative area of 0.9505. We conclude that $z = z_{0.05} = 1.645$.

7.57 Since the area under the standard normal curve to the right of $z_{0.01}$ is 0.01, we look in Table III for the z-value that corresponds to a cumulative area of 1 - 0.01 = 0.99. We find that the nearest associated z-value is $z = z_{0.01} = 2.33$.

7.59 Since the area under the standard normal curve to the right of $z_{0.975}$ is 0.975, we look in Table III for the z-value that corresponds to a cumulative area of 1 - 0.975 = 0.025. We find that the nearest associated z-value is $z = z_{0.975} = -1.96$.

MINITAB LAB ASSIGNMENTS

Note: The precise shape of the plots will depend upon your choice of data for the horizontal axis.

7.61(M)

```
MTB > # Exercise 7.61(M)
MTB > SET C1
DATA> -2.4:2.4/0.2
DATA> END
MTB > PDF C1 C2;
SUBC> NORMAL 0 1.
MTB > PLOT C2 C1
```

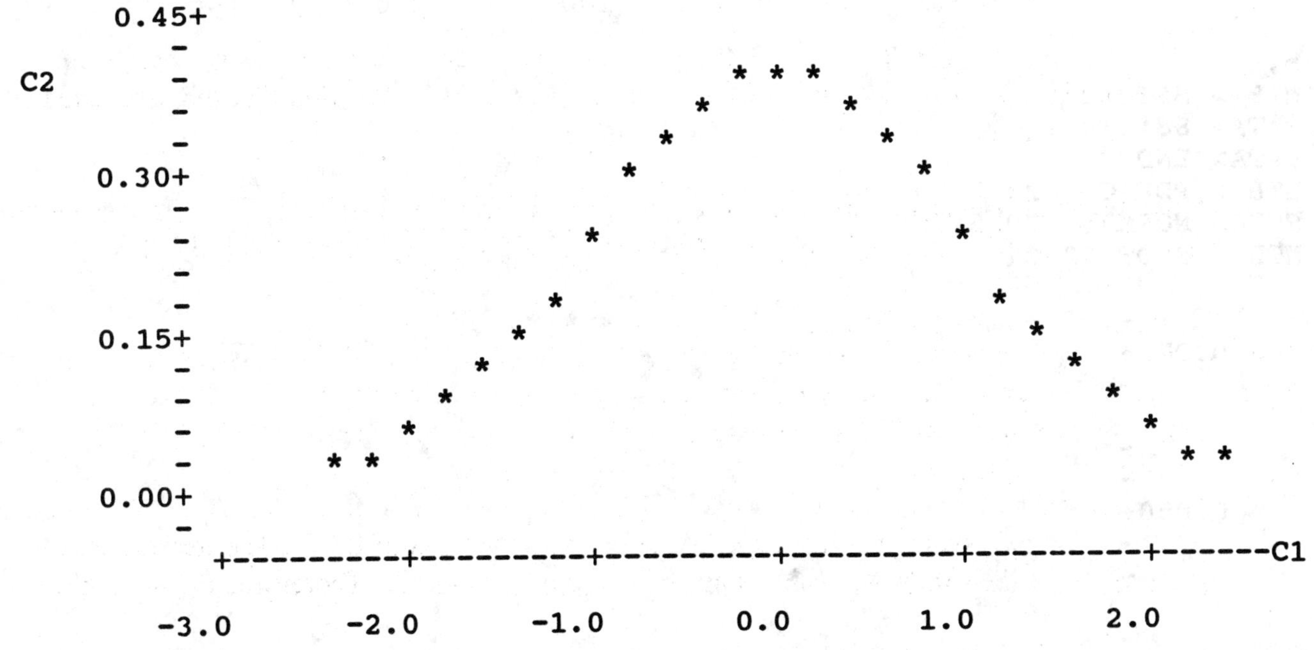

7.63(M) a.

```
MTB > # Exercise 7.63(M)
MTB > SET C1
DATA> 50:150/4
DATA> END
```

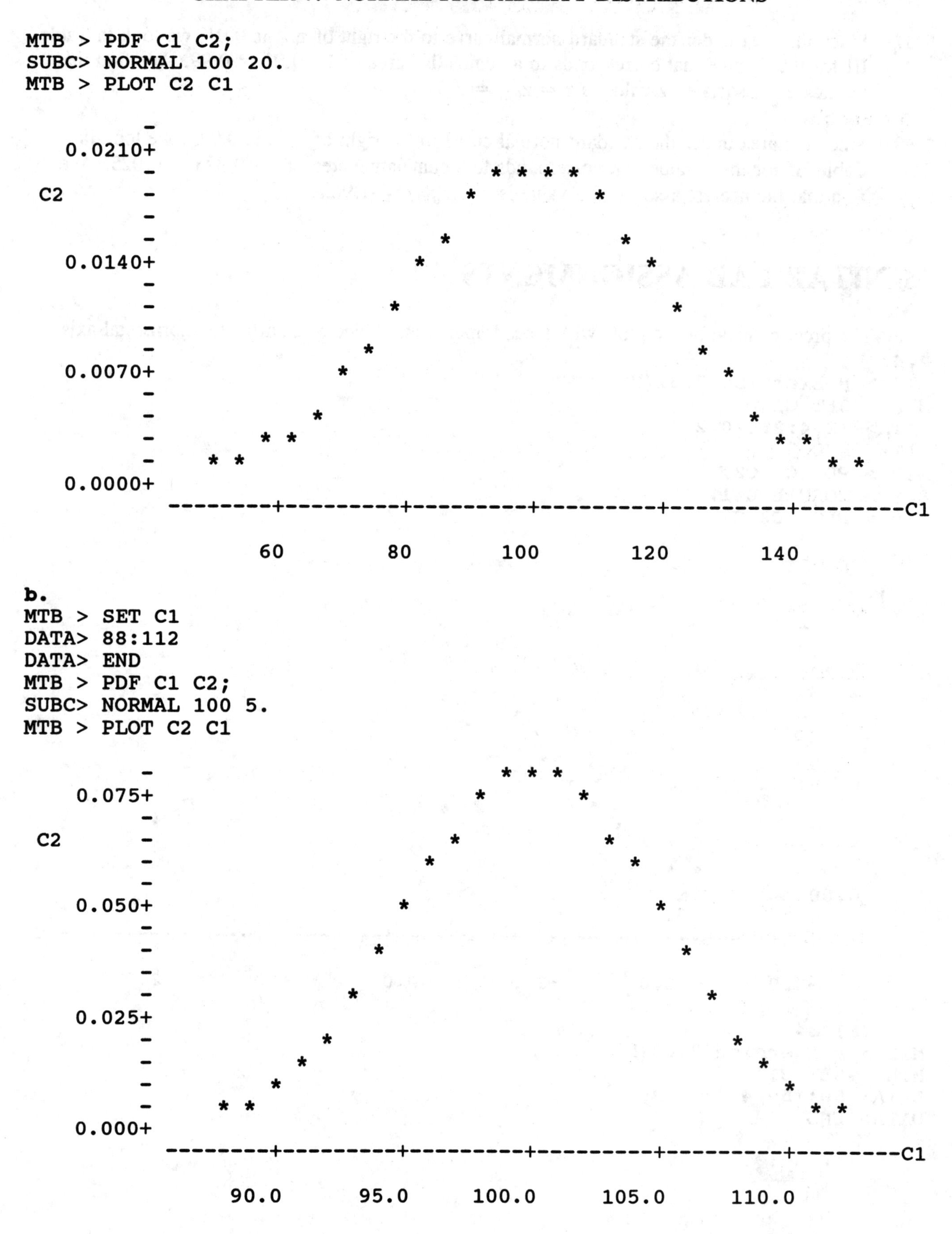
MTB > PDF C1 C2;
SUBC> NORMAL 100 20.
MTB > PLOT C2 C1
0.0210+
C2
0.0140+
0.0070+
0.0000+
C1
60
80
100
120
140
b.
MTB > SET C1
DATA> 88:112
DATA> END
MTB > PDF C1 C2;
SUBC> NORMAL 100 5.
MTB > PLOT C2 C1
0.075+
C2
0.050+
0.025+
0.000+
C1
90.0
95.0
100.0
105.0
110.0

EXERCISES for Section 7.3
Applications of Normal Distributions

7.65 For x = 68, the z-value is $z = \frac{x - \mu}{\sigma} = \frac{68 - 60}{4} = 2.$

7.67 For x = 56, the z-value is $z = \frac{x - \mu}{\sigma} = \frac{56 - 60}{4} = -1.$

7.69 For x = 54, the z-value is $z = \frac{x - \mu}{\sigma} = \frac{54 - 60}{4} = -1.5.$

7.71 For x = 47, the z-value is $z = \frac{x - \mu}{\sigma} = \frac{47 - 60}{4} = -3.25.$

7.73 The z-value for x = 110 is $z = \frac{x - \mu}{\sigma} = \frac{110 - 100}{10} = 1.$
Therefore, $P(x \leq 110) = P(z \leq 1) = 0.8413$.

7.75 The z-value for x = 95 is $z = \frac{x - \mu}{\sigma} = \frac{95 - 100}{10} = -0.5.$
Therefore, $P(x \geq 95) = P(z \geq -0.5) = 1 - 0.3085 = 0.6915$.

7.77 The z-value for x = 108 is $z = \frac{x - \mu}{\sigma} = \frac{108 - 100}{10} = 0.8$, and for x = 119 is
$z = \frac{x - \mu}{\sigma} = \frac{119 - 100}{10} = 1.9.$
Therefore, $P(108 \leq x \leq 119) = P(0.8 \leq z \leq 1.9) = 0.9713 - 0.7881 = 0.1832$.

7.79 The z-value for x = 87 is $z = \frac{x - \mu}{\sigma} = \frac{87 - 100}{10} = -1.3$, and for x = 102 is
$z = \frac{x - \mu}{\sigma} = \frac{102 - 100}{10} = 0.2.$
Therefore, $P(87 \leq x \leq 102) = P(-1.3 \leq z \leq 0.2) = 0.5793 - 0.0968 = 0.4825$.

7.81 The z-value for x = 71 is $z = \frac{x - \mu}{\sigma} = \frac{71 - 100}{10} = -2.9$, and for x = 93 is
$z = \frac{x - \mu}{\sigma} = \frac{93 - 100}{10} = -0.7.$
Therefore, $P(71 \leq x \leq 93) = P(-2.9 \leq z \leq -0.7) = 0.2420 - 0.0019 = 0.2401$.

7.83 The z-value for x = 38.2 is $z = \frac{x - \mu}{\sigma} = \frac{38.2 - 40.8}{2} = -1.3$, and for x = 44.6 is
$z = \frac{x - \mu}{\sigma} = \frac{44.6 - 40.8}{2} = 1.9.$

Therefore, $P(38.2 < x < 44.6) = P(-1.3 < z < 1.9) = P(z < 1.9) - P(z < -1.3) = 0.9713 - 0.0968 = 0.8745$.

7.85 The z-value for x = 45.1 is $z = \frac{x - \mu}{\sigma} = \frac{45.1 - 40.8}{2} = 2.15$.
Therefore, $P(x < 45.1) = P(z < 2.15) = 0.9842$.

7.87 The z-value for x = 37.9 is $z = \frac{x - \mu}{\sigma} = \frac{37.9 - 40.8}{2} = -1.45$, and for x = 40.3 is $z = \frac{x - \mu}{\sigma} = \frac{40.3 - 40.8}{2} = -0.25$. Therefore, $P(37.9 < x < 40.3) = P(-1.45 < z < -0.25) = P(z < -0.25) - P(z < -1.45) = 0.4013 - 0.0735 = 0.3278$.

7.89 We need to first find the z-value for which 78% of the distribution lies to the left. The z-value with area nearest to 0.7800 in Table III is z = 0.77, which means that P_{78} is 0.77 standard deviations above the mean. Therefore, $P_{78} = \mu + (0.77)\cdot\sigma = 280 + (0.77)\cdot(50) = 318.5$.

7.91 We need to first find the z-value for which 32% of the distribution lies to the left. The z-value with area nearest to 0.3200 in Table III is z = -0.47, which means that P_{32} is 0.47 standard deviations below the mean. Therefore, $P_{32} = \mu - (0.47)\cdot\sigma = 280 - (0.47)\cdot(50) = 256.5$.

7.93 We are given that the time required for a flight between two cities has approximately a normal distribution with $\mu = 54.8$ and $\sigma = 1.2$. The probability that a flight will take more than 56.6 minutes is $P(x > 56.6)$. The z-value for x = 56.6 is
$z = \frac{x - \mu}{\sigma} = \frac{56.6 - 54.8}{1.2} = 1.5$.
Therefore, $P(x > 56.6) = P(z > 1.5) = 1 - P(z \leq 1.5) = 1 - 0.9332 = 0.0668$.

7.95 We are given that the weight of the potatoes (in pounds) contained in a barrel has approximately a normal distribution with $\mu = 165$ and $\sigma = 0.8$. The probability that a barrel will contain at least 166 pounds of potatoes is $P(x \geq 166)$. The z-value for x = 166 is $z = \frac{x - \mu}{\sigma} = \frac{166 - 165}{0.8} = 1.25$.
Therefore, $P(x \geq 166) = P(z \geq 1.25) = 1 - 0.8944 = 0.1056 = 10.56\%$.

7.97 We need to first find the z-value for which 14% of the distribution lies to the left. The z-value with area nearest to 0.1400 in Table III is z = -1.08, which means that x is 1.08 standard deviations below the mean. Therefore, $x = \mu - (1.08)\cdot\sigma = 95 - (1.08)\cdot(12) = 82.04$.

7.99 We need to first find the z-value for which 1% of the distribution lies to the right and therefore, 99% of the distribution lies to the left. The z-value with area nearest to 0.9900 in Table III is z = 2.33, which means that x is 2.33 standard deviations above the mean. Therefore, $x = \mu + (2.33)\cdot\sigma = 7.4 + (2.33)\cdot(0.4) = 8.332$.

7.101 We need to first find the z-value for which 75% of the distribution lies to the right and, therefore, 25% of the distribution lies to the left. The z-value with area nearest to 0.2500 in Table III is $z = -0.67$, which means that x is 0.67 standard deviations below the mean. Therefore, $x = \mu - (0.67)\cdot\sigma = 172 - (0.67)\cdot(15.4) = 161.68$.

7.103 **a.** We are given that the distribution of costs has approximately a normal distribution with $\mu = 28{,}000$ and $\sigma = 2500$. We wish to find $P(x \le 25{,}000)$, the percentage of homes for which the supplies will cost no more than \$25,000. The z-value for x = 25,000 is

$$z = \frac{x - \mu}{\sigma} = \frac{25000 - 28000}{2500} = -1.2.$$

Therefore, $P(x \le 25{,}000) = P(z \le -1.2) = 0.1151$ which is 11.51%.

b. We wish to find $P(20{,}000 < x < 30{,}000)$, the percentage of homes for which the supplies will cost between \$20,000 and \$30,000. The z-value for x = 20,000 is

$$z = \frac{x - \mu}{\sigma} = \frac{20000 - 28000}{2500} = -3.2,$$ and for x = 30,000 is

$$z = \frac{x - \mu}{\sigma} = \frac{30000 - 28000}{2500} = 0.8.$$ Therefore, $P(20{,}000 < x < 30{,}000) = P(-3.2 < z < 0.8) = P(z < 0.8) - P(z < -3.2) = 0.7881 - 0.0007 = 0.7874$ which is 78.74%.

7.105 We need to first find the z-value for which 20% of the distribution lies to the right and therefore, 80% of the distribution lies to the left. The z-value with area nearest to 0.8000 in Table III is $z = 0.84$, which means that x is 0.84 standard deviations above the mean. Therefore, $x = \mu + (0.84)\cdot\sigma = 5 + (0.84)\cdot(0.2) = 5.168$.

MINITAB LAB ASSIGNMENTS

7.107(M) a.

```
MTB > # Exercise 7.107(M)
MTB > CDF 13.5;
SUBC> NORMAL 15.7 4.2.
   13.5000    0.3002
```

The proportion of ranges that last fewer than 13.5 years is 0.3002.

b.

```
MTB > CDF 17;
SUBC> NORMAL 15.7 4.2.
   17.0000    0.6215
```

The proportion of ranges that last more than 17 years is 1 - 0.6215 = 0.3785.

c.

```
Method 1:
MTB > SET C1
DATA> 10 20
DATA> END
```

```
MTB > CDF C1 C2;
SUBC> NORMAL 15.7 4.2.
MTB > PRINT C2

C2
  0.087368    0.847037

MTB > LET K1 = C2(2)-C2(1)
MTB > PRINT K1
K1        0.759669

Method 2:
MTB > CDF 20;
SUBC> NORMAL 15.7 4.2.
   20.0000      0.8470
MTB > CDF 10;
SUBC> NORMAL 15.7 4.2.
   10.0000      0.0874
```

The proportion of ranges that last between 10 and 20 years is 0.759669, using Method 1, or 0.8470 - 0.0874 = 0.7596, using Method 2.

7.109(M)

```
MTB > # Exercise 7.109(M)
MTB > INVCDF 0.05;
MTB > NORMAL 15.7 4.2.
    0.0500     8.7916
```

95% of them will last at least 8.7916 years.

EXERCISES for Section 7.4

Using Normal Distributions to Approximate Binomial Distributions

7.111 **a.** We are given n = 20 and p = ½ so $P(x) = C(20,x)\cdot(\frac{1}{2})^{x}\cdot(\frac{1}{2})^{(20-x)}$.

Therefore, $P(10) = C(20,10)\cdot(\frac{1}{2})^{10}\cdot(\frac{1}{2})^{10} = 184756\cdot\frac{1}{2^{20}} = 0.1762$.

b. By Table I, $P(10) = P(x \le 10) - P(x \le 9) = 0.588 - 0.412 = 0.176$.

c. First we note that $\mu = np = 20\cdot(½) = 10$, and $\sigma^2 = npq = 20\cdot(½)\cdot(½) = 5$, so $\sigma = \sqrt{5}$. Using the normal approximation, $P(10) \approx P(9.5 < x < 10.5)$, where x is normal with mean 10 and standard deviation $\sigma = \sqrt{5}$. The z-value for x = 9.5 is $z = \frac{x-\mu}{\sigma} = \frac{9.5-10}{\sqrt{5}} = -0.22$, and for x = 10.5 is $z = \frac{x-\mu}{\sigma} = \frac{10.5-10}{\sqrt{5}} = 0.22$.

Therefore, $P(10) \approx P(9.5 < x < 10.5) = P(-0.22 < z < 0.22) = 0.5871 - 0.4129 = 0.1742$.

7.113 First we note that $\mu = np = 15\cdot(0.6) = 9$, and $\sigma^2 = npq = 15\cdot(0.6)\cdot(0.4) = 3.6$, so $\sigma = 1.897$. Using the normal approximation, $P(9) \approx P(8.5 < x < 9.5)$, where x is normal with mean 9 and standard deviation $\sigma = 1.897$. The z-value for x = 8.5 is

$z = \frac{x - \mu}{\sigma} = \frac{8.5 - 9}{\sqrt{3.6}} = -0.26$, and for x = 9.5 is $z = \frac{x - \mu}{\sigma} = \frac{9.5 - 9}{\sqrt{3.6}} = 0.26$.
Therefore, P(9) ≈ P(8.5 < x < 9.5) = P(-0.26 < z < 0.26) = 0.6026 - 0.3974 = 0.2052. Using Table I, with n = 15 and p = 0.6, the probability of obtaining exactly 9 successes is P(9) = P(x ≤ 9) - P(x ≤ 8) = 0.597 - 0.390 = 0.207.

7.115 We note that n = 952 and p = 0.17. Therefore,

$$\mu = np = 952 \cdot (0.17) = 161.84 \text{ and } \sigma = \sqrt{npq} = \sqrt{952 \cdot (0.17) \cdot (0.83)} = 11.59.$$

Using the normal approximation, the probability that more than 175 moonlight during the school year is: P(176) +P(177) +· · · ≈ P(x > 175.5), where x is normal with mean 161.84 and standard deviation σ = 11.59. The z-value for x = 175.5 is

$z = \frac{x - \mu}{\sigma} = \frac{175.5 - 161.84}{11.59} = 1.18$. Therefore, P(x > 175.5) = P(z > 1.18) = 1 - 0.8810 = 0.1190.

7.117 We note that n = 1000 and p = 0.20. Therefore,

$\mu = np = 1000 \cdot (0.20) = 200 \text{ and } \sigma = \sqrt{npq} = \sqrt{1000 \cdot (0.20) \cdot (0.80)} = 12.65.$ Using the normal approximation, the probability that at least 220 are concerned with baggage is: P(220) +P(221) +· · · ≈ P(x > 219.5), where x is normal with mean 200 and standard deviation σ = 12.65. The z-value for x = 219.5 is $z = \frac{x - \mu}{\sigma} = \frac{219.5 - 200}{12.65} = 1.54$.
Therefore, P(x >219.5) = P(z > 1.54) = 1 - 0.9382 = 0.0618.

7.119 We note that n = 580 and p = 0.20. Therefore,

$\mu = np = 580 \cdot (0.20) = 116 \text{ and } \sigma = \sqrt{npq} = \sqrt{580 \cdot (0.20) \cdot (0.80)} = 9.63.$ Using the normal approximation, the probability that more than 125 were accompanied by dessert is: P(126) +P(127) +· · · ≈ P(x > 125.5), where x is normal with mean 116 and standard deviation σ = 9.63. The z-value for x = 125.5 is $z = \frac{x - \mu}{\sigma} = \frac{125.5 - 116}{9.63} = 0.99$.
Therefore, P(x > 125.5) = P(z > 0.99) = 1 - 0.8389 = 0.1611.

7.121 We note that n = 3375 and p = 0.90. Therefore,

$$\mu = np = 3375 \cdot (0.90) = 3037.5 \text{ and } \sigma = \sqrt{npq} = \sqrt{3375 \cdot (0.90) \cdot (0.10)} = 17.43.$$

Using the normal approximation, the probability that the number of cars equipped with this type of tire will be between 3,000 and 3,075, inclusive, is:
P(3000) +P(3001) +· · ·+P(3075) ≈ P(2999.5 < x <3075.5), where x is normal with mean 3037.5 and standard deviation σ = 17.43. The z-value for x = 2999.5 is

$z = \frac{x - \mu}{\sigma} = \frac{2999.5 - 3037.5}{17.43} = -2.18$, and for x = 3075.5 is $z = \frac{3075.5 - 3037.5}{17.43}$ = 2.18. Therefore, P(2999.5 < x <3075.5) = P(-2.18 < z < 2.18) = 0.9854 - 0.0146 = 0.9708.

MINITAB LAB ASSIGNMENTS

7.123(M) a.

```
MTB > # Exercise 7.123(M)
MTB > CDF 4;
SUBC> BINOMIAL 50 .12.
      K  P( X LESS OR = K)
   4.00              0.2680
```

Using the binomial distribution, the probability that there will be fewer than 5 overdue accounts is 0.2680.

b.

```
MTB > LET K1 = 50*.12 # Mean
MTB > LET K2 = SQRT(50*.12*.88) # standard deviation
MTB > CDF 4.5;  # note use of continuity correction
SUBC> NORMAL K1 K2.
    4.5000     0.2569
MTB > CDF -0.5;
SUBC> NORMAL K1 K2.
   -0.5000     0.0023
```

Using the normal approximation, the probability that there will be fewer than 5 overdue accounts is 0.2569 - 0.0023 = 0.2546.

REVIEW EXERCISES
CHAPTER 7

7.125 The area under the standard normal curve to the left of $z = 2.17$ is 0.9850.

7.126 The area under the standard normal curve to the left of $z = 1.95$ is 0.9744. Therefore, the area under the curve to the right of $z = 1.95$ is $1 - 0.9744 = 0.0256$.

7.127 **a.** $P(-1.98 < z < -0.95) = P(z < -0.95) - P(z < -1.98) = 0.1711 - 0.0239 = 0.1472$.
b. $P(-1.98 < z < 0.95) = P(z < 0.95) - P(z < -1.98) = 0.8289 - 0.0239 = 0.8050$.
c. $P(0.95 < z < 1.98) = P(z < 1.98) - P(z < 0.95) = 0.9761 - 0.8289 = 0.1472$.

7.128 The standard normal curve has a mean of 0 and a standard deviation of 1. The probability that its value lies within 2.5 standard deviations of the mean is $P(-2.5 < z < 2.5) = P(z < 2.5) - P(z < -2.5) = 0.9938 - 0.0062 = 0.9876$. This compares to "at least 0.84" ($= 1 - 1/(2.5)^2$), which is the value given by Chebyshev's Theorem.

7.129 Since the area to the right of z is 0.9726, we look in Table III for the z-value that corresponds to a cumulative area of $1 - 0.9726 = 0.0274$. We find that the associated z-value is $z = -1.92$.

7.130 Since the area between -z and z is 0.7198, and because of the symmetry of the standard normal distribution, the area to the left of -z is $(\frac{1}{2})\cdot(1 - 0.7198) = 0.1401$. We find from Table III that $-z = -1.08$ so $z = 1.08$.

7.131 **a.** We need to find the value of z for which 70% of the distribution lies to the left. We look in Table III for the z-value that corresponds to a cumulative area of 0.70. The nearest area to 0.70 in Table III is 0.6985, which is associated with z = 0.52. Therefore, $P_{70} = 0.52$.
b. We need to find the value of z for which 15% of the distribution lies to the left. We look in Table III for the z-value that corresponds to a cumulative area of 0.15. The nearest area to 0.15 in Table III is 0.1492, which is associated with z = -1.04. Therefore, $P_{15} = -1.04$.

7.132 **a.** Since the area under the standard normal curve to the right of $z_{0.02}$ is 0.02, we look in Table III for the z-value that corresponds to a cumulative area of 1 - 0.02 = 0.98. We find that the nearest associated z-value is $z = z_{0.02} = 2.05$.
b. Since the area under the standard normal curve to the right of $z_{0.08}$ is 0.08, we look in Table III for the z-value that corresponds to a cumulative area of 1 - 0.08 = 0.92. We find that the nearest associated z-value is $z = z_{0.08} = 1.41$.

7.133 We need to find the value of z for which 75% of the distribution lies to the left. We look in Table III for the z-value that corresponds to a cumulative area of 0.75. The nearest area to 0.75 in Table III is 0.7486, which is associated with z = 0.67. This means that the value of Q_3 is 0.67 standard deviations above the mean or $Q_3 = 150 + (0.67)(10) = 156.7$.

7.134 **a.** The z-value for x = 70.3 is $z = \frac{x - \mu}{\sigma} = \frac{70.3 - 78.4}{3.6} = -2.25.$
Therefore, $P(x \geq 70.3) = P(z \geq -2.25) = 1 - 0.0122 = 0.9878$.
b. The z-value for x = 73.9 is $z = \frac{x - \mu}{\sigma} = \frac{73.9 - 78.4}{3.6} = -1.25$, and for x = 89.2 is
$z = \frac{x - \mu}{\sigma} = \frac{89.2 - 78.4}{3.6} = 3$. Therefore, $P(73.9 \leq x \leq 89.2) = P(-1.25 \leq z \leq 3)$
$= P(z \leq 3) - P(z \leq -1.25) = 0.9987 - 0.1056 = 0.8931$.

7.135 We are given that the per capita consumption of yogurt in pounds has approximately a normal distribution with $\mu = 4.6$ and $\sigma = 1.3$. The proportion of the population that will consume less than one pound of yogurt is $P(x < 1)$. The z-value for x = 1 is
$z = \frac{x - \mu}{\sigma} = \frac{1 - 4.6}{1.3} = -2.77$. Therefore, $P(x < 1) = P(z < -2.77) = 0.0028$.

7.136 We need to first find the z-value for which 12% of the distribution lies to the left. The z-value with area 0.1200 in Table III is z = -1.175, (interpolate between -1.17 and -1.18) which means that x is -1.175 standard deviations below the mean. Therefore,
$x = \mu - (1.175)\cdot\sigma = 4.6 - (1.175)\cdot(1.3) = 3.07$ lbs.

7.137 We note that n = 1200 and $p = \frac{1}{6}$. Therefore,

$\mu = np = 1200\cdot\left(\frac{1}{6}\right) = 200$ *and* $\sigma = \sqrt{npq} = \sqrt{1200\cdot\left(\frac{1}{6}\right)\cdot\left(\frac{5}{6}\right)} = 12.91$. Using the normal approximation, the probability of tossing more than 225 ones in 1200 throws of a fair die is: P(226) +P(227) +· · · ≈ P(x > 225.5), where x is normal with mean 200 and standard deviation σ = 12.91. The z-value for x = 225.5 is

$$z = \frac{x - \mu}{\sigma} = \frac{225.5 - 200}{12.91} = 1.98.$$

Therefore, P(x > 225.5) = P(z > 1.98) = 1 - 0.9761 = 0.0239.

7.138 We are given that the time required to serve a customer a pizza has approximately a normal distribution with μ = 12.3 and σ = 1.7. The probability that a customer will wait longer than fifteen minutes is P(x > 15). The z-value for x = 15 is

$$z = \frac{x - \mu}{\sigma} = \frac{15 - 12.3}{1.7} = 1.59.$$

Therefore, P(x > 15) = P(z > 1.59) = 1 - 0.9441 = 0.0559.

7.139 We note that n = 500 and p = 0.15. Therefore,

$$\mu = np = 500\cdot(0.15) = 75 \text{ and } \sigma = \sqrt{npq} = \sqrt{500\cdot(0.15)\cdot(0.85)} = 7.98.$$

Using the normal approximation, the probability that the number of smokers will be between 60 and 80, inclusive, is: P(60) +P(61) +· · ·+P(80) ≈ P(59.5 < x < 80.5), where x is normal with mean 75 and standard deviation σ = 7.98. The z-value for x = 59.5 is

$$z = \frac{x - \mu}{\sigma} = \frac{59.5 - 75}{7.98} = -1.94,$$ and for x = 80.5 is $$z = \frac{80.5 - 75}{7.98} = 0.69.$$

Therefore, P(59.5 < x <80.5) = P(-1.94 < z < 0.69) = 0.7549 - 0.0262 = 0.7287.

7.140 We are given that the distribution of times has approximately a normal distribution with μ = 57.8 and σ = 9.6. We wish to find P(50 < x < 60), the percentage of returns completed that require between 50 and 60 minutes to complete. The z-value for x = 50 is

$$z = \frac{x - \mu}{\sigma} = \frac{50 - 57.8}{9.6} = -0.81,$$ and for x = 60 is $$z = \frac{x - \mu}{\sigma} = \frac{60 - 57.8}{9.6} = 0.23.$$

Therefore, P(50 < x < 60) = P(-0.81 < z < 0.23) = P(z < 0.23) - P(z < -0.81) = 0.5910 - 0.2090 = 0.3820 which is 38.20%.

7.141 We need to first find the z-value for which 10% of the distribution lies to the right and therefore, 90% of the distribution lies to the left. The z-value with area nearest to 0.9000 in Table III is z = 1.28, which means that x is 1.28 standard deviations above the mean. Therefore, x = μ + (1.28)·σ = 57.8 + (1.28)·(9.6) = 70.09 minutes.

7.142 We are given that the distribution of home prices has approximately a normal distribution with μ = 100,661 and σ = 24,000. We wish to find P(50,000 < x < 150,000), the percentage of homes that cost between \$50,000 and \$150,000. The z-value for x = 50,000 is

$$z = \frac{x - \mu}{\sigma} = \frac{50000 - 100661}{24000} = -2.11,$$ and for x = 150,000 is

$$z = \frac{x - \mu}{\sigma} = \frac{150000 - 100661}{24000} = 2.06.$$

Therefore, P(50,000 < x < 150,000) = P(-2.11 < z < 2.06) = P(z < 2.06) - P(z < -2.11) = 0.9803 - 0.0174 = 0.9629 which is 96.29%.

7.143 We note that n = 682 and p = 0.90. Therefore,

$$\mu = np = 682 \cdot (0.90) = 613.8 \text{ and } \sigma = \sqrt{npq} = \sqrt{682 \cdot (0.90) \cdot (0.10)} = 7.83.$$ Using the normal approximation, the probability that more than 600 were accompanied by dessert is: P(601) +P(602) + · · · ≈ P(x > 600.5), where x is normal with mean 613.8 and standard deviation σ = 7.83. The z-value for x = 600.5 is

$$z = \frac{x - \mu}{\sigma} = \frac{600.5 - 613.8}{7.83} = -1.70.$$

Therefore, P(x > 600.5) = P(z > -1.70) = 1 - 0.0446 = 0.9554.

7.144 We are given that the distribution of costs for a normal pregnancy and hospital delivery has approximately a normal distribution with μ = 4334 and σ = 125. We wish to find P(x > 4500), the proportion of time that the cost will exceed $4500. The z-value for x = 4500 is $z = \frac{x - \mu}{\sigma} = \frac{4500 - 4334}{125} = 1.33.$

Therefore, P(x > 4500) = P(z > 1.33) = 1 - P(z ≤ 1.33) = 1 - 0.9082 = 0.0918.

MINITAB LAB ASSIGNMENTS

7.145(M) a.

```
MTB > # Exercise 7.145(M)
MTB > CDF 24;
SUBC> BINOMIAL 50 0.60.
         K   P( X LESS OR = K)
     24.00                0.0573
```

Using the binomial distribution, we find the probability that at least 25 have a blood lead level that may impose this hazard is 1 - 0.0573 = 0.9427.

b.

```
MTB > LET K1 = 50*0.6 # mean
MTB > LET K2 = SQRT(50*0.6*0.4) # standard deviation
MTB > CDF 24.5;
SUBC> NORMAL k1 k2.
    24.5000     0.0562
```

Using the normal distribution, we find the probability at least 25 have a blood lead level that may impose this hazard is 1 - 0.0562 = 0.0.9438.

7.146(M) a.

```
MTB > # Exercise 7.146(M)
MTB > CDF 43.3;
SUBC> NORMAL 49.6 4.8.
   43.3000     0.0947
MTB > # P(x < 43.3) = 0.0947.
```

b.

```
MTB > INVCDF 0.75;
SUBC> NORMAL 49.6 4.8.
    0.7500    52.8376
MTB > # The 3rd quartile is 52.8376.
```

7.147(M) a.

```
MTB > # Exercise 7.147(M)
MTB > CDF 3.57;
SUBC> NORMAL 0 1.
    3.5700    0.9998
MTB > # P(z ≤ 3.57) = 0.9998.
```

b.

```
MTB > INVCDF 0.91;
SUBC> NORMAL 0 1.
    0.9100    1.3408
MTB > # The 91 st percentile for the standard normal distribution is
MTB > # 1.3408.
```

7.148(M) a.

```
MTB > # Exercise 7.148(M)
MTB > CDF -0.62;
SUBC> NORMAL 0 1.
   -0.6200    0.2676
MTB > # P(z < -0.62) = 0.2676
```

b.

```
MTB > CDF 1.86;
SUBC> NORMAL 0 1.
    1.8600    0.9686
MTB > # P(z > 1.86) = 1 - 0.9686 = 0.0314.
```

c.

```
MTB > SET C1
DATA> -1.64 1.11
DATA> END
MTB > CDF C1 C2;
SUBC> NORMAL 0 1.
MTB > PRINT C2

C2
  0.050503   0.866500

MTB > # P(-1.64 <= z <= 1.11) = 0.866500 - 0.050503 = 0.815997.
```

7.149(M)

```
MTB > # Exercise 7.149(M)
MTB > INVCDF 0.83;
SUBC> NORMAL 154 12.3.
    0.8300  165.7362
MTB > # the 83 rd percentile for a normal distribution with mean 154
MTB > # and standard deviation 12.3 is 165.7362.
```

7.150(M)

```
MTB > # Exercise 7.150(M)
MTB > INVCDF 0.96;
SUBC> NORMAL 0 1.
    0.9600    1.7507
MTB > # The value of z(0.04) is 1.7507.
```

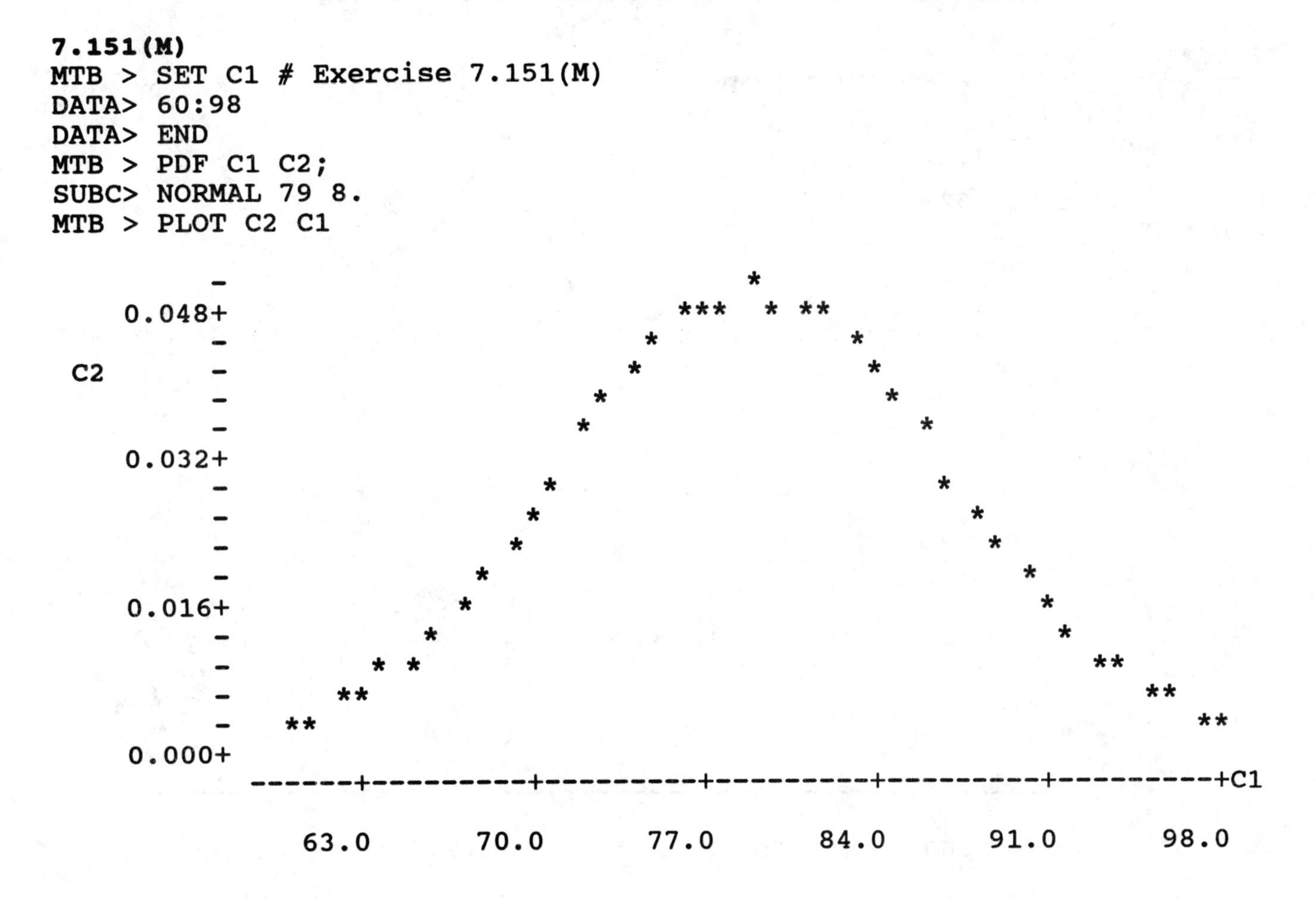

7.151(M)

```
MTB > SET C1 # Exercise 7.151(M)
DATA> 60:98
DATA> END
MTB > PDF C1 C2;
SUBC> NORMAL 79 8.
MTB > PLOT C2 C1
```

7.152(M)

```
MTB > SET C1 # Exercise 7.152(M)
DATA> 680:820/6
DATA> END
MTB > PDF C1 C2;
SUBC> NORMAL 750 32.
MTB > PLOT C2 C1
```

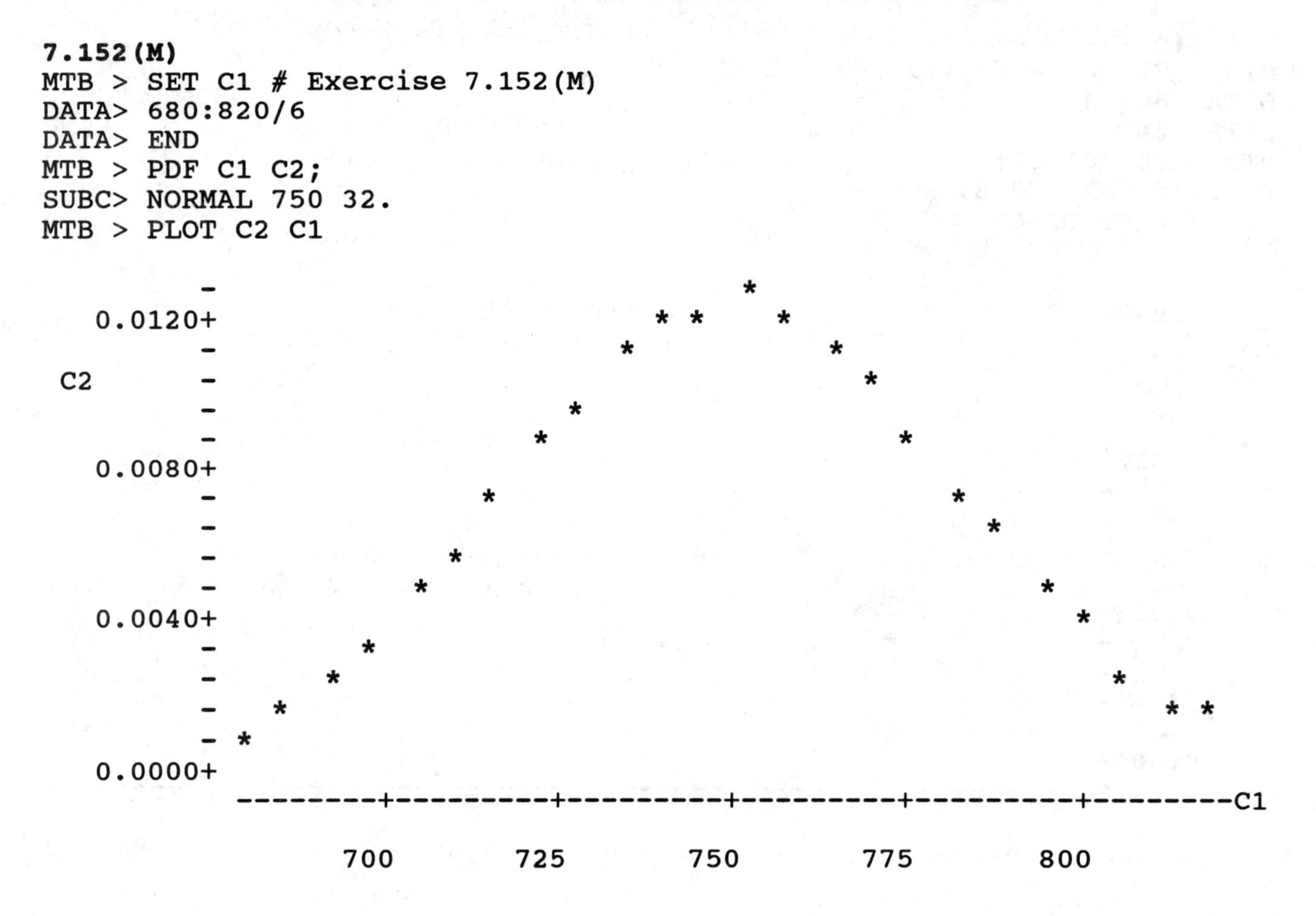

CHAPTER 8
SAMPLING METHODS AND SAMPLING DISTRIBUTIONS

EXERCISES for Section 8.1
Random Sampling

8.1 The number of possible samples of size n = 3 from a population of size N = 8 is C(8,3) = 56.

8.3 The number of possible samples of size n = 2 from a population of size N = 10 is C(10,2) = 45.

8.5 The number of possible samples of size n = 5 from a population of size N = 100 is C(100,5) = 75,287,520.

8.7 The number of possible samples of size n = 4 from a population of size N = 25 is C(25,4) = 12650. Since the samples are random, each possible sample of size n has the same probability of being selected. Therefore, we may apply the theoretical concept and conclude that the probability of each possible sample is $p = \frac{1}{C(25,4)} = \frac{1}{12650}$.

8.9 **a.** The number of possible samples of size n = 2 from a population of size N = 4 is C(4,2) = 6.
b. The 6 possible samples are: WX WY WZ XY XZ YZ
c. Since the samples are random, each possible sample of size n has the same probability of being selected. Therefore, we may apply the theoretical concept and conclude that the probability of each possible sample is $p = \frac{1}{C(4,2)} = \frac{1}{6}$.

8.11 **a.** We note that W is contained in 3 of the 6 samples listed in Exercise 8.9. Since the samples are random, each of the 6 has the same probability of being selected. Therefore, we may apply the theoretical concept and conclude that the probability that the sample contains W is $p = \frac{3}{6} = \frac{1}{2}$.

b. X is also contained in 3 of the 6 samples. As above, we may conclude that the probability that the sample contains X is ½.
c. Y is contained in 3 of the 6 samples. As above, we may conclude that the probability that the sample contains Y is ½.
d. Z is contained in 3 of the 6 samples. As above, we may conclude that the probability that the sample contains Z is ½.

8.13 **a.** Assign a number from 1 to 7 to each faculty member. Use MINITAB to select 2 numbers at random. (Other answers are possible.)
b. There are C(7,2) = 21 possible samples and, C(2,2) = 1 sample which contains the two department members who are are married to each other. Therefore, p=1/21.

MINITAB LAB ASSIGNMENTS

Note: Results for the following MINITAB problems will vary.

8.15(M)

```
MTB > # Exercise 8.15(M)
MTB > SET C1
DATA> 1:109
DATA> END
MTB > SAMPLE 25 C1 C2
MTB > PRINT C2

C2
    20     8    92    36    40    69     2    11   103    68    21
    34    90    43    75   102    86    70    76    87    26    78
     9    59    37
```

8.17(M)

```
MTB > # Exercise 8.17(M)
MTB > SET C1
DATA> 510:749
DATA> END
MTB > SAMPLE 20 C1 C2
MTB > PRINT C2

C2
   625   742   567   587   748   665   585   601   551   613   692
   619   732   712   589   649   663   646   726   526
```

EXERCISES for Section 8.2

Other Scientific Sampling Methods

8.19 Stratified random sample

8.21 Cluster sample

8.23 Systematic sample

EXERCISES for Section 8.3
Sampling Distributions

8.25 **a,b.** We list the 16 samples of two digits and the sample means in the following table:

Sample	1,1	1,3	1,5	1,7	3,1	3,3	3,5	3,7
Mean $\bar{x}$	1	2	3	4	2	3	4	5

Sample	5,1	5,3	5,5	5,7	7,1	7,3	7,5	7,7
Mean $\bar{x}$	3	4	5	6	4	5	6	7

c. Distribution of $\bar{x}$:

Mean	Probability
1	1/16
2	2/16
3	3/16
4	4/16
5	3/16
6	2/16
7	1/16

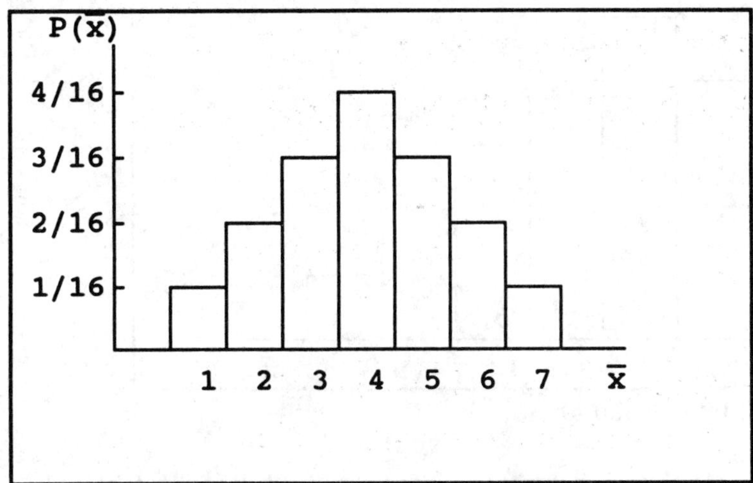

d. Histogram of the sampling distribution

8.27 **a,b.** We list the 25 samples of two digits and the sample means in the following table:

Sample	1,1	1,2	1,3	1,4	1,5	2,1	2,2	2,3	2,4	2,5	3,1	3,2	3,3
Mean $\bar{x}$	1.0	1.5	2.0	2.5	3.0	1.5	2.0	2.5	3.0	3.5	2.0	2.5	3.0

Sample	3,4	3,5	4,1	4,2	4,3	4,4	4,5	5,1	5,2	5,3	5,4	5,5
Mean $\bar{x}$	3.5	4.0	2.5	3.0	3.5	4.0	4.5	3.0	3.5	4.0	4.5	5.0

c. Distribution of $\bar{x}$:

Mean $\bar{x}$	Probability
1.0	1/25
1.5	2/25
2.0	3/25
2.5	4/25
3.0	5/25
3.5	4/25
4.0	3/25
4.5	2/25
5.0	1/25

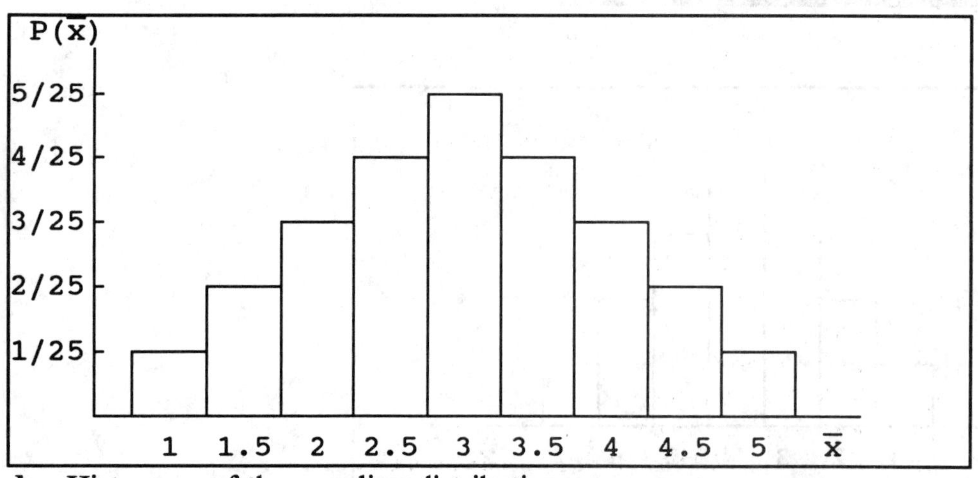

d. Histogram of the sampling distribution.

8.29 We list the 25 samples of two digits and the sample ranges in the following table:

Sample	1,1	1,2	1,3	1,4	1,5	2,1	2,2	2,3	2,4	2,5	3,1	3,2	3,3
Range	0	1	2	3	4	1	0	1	2	3	2	1	0

Sample	3,4	3,5	4,1	4,2	4,3	4,4	4,5	5,1	5,2	5,3	5,4	5,5
Range	1	2	3	2	1	0	1	4	3	2	1	0

Distribution of Sample Range:

Range	Frequency	Probability
0	5	0.20
1	8	0.32
2	6	0.24
3	4	0.16
4	2	0.08

MINITAB LAB ASSIGNMENTS

NOTE: Results on the following MINITAB problems will vary.

8.31(M)

```
MTB > # Exercise 8.31(M)
MTB > LET K1 = 1
MTB > STORE
STOR> NOECHO
STOR> RANDOM 2 C1;
STOR> INTEGER 1 5.
STOR> LET C2(K1) = MEAN(C1)
STOR> LET K1 = K1 + 1
STOR> END
MTB > EXECUTE 100 # It will take some time to do this
MTB > HISTOGRAM C2

Histogram of C2   N = 100

Midpoint   Count
     1.0       5  *****
     1.5       8  ********
     2.0      13  *************
     2.5      13  *************
     3.0      30  ******************************
     3.5      11  ***********
     4.0      10  **********
     4.5       8  ********
     5.0       2  **
```

The expected counts, according to Exercise 8.27, are 4, 8, 12, 16, 20, 16, 12, 8, and 4 (we multiply the probabilities found in 8.27 c. by 100). These compare favorably with the observed counts.

EXERCISES for Section 8.4

The Central Limit Theorem

8.33 Statement c; since the population has a normal distribution, the sampling distribution of the sample mean will be exactly normal, regardless of the size of n.

8.35 **a.** The standard deviation of the sampling distribution of the sample mean is $\sigma_{\bar{x}}$, where $\sigma_{\bar{x}} = \frac{\sigma}{\sqrt{n}} = \frac{18}{\sqrt{144}} = 1.5$.

b. For n = 81, $\sigma_{\bar{x}} = \frac{\sigma}{\sqrt{n}} = \frac{18}{\sqrt{81}} = 2$.

8.37 We are given $\sigma_{\bar{x}} = 2$ and $\sigma = 48$ and wish to determine n. We have

$2 = \sigma_{\bar{x}} = \frac{\sigma}{\sqrt{n}} = \frac{48}{\sqrt{n}} \rightarrow \sqrt{n} = 24$, so n = 576.

8.39 **a.** Since n is large (n = 81 > 30), by the Central Limit theorem, we would expect the histogram of the sample means to be approximately normal.

b. The expected mean is $\mu_{\bar{x}} = \mu = 79$, the mean of the population.

c. The estimated value for the standard deviation of the histogram of sample means is

$\sigma_{\bar{x}} = \frac{\sigma}{\sqrt{n}} = \frac{27}{\sqrt{81}} = 3$.

8.41 **a.** The mean of the random variable is $\mu = \Sigma x P(x) = 2 \cdot (0.5) + 4 \cdot (0.5) = 3$.
Therefore, $\sigma^2 = \Sigma(x - \mu)^2 \cdot P(x) = (2 - 3)^2 \cdot (0.5) + (4 - 3)^2 \cdot (0.5) = 1$, so $\sigma = 1$.

b.

$\bar{x}$	$P(\bar{x})$	$\bar{x} \cdot P(\bar{x})$	$(\bar{x} - 3)^2 \cdot P(\bar{x})$
2.0	1/16	2/16	1/16
2.5	4/16	10/16	1/16
3.0	6/16	18/16	0/16
3.5	4/16	14/16	1/16
4.0	1/16	4/16	1/16
		48/16 = 3	4/16 = ¼

Therefore, using the sampling distribution of $\bar{x}$, we find $\mu_{\bar{x}} = \Sigma \bar{x} \cdot P(\bar{x}) = 3$ and $\sigma_{\bar{x}}^2 = \Sigma(\bar{x} - 3)^2 \cdot P(\bar{x}) = ¼$, so $\sigma_{\bar{x}} = ½$.

c. Checking the results by using the formulas we have $\mu_{\bar{x}} = \mu = 3$, and

$\sigma_{\bar{x}} = \frac{\sigma}{\sqrt{n}} = \frac{1}{\sqrt{4}} = \frac{1}{2}$.

MINITAB LAB ASSIGNMENTS

Note: The answers to the following MINITAB problems will vary.

8.43(M) a,b.

```
MTB > # Exercise 8.43(M)
MTB > LET K1 = 1
MTB > STORE
STOR> NOECHO
STOR> RANDOM 25 C1;
STOR> NORMAL 50 12.
STOR> LET C2(K1) = MEAN(C1)
STOR> LET K1 = K1 + 1
STOR> END
MTB > EXECUTE 100
```

c.

```
Histogram of C2    N = 100

Midpoint   Count
      44       2  **
      45       1  *
      46       7  *******
      47       8  ********
      48      13  *************
      49      18  ******************
      50      12  ************
      51      15  ***************
      52       9  *********
      53       7  *******
      54       3  ***
      55       3  ***
      56       1  *
      57       1  *

MTB > MEAN C2
   MEAN     =       49.820
MTB > STDEV C2
   ST.DEV.  =       2.6252
```

d. We expect the mean of the sample means in C2 to be $\mu_{\bar{x}} = \mu = 50$, which compares favorably to our empirical value of 49.820. The estimated value for the standard deviation is $\sigma_{\bar{x}} = \frac{\sigma}{\sqrt{n}} = \frac{12}{\sqrt{25}} = 2.4$, compared to our empirical value of 2.6252. We also expect the distribution to have the shape of a normal curve.

8.45(M) a.,b.

```
MTB > # Exercise 8.45(M)
MTB > READ C3 C4
DATA> 1 0.2
DATA> 2 0.2
DATA> 3 0.2
DATA> 4 0.2
DATA> 5 0.2
DATA> END
      5 ROWS READ
MTB > LET K1 = 1
MTB > STORE
STOR> NOECHO
STOR> RANDOM 60 C1;
STOR> DISCRETE C3 C4.
STOR> LET C2(K1) = MEAN(C1)
STOR> LET K1 = K1 + 1
STOR> END
MTB > EXECUTE 100
```

c.

```
MTB > HISTOGRAM C2

Histogram of C2   N = 100

Midpoint   Count
     2.5       1  *
     2.6       2  **
     2.7       2  **
     2.8       9  *********
     2.9      16  ****************
     3.0      24  ************************
     3.1      21  *********************
     3.2      14  **************
     3.3      10  **********
     3.4       1  *

MTB > MEAN C2
   MEAN     =       3.0167
MTB > STDEV C2
   ST.DEV. =      0.17429
```

d. We expect the mean of the sample means in C2 to be $\mu_{\bar{x}} = \mu = 3$, which compares favorably to our empirical value of 3.0167. The estimated value for the standard deviation is

$\sigma_{\bar{x}} = \frac{\sigma}{\sqrt{n}} = \frac{\sqrt{2}}{\sqrt{60}} = 0.1826$, compared to our empirical value of 0.17429. We also expect the distribution to have approximately the shape of a normal curve.

EXERCISES for Section 8.5
Applications of the Central Limit Theorem

8.47 Since the sample size of n = 36 is large, the sampling distribution of $\overline{x}$ will, by the Central Limit Theorem, be approximately normal. The mean of the sampling distribution is $\mu_{\overline{x}} = \mu$ = 450 and the standard deviation is $\sigma_{\overline{x}} = \frac{\sigma}{\sqrt{n}} = \frac{12}{\sqrt{36}} = 2$. We wish to find the probability that the sample mean will be at least 454 grams, which is $P(\overline{x} \geq 454)$. The z-value for 454 is: $z = \frac{\overline{x} - \mu}{\sigma_{\overline{x}}} = \frac{454 - 450}{2} = 2$.
Therefore, $P(\overline{x} \geq 454) = P(z \geq 2) = 1 - 0.9772 = 0.0228$.

8.49 We note that n = 52, so the sample size is large, and the sampling distribution of $\overline{x}$ will, by the Central Limit Theorem, be approximately normal. The mean of the sampling distribution is $\mu_{\overline{x}} = \mu = 174$ and the standard deviation is $\sigma_{\overline{x}} = \frac{\sigma}{\sqrt{n}} = \frac{10}{\sqrt{52}} = 1.387$. The probability that the average number of calories consumed per steak will be less than 175 is $P(\overline{x} < 175)$. The z-value for 175 is: $z = \frac{\overline{x} - \mu}{\sigma_{\overline{x}}} = \frac{175 - 174}{1.387} = 0.72$.
Therefore, $P(\overline{x} < 175) = P(z < 0.72) = 0.7642$.

8.51 Since the sample size of n = 35 is large, the sampling distribution of $\overline{x}$ will, by the Central Limit Theorem, be approximately normal. The mean of the sampling distribution is $\mu_{\overline{x}} = \mu$ = 4334 and the standard deviation is $\sigma_{\overline{x}} = \frac{\sigma}{\sqrt{n}} = \frac{125}{\sqrt{35}} = 21.13$. The probability that the average cost exceeds $4,350 is $P(\overline{x} > 4350)$. The z-value for 4350 is:
$z = \frac{\overline{x} - \mu}{\sigma_{\overline{x}}} = \frac{4350 - 4334}{21.13} = 0.76$.
Therefore, $P(\overline{x} > 4350) = P(z > 0.76) = 1 - 0.7764 = 0.2236$.

8.53 Since the sample size of n = 100 is large, the sampling distribution of $\overline{x}$ will, by the Central Limit Theorem, be approximately normal. The mean of the sampling distribution is $\mu_{\overline{x}} = \mu$ = 7.4 and the standard deviation is $\sigma_{\overline{x}} = \frac{\sigma}{\sqrt{n}} = \frac{2.4}{\sqrt{100}} = 0.24$. The caterer will run out of shrimp at the affair if $\Sigma x > 50$ pounds = 800 ounces, which is equivalent to $\overline{x} > 8$.
The z-value for 8 is: $z = \frac{\overline{x} - \mu}{\sigma_{\overline{x}}} = \frac{8 - 7.4}{0.24} = 2.5$.
Therefore, $P(\overline{x} > 8) = P(z > 2.5) = 1 - 0.9938 = 0.0062$.

8.55 Since the sample size of n = 250 is large, the sampling distribution of $\bar{x}$ will, by the Central Limit Theorem, be approximately normal. The standard deviation of the sampling distribution is $\sigma_{\bar{x}} = \frac{\sigma}{\sqrt{n}} = \frac{45}{\sqrt{250}} = 2.846$.
The probability that the estimate will be in error by more than five cents is $P(|\bar{x} - \mu| > 5) = 1 - P(\mu - 5 \le \bar{x} \le \mu + 5)$. The z-value for $\mu - 5$ is:
$z = \frac{\bar{x} - \mu}{\sigma_{\bar{x}}} = \frac{(\mu - 5) - \mu}{2.846} = -1.76$, and the z-value for $\mu + 5$ is:
$z = \frac{\bar{x} - \mu}{\sigma_{\bar{x}}} = \frac{(\mu + 5) - \mu}{2.846} = 1.76$.
Therefore, $P(\mu - 5 \le \bar{x} \le \mu + 5) = P(-1.76 \le z \le 1.76) = 0.9608 - 0.0392 = 0.9216$, so $P(|\bar{x} - \mu| > 5) = 1 - P(\mu - 5 \le \bar{x} \le \mu + 5) = 1 - 0.9216 = 0.0784$.

REVIEW EXERCISES
CHAPTER 8

8.57 Since the sample size of n = 49 is large, the sampling distribution of $\bar{x}$ will, by the Central Limit Theorem, be approximately normal. The mean of the sampling distribution is $\mu_{\bar{x}} = \mu = 24.8$ and the standard deviation is $\sigma_{\bar{x}} = \frac{\sigma}{\sqrt{n}} = \frac{2.8}{\sqrt{49}} = 0.4$.
a. The probability that that the sample mean will be more than 26 is $P(\bar{x} > 26)$, and the z-value for 26 is: $z = \frac{\bar{x} - \mu}{\sigma_{\bar{x}}} = \frac{26 - 24.8}{0.4} = 3$.
Therefore, $P(\bar{x} > 26) = P(z > 3) = 1 - 0.9987 = 0.0013$.
b. The probability that that the sample mean will be less than 24 is $P(\bar{x} < 24)$, and the z-value for 24 is: $z = \frac{\bar{x} - \mu}{\sigma_{\bar{x}}} = \frac{24 - 24.8}{0.4} = -2$.
Therefore, $P(\bar{x} < 24) = P(z < -2) = 0.0228$.
c. The probability that that the sample mean will be between 24.2 and 25.5 is $P(24.2 < \bar{x} < 25.5)$. The z-value for 24.2 is: $z = \frac{\bar{x} - \mu}{\sigma_{\bar{x}}} = \frac{24.2 - 24.8}{0.4} = -1.5$, and for 25.5 is: $z = \frac{\bar{x} - \mu}{\sigma_{\bar{x}}} = \frac{25.5 - 24.8}{0.4} = 1.75$. Therefore,
$P(24.2 < \bar{x} < 25.5) = P(-1.5 < z < 1.75) = 0.9599 - 0.0668 = 0.8931$.

8.58 Since the population is normal, the sampling distribution of $\bar{x}$ will also be normal. The mean of the sampling distribution is $\mu_{\bar{x}} = \mu = 76$ and the standard deviation is
$\sigma_{\bar{x}} = \frac{\sigma}{\sqrt{n}} = \frac{10}{\sqrt{16}} = 2.5$.
a. The z-value for 75 is: $z = \frac{\bar{x} - \mu}{\sigma_{\bar{x}}} = \frac{75 - 76}{2.5} = -0.4$.
Therefore, $P(\bar{x} > 75) = P(z > -0.4) = 1 - 0.3446 = 0.6554$.
b. The z-value for 72.5 is: $z = \frac{\bar{x} - \mu}{\sigma_{\bar{x}}} = \frac{72.5 - 76}{2.5} = -1.4$.
Therefore, $P(\bar{x} < 72.5) = P(z < -1.4) = 0.0808$.

c. The z-value for 70 is $z = \frac{\bar{x} - \mu}{\sigma_{\bar{x}}} = \frac{70 - 76}{2.5} = -2.4$, and for 80 is
$z = \frac{\bar{x} - \mu}{\sigma_{\bar{x}}} = \frac{80 - 76}{2.5} = 1.6$.
Therefore, $P(70 < \bar{x} < 80) = P(-2.4 < z < 1.6) = 0.9452 - 0.0082 = 0.9370$.

8.59 Since the sample size of n = 73 is large, the sampling distribution of $\bar{x}$ will, by the Central Limit Theorem, be approximately normal. The mean of the sampling distribution is $\mu_{\bar{x}} = \mu = 7.80$ and the standard deviation is $\sigma_{\bar{x}} = \frac{\sigma}{\sqrt{n}} = \frac{1.40}{\sqrt{73}} = 0.1639$. We wish to find the probability that the average fare for that day exceeded \$8.00, which is $P(\bar{x} > 8)$. The z-value for 8 is: $z = \frac{\bar{x} - \mu}{\sigma_{\bar{x}}} = \frac{8 - 7.80}{0.1639} = 1.22$.
Therefore, $P(\bar{x} > 8) = P(z > 1.22) = 1 - 0.8888 = 0.1112$.

8.60 **a.** The standard deviation of the sampling distribution of the sample mean is $\sigma_{\bar{x}}$, where $\sigma_{\bar{x}} = \frac{\sigma}{\sqrt{n}} = \frac{22}{\sqrt{121}} = 2$.

b. For n = 400, $\sigma_{\bar{x}} = \frac{\sigma}{\sqrt{n}} = \frac{22}{\sqrt{400}} = 1.1$.

8.61 The standard deviation of the sampling distribution of the sample mean is $\sigma_{\bar{x}}$, where $\sigma_{\bar{x}} = \frac{\sigma}{\sqrt{n}}$. If the size of a random sample is doubled, the new value for the standard deviation is $\sigma_{\bar{x}} = \frac{\sigma}{\sqrt{2n}} = \frac{1}{\sqrt{2}} \cdot \frac{\sigma}{\sqrt{n}} = (0.707) \cdot \frac{\sigma}{\sqrt{n}}$. Therefore, the standard deviation of the sampling distribution of the sample mean is reduced to 70.7% of the original value.

8.62 Since n is large (n = 100 > 30), by the Central Limit theorem, we would expect the relative frequency histogram of the sample means to be approximately normal. The expected mean is $\mu_{\bar{x}} = \mu = 88.5$, which is the mean of the population. The expected standard deviation of the histogram of sample means is $\sigma_{\bar{x}} = \frac{\sigma}{\sqrt{n}} = \frac{12}{\sqrt{100}} = 1.2$.

8.63 The number of possible samples of size n = 5 from a population of size N = 50 is C(50,5) = 2,118,760

8.64 Cluster sampling

8.65 Systematic sampling

8.66 Stratified random sampling

8.67 Random sampling

8.68 Since the sample size of n = 36 is large, the sampling distribution of $\bar{x}$ will, by the Central Limit Theorem, be approximately normal. The mean of the sampling distribution is $\mu_{\bar{x}} = \mu = 6.4$ and the standard deviation is $\sigma_{\bar{x}} = \frac{\sigma}{\sqrt{n}} = \frac{0.6}{\sqrt{36}} = 0.1$. The boxes will contain at least 223.2 ounces of oranges if $\Sigma x \geq 223.2$, which is equivalent to $\bar{x} \geq 6.2$. (Divide by 36.)

The z-value for 6.2 is: $z = \frac{\bar{x}-\mu}{\sigma_{\bar{x}}} = \frac{6.2-6.4}{0.1} = -2$.

Therefore, $P(\bar{x} \geq 6.2) = P(z \geq -2) = 1 - 0.0228 = 0.9772$.

8.69 Since the sample size of n = 125 is large, the sampling distribution of $\bar{x}$ will, by the Central Limit Theorem, be approximately normal. The standard deviation of the sampling distribution is $\sigma_{\bar{x}} = \frac{\sigma}{\sqrt{n}} = \frac{35}{\sqrt{125}} = 3.13$. The probability that the estimate will be in error by less than \$7 is $P(\mu - 7 < \bar{x} < \mu + 7)$.

The z-value for $\mu - 7$ is: $z = \frac{\bar{x}-\mu}{\sigma_{\bar{x}}} = \frac{(\mu-7)-\mu}{3.13} = -2.24$,

and the z-value for $\mu + 7$ is: $z = \frac{\bar{x}-\mu}{\sigma_{\bar{x}}} = \frac{(\mu+7)-\mu}{3.13} = 2.24$. Therefore,

$P(\mu - 7 < \bar{x} < \mu + 7) = P(-2.24 < z < 2.24) = 0.9875 - 0.0125 = 0.9750$.

8.70 Since the sample size of n = 39 is large, the sampling distribution of $\bar{x}$ will, by the Central Limit Theorem, be approximately normal. The standard deviation of the sampling distribution is $\sigma_{\bar{x}} = \frac{\sigma}{\sqrt{n}} = \frac{25}{\sqrt{39}} = 4.003$. The probability that the estimate will be in error by more than five is $P(|\bar{x} - \mu| > 5) = 1 - P(\mu - 5 \leq \bar{x} \leq \mu + 5)$.

The z-value for $\mu - 5$ is: $z = \frac{\bar{x}-\mu}{\sigma_{\bar{x}}} = \frac{(\mu-5)-\mu}{4.003} = -1.25$, and the z-value for $\mu + 5$ is: $z = \frac{\bar{x}-\mu}{\sigma_{\bar{x}}} = \frac{(\mu+5)-\mu}{4.003} = 1.25$. Therefore,

$P(\mu - 5 \leq \bar{x} \leq \mu + 5) = P(-1.25 \leq z \leq 1.25) = 0.8944 - 0.1056 = 0.7888$, so

$P(|\bar{x} - \mu| > 5) = 1 - P(\mu - 5 \leq \bar{x} \leq \mu + 5) = 1 - 0.7888 = 0.2112$.

8.71 Since the sample size of n = 33 is large, the sampling distribution of $\bar{x}$ will, by the Central Limit Theorem, be approximately normal. The mean of the sampling distribution is $\mu_{\bar{x}} = \mu = 55$ and the standard deviation is $\sigma_{\bar{x}} = \frac{\sigma}{\sqrt{n}} = \frac{19}{\sqrt{33}} = 3.31$. The probability that the average cost is as large as \$65.48 is $P(\bar{x} \geq 65.48)$. The z-value for 65.48 is:

$z = \frac{\bar{x}-\mu}{\sigma_{\bar{x}}} = \frac{65.48-55}{3.31} = 3.17$. Therefore, $P(\bar{x} \geq 65.48) = P(z \geq 3.17) = 1 - 0.9992 = 0.0008$; yes, we would doubt the laboratory's claim.

8.72 **a.** The number of possible samples of size n = 2 from a population of size N = 7 is C(7,2) = 21.

b. Denote Airedale by A, Boston Terrier by B, Collie by C, Dalmatian by D, Elkhound by E, Fox Terrier by F, and Great Dane by G. The 21 possible samples are: AB AC AD AE AF AG BC BD BE BF BG CD CE CF CG DE DF DG EF EG FG.

c. Place the names of each dog on a slip of paper, place the 7 slips in a hat, and then draw two names from the hat. (Other answers are possible)
d. Since the samples are random, each possible sample of size n has the same probability of being selected. Therefore, we may apply the theoretical concept and conclude that the probability of each possible sample (including the sample AF) is $p = \frac{1}{C(7,2)} = \frac{1}{21}$.

8.73 **a.** In order to find the mean and the standard deviation of x, we create a table:

x	P(x)	x•P(x)	x^2•P(x)
3	1/6	3/6	9/6
4	1/6	4/6	16/6
5	1/6	5/6	25/6
6	1/6	6/6	36/6
7	1/6	7/6	49/6
8	1/6	8/6	64/6
		33/6 = 5.5	199/6

We have $\mu = \Sigma x \cdot P(x) = 5.5$, and $\sigma = \sqrt{\Sigma X^2 \cdot P(X) - \mu^2} = \sqrt{\frac{199}{6} - (5.5)^2} = 1.708$.
b, c. The following table lists all possible samples of two independent observations from this population and their sample means:

Sample	Mean	Sample	Mean	Sample	Mean
3,3	3.0	5,3	4.0	7,3	5.0
3,4	3.5	5,4	4.5	7,4	5.5
3,5	4.0	5,5	5.0	7,5	6.0
3,6	4.5	5,6	5.5	7,6	6.5
3,7	5.0	5,7	6.0	7,7	7.0
3,8	5.5	5,8	6.5	7,8	7.5
4,3	3.5	6,3	4.5	8,3	5.5
4,4	4.0	6,4	5.0	8,4	6.0
4,5	4.5	6,5	5.5	8,5	6.5
4,6	5.0	6,6	6.0	8,6	7.0
4,7	5.5	6,7	6.5	8,7	7.5
4,8	6.0	6,8	7.0	8,8	8.0

d, e. The following table shows the sampling distribution of the sample mean by listing the possible values of $\bar{x}$ and the associated probabilities. Also listed are calculations needed to find the mean and the standard deviation for the sampling distribution.

$\bar{x}$	$P(\bar{x})$	$\bar{x} \cdot P(\bar{x})$	$(\bar{x})^2 \cdot P(\bar{x})$
3.0	1/36	3/36	9/36
3.5	2/36	7/36	49/72
4.0	3/36	12/36	48/36
4.5	4/36	18/36	81/36
5.0	5/36	25/36	125/36
5.5	6/36	33/36	363/72
6.0	5/36	30/36	180/36
6.5	4/36	26/36	169/36
7.0	3/36	21/36	147/36
7.5	2/36	15/36	225/72
8.0	1/36	8/36	64/36
SUM:	36/36 =1	$\mu_{\bar{x}}$ = 198/36 =5.5	2283/72

We note that $\mu_{\bar{x}} = 5.5$ and that

$$\sigma_{\bar{x}} = \sqrt{\Sigma (\bar{x})^2 \cdot P(\bar{x}) - \mu^2} = \sqrt{\frac{2283}{72} - (5.5)^2} = 1.208.$$

f. In part a, we found that $\mu = 5.5$ which confirms $\mu = \mu_{\bar{x}} = 5.5$ as found in part e. Also, using the results from part a, $\sigma_{\bar{x}} = \frac{\sigma}{\sqrt{n}} = \frac{1.708}{\sqrt{2}} = 1.208$. This agrees with the value found in part e.

MINITAB LAB ASSIGNMENTS

Note: The answers to the following MINITAB problems will vary.

8.74(M) a.,b.

```
MTB > # Exercise 8.74(M)
MTB > READ C3 C4
DATA> 3 1
DATA> 4 1
DATA> 5 1
DATA> 6 1
DATA> 7 1
DATA> 8 1
DATA> END
      6 ROWS READ
MTB > LET C4 = C4/6
MTB > LET K1 = 1
MTB > STORE
STOR> NOECHO
STOR> RANDOM 2 C1;
STOR> DISCRETE C3 C4.
STOR> LET C2(K1) = MEAN(C1)
STOR> LET K1 = K1 + 1
STOR> END
MTB > EXECUTE 100
```

c.

```
MTB > HIST C2

Histogram of C2   N = 100

Midpoint   Count
     3.0       2  **
     3.5       8  ********
     4.0      11  ***********
     4.5      12  ************
     5.0      14  **************
     5.5      14  **************
     6.0      14  **************
     6.5      14  **************
     7.0       6  ******
     7.5       4  ****
     8.0       1  *
```

d.

```
MTB > MEAN C2
   MEAN    =       5.3400
MTB > STDEV C2
   ST.DEV. =       1.1696
```

We expect the mean of the sample means in C2 to be $\mu_{\bar{x}} = \mu = 5.5$, which compares favorably to our empirical value of 5.34. The estimated value for the standard deviation is

$\sigma_{\bar{x}} = \dfrac{\sigma}{\sqrt{n}} = \dfrac{1.708}{\sqrt{2}} = 1.208$, compared to our empirical value of 1.1696.

8.75(M)a,b.

```
MTB > # Exercise 8.75(M)
MTB > # C3 and C4 were retained from 8.74(M)
MTB > LET K1 = 1
MTB > STORE
STOR> NOECHO
STOR> RANDOM 32 C1;
STOR> DISCRETE C3 C4.
STOR> LET C2(K1) = MEAN(C1)
STOR> LET K1 = K1 + 1
STOR> END
MTB > EXECUTE 100
```

c.

```
MTB > HIST C2
Histogram of C2   N = 100

Midpoint   Count
     4.8       3  ***
     5.0       6  ******
     5.2      19  *******************
     5.4      22  **********************
     5.6      25  *************************
     5.8      15  ***************
     6.0       5  *****
     6.2       3  ***
     6.4       2  **
```

d.

```
MTB > MEAN C2
   MEAN    =      5.5012
MTB > STDEV C2
   ST.DEV. =    0.32167
```

We expect the mean of the sample means in C2 to be $\mu_{\bar{x}} = \mu = 5.5$, which compares favorably to our empirical value of 5.5012. The estimated value for the standard deviation is

$\sigma_{\bar{x}} = \frac{\sigma}{\sqrt{n}} = \frac{1.708}{\sqrt{32}} = 0.302$, compared to our empirical value of 0.32167.

8.76(M)

```
MTB > # Exercise 8.76(M)
MTB > LET K1 = 1
MTB > STORE
STOR> NOECHO
STOR> RANDOM 20 C1;
STOR> NORMAL 500 75.
STOR> LET C2(K1) = MEAN(C1)
STOR> LET K1 = K1 + 1
STOR> END
MTB > EXECUTE 100
MTB > HISTOGRAM C2
Histogram of C2   N = 100
```

```
Midpoint   Count
     460       1  *
     470       3  ***
     480      14  **************
     490      19  *******************
     500      19  *******************
     510      25  *************************
     520      13  *************
     530       4  ****
     540       1  *
     550       1  *

MTB > MEAN C2
   MEAN     =        501.60
MTB > STDEV C2
   ST.DEV.  =        16.646
```

We expect the mean of the sample means in C2 to be $\mu_{\bar{x}} = \mu = 500$, which compares favorably to our empirical value of 501.60. The estimated value for the standard deviation is $\sigma_{\bar{x}} = \frac{\sigma}{\sqrt{n}} = \frac{75}{\sqrt{20}} = 16.771$, compared to our empirical value of 16.646. We also expect the histogram to approximate the shape of a normal curve.

8.77(M)

```
MTB > # Exercise 8.77(M)
MTB > SET C1
DATA> 1:100
DATA> END
MTB > SAMPLE 50 C1 C2
MTB > PRINT C2

C2
    54    56    93    52    57    63    81    33    70    44    84
    32    21     2     3    83   100    41    24    26     6    49
    88    66    23    60     1    58    39    43    61    31    20
    79    37    99    67    78     7    91    47    10    97    62
    27    30    95    74    28    73
```

8.78(M)

```
MTB > # Exercise 8.78(M)
MTB > SET C1
DATA> 1:308
DATA> END
MTB > SAMPLE 30 C1 C2
MTB > PRINT C2

C2
   290   192   181    78   133   278   224   282     1   168    24
   138    33   121   298   155   189    88   279   300   230   122
   199    52     9    89   119   246    38   217
```

CHAPTER 9
ESTIMATING MEANS, PROPORTIONS, AND VARIANCES: SINGLE SAMPLE

EXERCISES for Section 9.1 & 9.2
Basic Concepts of Estimating Population Parameters
Confidence Interval for a Mean: Large Sample

9.1 **a.** For a 91% confidence level, the subscript for the z-value is (1 - 0.91)/2 = 0.045. Therefore, the desired z-value will have a cumulative area to the left of it equal to (1 - 0.045) = 0.955. From Table III, we find that the z-value is $z_{0.045} = 1.70$.
b. For a 94% confidence level, the subscript for the z-value is (1 - 0.94)/2 = 0.03. Therefore, the desired z-value will have a cumulative area to the left of it equal to (1 - 0.03) = 0.97. From Table III, we find that the z-value is $z_{0.03} = 1.88$.
c. For a 96% confidence level, the subscript for the z-value is (1 - 0.96)/2 = 0.02. Therefore, the desired z-value will have a cumulative area to the left of it equal to (1 - 0.02) = 0.98. From Table III, we find that the z-value is $z_{0.02} = 2.05$.
d. For a 97% confidence level, the subscript for the z-value is (1 - 0.97)/2 = 0.015. Therefore, the desired z-value will have a cumulative area to the left of it equal to (1 - 0.015) = 0.985. From Table III, we find that the z-value is $z_{0.015} = 2.17$.

9.3 A point estimate is a single value; a confidence interval estimate consists of an interval of values with an associated level of confidence.

9.5 We are given $\bar{x} = 75.9$, s= 8, and n = 100. For a 90% confidence level, the subscript for the z-value is (1 - 0.90)/2 = 0.05. From Table III, we find that the z-value is $z_{\alpha/2} = z_{0.05} = 1.645$. Therefore, we have: $\bar{x} \pm z_{\alpha/2}\frac{\sigma}{\sqrt{n}} \Rightarrow 75.9 \pm 1.645\frac{8}{\sqrt{100}}$, where we replace σ with s = 8, $\Rightarrow 75.9 \pm 1.32 \Rightarrow 74.58 < \mu < 77.22$.

9.7 **a.** We are given $\bar{x} = 25.4$, $\sigma = 12$, and n = 36. For a 95% confidence level, the subscript for the z-value is (1 - 0.95)/2 = 0.025. From Table III, we find that the z-value is $z_{\alpha/2} = z_{0.025} = 1.960$. Therefore the endpoints are given by:

$$\bar{x} \pm z_{\alpha/2}\frac{\sigma}{\sqrt{n}} \Rightarrow 25.4 \pm 1.960\frac{12}{\sqrt{36}} \Rightarrow 25.4 \pm 3.92 \Rightarrow 21.48 < \mu < 29.32.$$

b. For a 99% confidence level, the subscript for the z-value is (1 - 0.99)/2 = 0.005. From Table III, we find that the z-value is $z_{\alpha/2} = z_{0.005} = 2.576$. Therefore:

$$\bar{x} \pm z_{\alpha/2}\frac{\sigma}{\sqrt{n}} \Rightarrow 25.4 \pm 2.576\frac{12}{\sqrt{36}} \Rightarrow 25.4 \pm 5.15 \Rightarrow 20.25 < \mu < 30.55.$$

c. The width of the confidence interval is increased from a width of 7.84 to a width of 10.30 when the level of confidence is increased from 95% to 99%.

9.9 **a.** We are given $\bar{x} = 98.84$, $\sigma = 10$, and $n = 100$. For a 95% confidence level, the subscript for the z-value is (1 - 0.95)/2 = 0.025. From Table III, we find that the z-value is $z_{\alpha/2} = z_{0.025} = 1.960$. Therefore the endpoints are given by:

$$\bar{x} \pm z_{\alpha/2}\frac{\sigma}{\sqrt{n}} \Rightarrow 98.84 \pm 1.960\frac{10}{\sqrt{100}} \Rightarrow 98.84 \pm 1.96 \Rightarrow 96.88 < \mu < 100.80.$$

b. We are given $\bar{x} = 98.84$, $\sigma = 20$, and $n = 100$. For a 95% confidence level, the subscript for the z-value is (1 - 0.95)/2 = 0.025. From Table III, we find that the z-value is $z_{\alpha/2} = z_{0.025} = 1.960$. Therefore the endpoints are given by:

$$\bar{x} \pm z_{\alpha/2}\frac{\sigma}{\sqrt{n}} \Rightarrow 98.84 \pm 1.960\frac{20}{\sqrt{100}} \Rightarrow 98.84 \pm 3.92 \Rightarrow 94.92 < \mu < 102.76.$$

c. The width of the confidence interval is directly proportional to the standard deviation.

9.11 We are given $\bar{x} = 13.36$, $s = 0.22$ and $n = 31$. For a 99% confidence level, the subscript for the z-value is (1 - 0.99)/2 = 0.005. From Table III, we find that the z-value is $z_{\alpha/2} = z_{0.005} = 2.576$. Therefore we find the endpoints of the confidence interval

using: $\bar{x} \pm z_{\alpha/2}\frac{\sigma}{\sqrt{n}} \Rightarrow 13.36 \pm 2.576\frac{0.22}{\sqrt{31}}$, where we replace σ with $s = 0.22$,

$\Rightarrow 13.36 \pm 0.102 \Rightarrow 13.258 < \mu < 13.462.$

9.13 We are given $\bar{x} = 850$, $s = 80$, and $n = 52$. For a 90% confidence level, the subscript for the z-value is (1 - 0.90)/2 = 0.05. From Table III, we find that the z-value is $z_{\alpha/2} = z_{0.05} =$ 1.645. Therefore, the endpoints are given by: $\bar{x} \pm z_{\alpha/2}\frac{\sigma}{\sqrt{n}} \Rightarrow 850 \pm 1.645\frac{80}{\sqrt{52}}$, where we replace σ with $s = 80$, $\Rightarrow 850 \pm 18.2 \Rightarrow 831.8 < \mu < 868.2.$

9.15 We are given $\bar{x} = 577$, $s = 173$ and $n = 924$. For a 99% confidence level, the subscript for the z-value is (1 - 0.99)/2 = 0.005. From Table III, we find that the z-value is $z_{\alpha/2} = z_{0.005}$ = 2.576. Therefore the endpoints are given by: $\bar{x} \pm z_{\alpha/2}\frac{\sigma}{\sqrt{n}} \Rightarrow 577 \pm 2.576\frac{173}{\sqrt{924}}$, where we replace σ with $s = 173$, $\Rightarrow 577 \pm 14.66 \Rightarrow \$562.34 < \mu < \$591.66.$

9.17 We are given $\bar{x} = 97.8$, $s = 3.1$, and $n = 48$. For a 90% confidence level, the subscript for the z-value is (1 - 0.90)/2 = 0.05. From Table III, we find that the z-value is

$z_{\alpha/2} = z_{0.05} = 1.645$. Therefore, the endpoints are given by: $\bar{x} \pm z_{\alpha/2}\frac{\sigma}{\sqrt{n}} \Rightarrow$

$97.8 \pm 1.645\frac{3.1}{\sqrt{48}}$, where we replace σ with $s = 3.1$, $\Rightarrow 97.8 \pm 0.74 \Rightarrow$
$97.06 < \mu < 98.54.$

MINITAB LAB ASSIGNMENTS

9.19(M)

```
MTB > # Exercise 9.19(M)
MTB > # The data from Exercise 9.18 were read into C1 using SET C1
MTB > STDEV C1
   ST.DEV. =       0.35503
MTB > ZINTERVAL 99 0.35503 C1

THE ASSUMED SIGMA =0.355

              N      MEAN     STDEV   SE MEAN     99.0 PERCENT C.I.
C1           41    9.8506    0.3550    0.0554   (  9.7075,   9.9937)
```

9.21(M)

```
MTB > # Exercise 9.21(M)
MTB > ZINTERVAL 90 1.6 C1

THE ASSUMED SIGMA =1.60

              N      MEAN     STDEV   SE MEAN     90.0 PERCENT C.I.
FAT          40    19.990     1.645     0.253   (  19.573,   20.407)
```

EXERCISES for Section 9.3

Student's t Probability Distributions

9.23 From Table IV, with df = 23, we find $t_{0.10} = 1.319$.

9.25 From Table IV, with df = 14, we find $t_{0.005} = 2.977$.

9.27 From Table IV, with df = 11, we find $t_{0.025} = 2.201$.

9.29 From Table IV, with df = 29, we find $t_{0.01} = 2.462$.

9.31 From Table IV, with df = ∞, we find $t_{0.05} = 1.645$.

9.33 The 95th percentile is the value of t for which the cumulative area to the left is 0.95. Therefore, the area under the curve to the right of t is 1 - 0.95 = 0.05. From Table IV, with df = 26, we find $P_{95} = t_{0.05} = 1.706$.

9.35 The 99th percentile is the value of t for which the cumulative area to the left is 0.99. Therefore, the area under the curve to the right of t is 1 - 0.99 = 0.01. From Table IV, with df = 15, we find $P_{99} = t_{0.01} = 2.602$.

9.37 The 90th percentile is the value of t for which the cumulative area to the left is 0.90. Therefore, the area under the curve to the right of t is 1 - 0.90 = 0.10. From Table IV with df = 3, we find $P_{90} = t_{0.10} = 1.638$.

9.39 The 5th percentile is the value of t for which the cumulative area to the left is 0.05. Therefore, the area under the curve to the right of t is 1 - 0.05 = 0.95. From Table IV, with df = 12 and using the symmetry of the Student's t distribution, we find $P_5 = -t_{0.05} = -1.782$.

9.41 The first percentile is the value of t for which the cumulative area to the left is 0.01. Therefore, the area under the curve to the right of t is 1 - 0.01 = 0.99. From Table IV, with df = ∞ and using the symmetry of the Student's t distribution (in this case the standard normal curve), we find $P_1 = -t_{0.01} = -2.326$.

9.43 The 10th percentile is the value of t for which the cumulative area to the left is 0.10. Therefore, the area under the curve to the right of t is 1 - 0.10 = 0.90. From Table IV, with df = 17 and using the symmetry of the Student's t distribution, we find $P_{10} = -t_{0.10} = -1.333$.

9.45 We note in Table IV that, with df = 13, $t_{0.01} = 2.650$. Therefore, $P(t > 2.650) = 0.01$.

9.47 We note in Table IV that, with df = 5, $t_{0.10} = 1.476$. Therefore, $P(t > 1.476) = 0.10$, so we have $P(t < 1.476) = 1 - 0.10 = 0.90$.

9.49 We note in Table IV that, with df = 19, $t_{0.025} = 2.093$. Therefore, $P(t < -2.093) = P(t > 2.093) = 0.025$.

9.51 We note in Table IV that, with df = 25, $t_{0.05} = 1.708$ and $t_{0.005} = 2.787$. Therefore, $P(-1.708 < t < 2.787) = P(t < 2.787) - P(t < -1.708) = 0.995 - 0.05 = 0.945$.

9.53 We note in Table IV that, with df = 11, $t_{0.01} = 2.718$. Therefore, $P(t > 2.718) = 0.01$, so we conclude that c = 2.718.

9.55 Since $P(t > c) = 0.90$, it follows that $P(t < c) = 0.10$. We note in Table IV that, with df = 3, $t_{0.10} = 1.638$. Therefore, $P(t > 1.638) = P(t < -1.638) = 0.10$, so we conclude that c = -1.638.

9.57 Since $P(-c < t < c) = 0.99$, $P(t > c) = ½(1 - 0.99) = 0.005$. Therefore, $c = t_{0.005}$. We note in Table IV that, with df = 26, $t_{0.005} = 2.779$, so we conclude that c = 2.779.

MINITAB LAB ASSIGNMENTS

9.59(M) a.

```
MTB > # Exercise 9.59(M) a.
MTB > CDF 2.22;
SUBC> T 48.
    2.2200    0.9844
MTB > # The answer is P(t > 2.22) = 1 - 0.9844 = 0.0156.
```

b.

```
MTB > # Exercise 9.59(M) b.
MTB > INVCDF 0.68;
SUBC> T 48.
    0.6800    0.4707
```

EXERCISES for Section 9.4

Confidence Interval for a Mean: Small Sample

9.61 **a.** For a confidence level of 80%, the subscript for the t-value is $\alpha/2 = (1 - 0.80)/2 = 0.10$. We use Table IV to find $t_{\alpha/2}$ based on df = (n -1) = 15 - 1 = 14. The desired t-value is $t_{\alpha/2} = t_{0.10} = 1.345$.
b. For a confidence level of 90%, the subscript for the t-value is $\alpha/2 = (1 - 0.90)/2 = 0.05$. We use Table IV to find $t_{\alpha/2}$ based on df = (n -1) = 22 - 1 = 21. The desired t-value is $t_{\alpha/2} = t_{0.05} = 1.721$.
c. For a confidence level of 99%, the subscript for the t-value is $\alpha/2 = (1 - 0.99)/2 = 0.005$. We use Table IV to find $t_{\alpha/2}$ based on df = (n -1) = 29 - 1 = 28. The desired t-value is $t_{\alpha/2} = t_{0.005} = 2.763$.
d. For a confidence level of 95%, the subscript for the t-value is $\alpha/2 = (1 - 0.95)/2 = 0.025$. We use Table IV to find $t_{\alpha/2}$ based on df = (n -1) = 9 - 1 = 8. The desired t-value is $t_{\alpha/2} = t_{0.025} = 2.306$.

9.63 We are given $\bar{x} = 575$, s = 95, and n = 25. For a 99% confidence level, the subscript for the t-value is (1 - 0.99)/2 = 0.005. From Table IV, using df = n - 1 = 24, we find that the t-value is $t_{\alpha/2} = t_{0.005} = 2.797$. Therefore, we find the endpoints of the confidence interval using: $\bar{x} \pm t_{\alpha/2}\frac{s}{\sqrt{n}} \Rightarrow 575 \pm 2.797\frac{95}{\sqrt{25}} \Rightarrow 575 \pm 53.1 \Rightarrow$
$521.9 < \mu < 628.1$.

9.65 We first calculate the sample mean and standard deviation. $\bar{x} = \Sigma x / n = 125/5 = 25$,

while $SS(x) = \Sigma x^2 - \frac{(\Sigma x)^2}{n} = 3163 - \frac{(125)^2}{5} = 38.$ Therefore the standard deviation is

$s = \sqrt{\frac{SS(x)}{n - 1}} = \sqrt{\frac{38}{4}} = 3.082.$ For a 90% confidence level, the subscript for the t-value is (1 - 0.90)/2 = 0.05. From Table IV, using df = 5 - 1 = 4, we find that the t-value is $t_{\alpha/2} = t_{0.05} = 2.132$. Therefore, we find the endpoints of the confidence interval using:

$$\bar{x} \pm t_{\alpha/2}\frac{s}{\sqrt{n}} \Rightarrow 25 \pm 2.132\frac{3.082}{\sqrt{5}} \Rightarrow 25 \pm 2.9 \Rightarrow 22.1 < \mu < 27.9.$$

9.67 We are given $\bar{x} = 28.9$, s = 2.8, and n = 27. For a 99% confidence level, the subscript for the t-value is (1 - 0.99)/2 = 0.005. From Table IV, using df = n - 1 = 26, we find that the t-value is $t_{\alpha/2} = t_{0.005} = 2.779$. Therefore, we find the endpoints of the confidence interval using:

$$\bar{x} \pm t_{\alpha/2}\frac{s}{\sqrt{n}} \Rightarrow 28.9 \pm 2.779\frac{2.8}{\sqrt{27}} \Rightarrow 28.9 \pm 1.50 \Rightarrow 27.40 < \mu < 30.40.$$

9.69 We are given $\bar{x} = 34.3$, s = 3.9, and n = 21. For a 95% confidence level, the subscript for the t-value is (1 - 0.95)/2 = 0.025. From Table IV, using df = n - 1 = 20, we find that the t-value is $t_{\alpha/2} = t_{0.025} = 2.086$. Therefore, we find the endpoints of the confidence interval using:

$$\bar{x} \pm t_{\alpha/2}\frac{s}{\sqrt{n}} \Rightarrow 34.3 \pm 2.086\frac{3.9}{\sqrt{21}} \Rightarrow 34.3 \pm 1.78 \Rightarrow 32.52 < \mu < 36.08.$$

9.71 We first calculate the sample mean and standard deviation. $\bar{x} = \Sigma x/n = 88.8/10 = 8.88$,

while $SS(x) = \Sigma x^2 - \frac{(\Sigma x)^2}{n} = 789.2 - \frac{(88.8)^2}{10} = 0.656.$ Therefore the standard

deviation is $s = \sqrt{\frac{SS(x)}{n-1}} = \sqrt{\frac{0.656}{9}} = 0.270.$ For a 90% confidence level, the subscript for the t-value is (1 - 0.90)/2 = 0.05. From Table IV, using df = 10 - 1 = 9, we find that the t-value is $t_{\alpha/2} = t_{0.05} = 1.833$. Therefore, we find the endpoints of the confidence interval using:

$$\bar{x} \pm t_{\alpha/2}\frac{s}{\sqrt{n}} \Rightarrow 8.88 \pm 1.833\frac{0.270}{\sqrt{10}} \Rightarrow 8.88 \pm 0.16 \Rightarrow 8.72 < \mu < 9.04.$$

9.73 We are given $\bar{x}$ = \$115, s = \$25, and n = 15. For a 95% confidence level, the subscript for the t-value is (1 - 0.95)/2 = 0.025. From Table IV, using df = n - 1 = 14, we find that the t-value is $t_{\alpha/2} = t_{0.025} = 2.145$. Therefore, we find the endpoints of the confidence interval using:

$$\bar{x} \pm t_{\alpha/2}\frac{s}{\sqrt{n}} \Rightarrow 115 \pm 2.145\frac{25}{\sqrt{15}} \Rightarrow 115 \pm 13.85 \Rightarrow \$101.15 < \mu < \$128.85.$$

MINITAB LAB ASSIGNMENTS

9.75(M)

```
MTB > # Exercise 9.75(M)
MTB > # The data were entered into C1 using SET C1
MTB > TINTERVAL 95 C1

                N      MEAN     STDEV   SE MEAN     95.0 PERCENT C.I.
C1             15    16.900     1.326     0.342   (  16.166,   17.634)
```

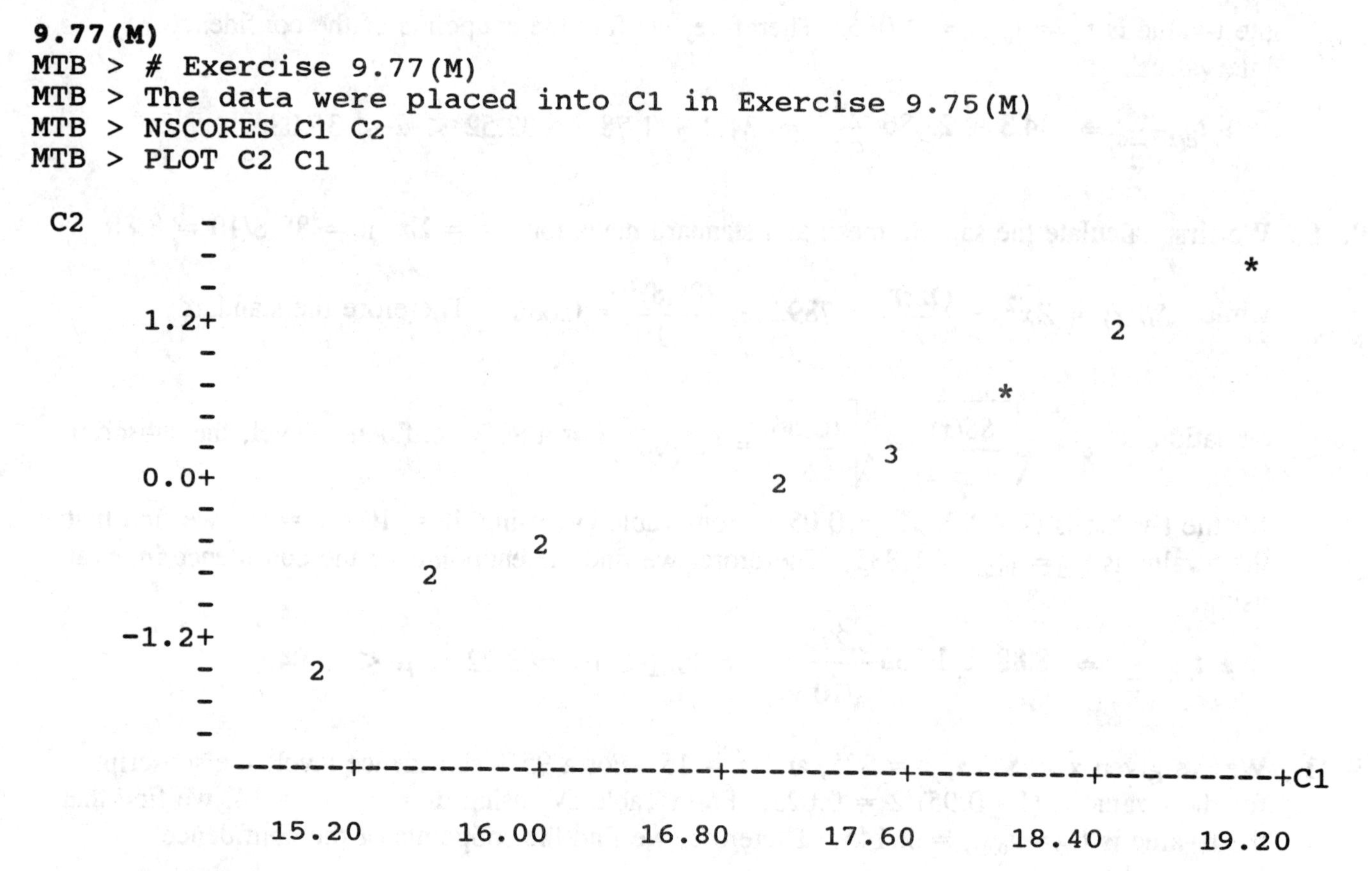


```
MTB > CORRELATION C1 C2

Correlation of C1 and C2 = 0.987
```

We note that the normal scores plot appears reasonably straight and that the correlation coefficient of 0.987 is close to 1. Therefore, it is reasonable to assume that the data are normally distributed.

EXERCISES for Section 9.5

Confidence Interval for a Proportion: Large Sample

9.79 We are given n = 400 and x = 120 so the sample proportion is $\hat{p} = x/n = 120/400 = 0.30$. For a 90% confidence level, the subscript for the z-value is (1 - 0.90)/2 = 0.05. From Table III, we find that the z-value is $z_{\alpha/2} = z_{0.05} = 1.645$. Therefore, a 90% confidence interval for p is given by:

$$\hat{p} \pm z_{\alpha/2}\sqrt{\frac{\hat{p}(1-\hat{p})}{n}} \Rightarrow 0.30 \pm 1.645\sqrt{\frac{(0.30)(1-0.30)}{400}} \Rightarrow 0.30 \pm 0.038.$$

Thus, with 90% confidence, we estimate that the proportion of successes in the population is some value between 0.262 and 0.338.

9.81 We are given n = 600 and x = 90 so the sample proportion is $\hat{p}$ = x/n = 90/600 = 0.15. For a 90% confidence level, the subscript for the z-value is (1 - 0.90)/2 = 0.05. From Table III, we find that the z-value is $z_{\alpha/2} = z_{0.05}$ = 1.645. Therefore, a 90% confidence interval for p is given by:

$$\hat{p} \pm z_{\alpha/2}\sqrt{\frac{\hat{p}(1-\hat{p})}{n}} \Rightarrow 0.15 \pm 1.645\sqrt{\frac{(0.15)(1-0.15)}{600}} \Rightarrow 0.15 \pm 0.024.$$

Thus, with 90% confidence, we estimate that the probability of tossing an ace with this die is some value between 0.126 and 0.174.

9.83 We are given n = 1200 and x = 780 so the sample proportion is $\hat{p}$ = x/n = 780/1200 = 0.65. For a 95% confidence level, the subscript for the z-value is (1 - 0.95)/2 = 0.025. From Table III, we find that the z-value is $z_{\alpha/2} = z_{0.025}$ = 1.960. Therefore, a 95% confidence interval for p is given by:

$$\hat{p} \pm z_{\alpha/2}\sqrt{\frac{\hat{p}(1-\hat{p})}{n}} \Rightarrow 0.65 \pm 1.960\sqrt{\frac{(0.65)(1-0.65)}{1200}} \Rightarrow 0.65 \pm 0.027.$$

Thus, with 95% confidence, we estimate that the proportion of the company's policyholders who would add the $500 deductible is some value between 0.623 and 0.677.

9.85 We are given n = 200 and x = 184 so the sample proportion is $\hat{p}$ = x/n = 184/200 = 0.92. For a 95% confidence level, the subscript for the z-value is (1 - 0.95)/2 = 0.025. From Table III, we find that the z-value is $z_{\alpha/2} = z_{0.025}$ = 1.960. Therefore, a 95% confidence interval for p is given by:

$$\hat{p} \pm z_{\alpha/2}\sqrt{\frac{\hat{p}(1-\hat{p})}{n}} \Rightarrow 0.92 \pm 1.960\sqrt{\frac{(0.92)(1-0.92)}{200}} \Rightarrow 0.92 \pm 0.038.$$

Thus, with 95% confidence, we estimate that the percentage of all employees for this company who do not smoke is some value between 88.2% and 95.8%.

9.87 We are given n = 250 and x = 35 so the sample proportion is $\hat{p}$ = x/n = 35/250 = 0.14. For a 99% confidence level, the subscript for the z-value is (1 - 0.99)/2 = 0.005. From Table III, we find that the z-value is $z_{\alpha/2} = z_{0.005}$ = 2.576. Therefore, a 99% confidence interval for p is given by:

$$\hat{p} \pm z_{\alpha/2}\sqrt{\frac{\hat{p}(1-\hat{p})}{n}} \Rightarrow 0.14 \pm 2.576\sqrt{\frac{(0.14)(1-0.14)}{250}} \Rightarrow 0.14 \pm 0.057.$$

Thus, with 99% confidence, we estimate that the probability that a package wrapped by this machine will contain excessive wrinkles is some value between 0.083 and 0.197.

9.89 We are given n = 1000 and that the sample proportion is $\hat{p}$ = x/n = 0.16. For a 90% confidence level, the subscript for the z-value is (1 - 0.90)/2 = 0.05. From Table III, we find that the z-value is $z_{\alpha/2} = z_{0.05}$ = 1.645. Therefore, a 90% confidence interval for p is given by:

$$\hat{p} \pm z_{\alpha/2}\sqrt{\frac{\hat{p}(1-\hat{p})}{n}} \Rightarrow 0.16 \pm 1.645\sqrt{\frac{(0.16)(1-0.16)}{1000}} \Rightarrow 0.16 \pm 0.019.$$

Thus, with 90% confidence, we estimate that the percentage of all employers that let their workers stay home on their birthday is some value between 14.1% and 17.9%.

9.91 We are given n = 11,242 and x = 2,467 so the sample proportion is $\hat{p} = x/n = 2467/11242$. For a 95% confidence level, the subscript for the z-value is (1 - 0.95)/2 = 0.025. From Table III, we find that the z-value is $z_{\alpha/2} = z_{0.025} = 1.960$. Therefore, a 95% confidence interval for p is given by:

$$\hat{p} \pm z_{\alpha/2}\sqrt{\frac{\hat{p}(1-\hat{p})}{n}} \Rightarrow \frac{2467}{11242} \pm 1.960\sqrt{\frac{\left(\frac{2467}{11242}\right)\left(1-\frac{2467}{11242}\right)}{11242}} \Rightarrow 0.219 \pm 0.008.$$

Thus, with 95% confidence, we estimate that the outpatients at these health centers who suffer from depressive symptoms is some value between 21.1% and 22.7%.

EXERCISES for Section 9.6

Determining the Required Sample Size

9.93 We are given that the range of the population is about 50, so we may estimate the population standard deviation σ as R/4 = 12.5. We wish to be 99% confident that the maximum error of the estimate is E = 2. The required sample size is

$$n = \left(\frac{\sigma z_{\alpha/2}}{E}\right)^2 = \left(\frac{(12.5)(2.576)}{2}\right)^2 = 259.21,$$ which we round up to 260.

9.95 We wish to be 95% confident that the maximum error of the estimate is E = 0.04, and we are given that the true proportion is somewhere between 0.65 and 0.80, so we use p = 0.65.

The required sample size is $$n = pq\left(\frac{z_{\alpha/2}}{E}\right)^2 = (0.65)(1-0.65)\left(\frac{1.960}{0.04}\right)^2 = 546.23,$$ which we round up to 547.

9.97 We wish to be 90% confident that the maximum error of the estimate is E = 5% = 0.05, and we are given that the true proportion is approximately 30%, so we use p = 0.30. The required sample size is

$$n = pq\left(\frac{z_{\alpha/2}}{E}\right)^2 = (0.30)(1-0.30)\left(\frac{1.645}{0.05}\right)^2 = 227.31,$$ which we round up to 228.

Determining the Required Sample Size

9.99 We wish to be 90% confident that the maximum error of the estimate is E = 0.05, and no information is given concerning the population proportion, so we use p = 0.5. The required sample size is

$$n = pq\left(\frac{z_{\alpha/2}}{E}\right)^2 = (0.50)(1-0.50)\left(\frac{1.645}{0.05}\right)^2 = 270.6, \text{ which we round up to 271.}$$

9.101 We wish to determine the required sample size and be 90% confident that the maximum error of the estimate is E = 0.05, when it is believed that about 25% of the plants have sustained damage. Therefore, we use p = 0.25 and find the required sample size is

$$n = pq\left(\frac{z_{\alpha/2}}{E}\right)^2 = (0.25)(1-0.25)\left(\frac{1.645}{0.05}\right)^2 = 202.95, \text{ which we round up to 203.}$$

9.103 We are given that the population standard deviation is σ =1.5, and we wish to be 95% confident that the maximum error of the estimate is E = 0.1. The required sample size is

$$n = \left(\frac{\sigma z_{\alpha/2}}{E}\right)^2 = \left(\frac{(1.5)(1.960)}{0.1}\right)^2 = 864.36, \text{ which we round up to 865.}$$

9.105 We wish to be 95% confident that the maximum error of the estimate is E = 2% = 0.02. Since we are to assume that we have no prior information about the actual percentage, we use p = 0.5. The required sample size is

$$n = pq\left(\frac{z_{\alpha/2}}{E}\right)^2 = (0.5)(1-0.5)\left(\frac{1.960}{0.02}\right)^2 = 2401.$$

EXERCISES for Section 9.7
Chi-Square Probability Distributions

9.107 From Table V, with df = 17, we find $X^2_{0.01}$ = 33.41.

9.109 From Table V, with df = 13, we find $X^2_{0.005}$ = 29.82.

9.111 From Table V, with df = 25, we find $X^2_{0.025}$ = 40.65.

9.113 From Table V, with df = 50, we find $X^2_{0.05}$ = 67.50.

9.115 From Table V, with df = 90, we find $X^2_{0.995}$ = 59.20.

9.117 The 10th percentile is the value of X^2 for which the cumulative area to the left is 0.10. Therefore, the area under the curve to the right of X^2 is 1 - 0.10 = 0.90. From Table V, with df = 11, we find $P_{10} = X^2_{0.90}$ = 5.58.

9.119 The 1st percentile is the value of X^2 for which the cumulative area to the left is 0.01. Therefore, the area under the curve to the right of X^2 is 1 - 0.01 = 0.99. From Table V, with df = 80, we find $P_1 = X^2_{0.99} = 53.54$.

9.121 The 5th percentile is the value of X^2 for which the cumulative area to the left is 0.05. Therefore, the area under the curve to the right of X^2 is 1 - 0.05 = 0.95. From Table V, with df = 14, we find $P_5 = X^2_{0.95} = 6.57$.

9.123 The 90th percentile is the value of X^2 for which the cumulative area to the left is 0.90. Therefore, the area under the curve to the right of X^2 is 1 - 0.90 = 0.10. From Table V, with df = 9, we find $P_{90} = X^2_{0.10} = 14.68$.

9.125 The 99th percentile is the value of X^2 for which the cumulative area to the left is 0.99. Therefore, the area under the curve to the right of X^2 is 1 - 0.99 = 0.01. From Table V, with df = 16, we find $P_{99} = X^2_{0.01} = 32.00$.

9.127 The 95th percentile is the value of X^2 for which the cumulative area to the left is 0.95. Therefore, the area under the curve to the right of X^2 is 1 - 0.95 = 0.05. From Table V, with df = 21, we find $P_{95} = X^2_{0.05} = 32.67$.

9.129 We note in Table V that, with df = 18, $X^2_{0.975} = 8.23$. Therefore, $P(X^2 > 8.23) = 0.975$.

9.131 We note in Table V that, with df = 22, $X^2_{0.01} = 40.29$. Therefore, $P(X^2 > 40.29) = 0.01$, so we have $P(X^2 < 40.29) = 1 - 0.01 = 0.99$.

9.133 We note in Table V that, with df = 80, $X^2_{0.975} = 57.15$ and $X^2_{0.10} = 96.58$. Therefore, $P(57.15 < X^2 < 96.58) = P(X^2 > 57.15) - P(X^2 > 96.58) = 0.975 - 0.10 = 0.875$.

9.135 We note in Table V that, with df = 9, $X^2_{0.01} = 21.67$. Therefore, $P(X^2 > 21.67) = 0.01$, so we conclude that c = 21.67.

9.137 If $P(X^2 < c) = 0.005$, then $P(X^2 > c) = 1 - 0.005 = 0.995$. From Table V, with df = 30, $X^2_{0.995} = 13.79$. Therefore, $P(X^2 > 13.79) = 0.995$, so we conclude that c = 13.79.

MINITAB LAB ASSIGNMENTS

9.139(M)

```
MTB > # Exercise 9.139(M)
MTB > # Part 1:
MTB > INVCDF 0.85;
SUBC> CHIS 42.
    0.8500    51.4746

MTB > # Part 2:
MTB > INVCDF 0.15;
SUBC> CHIS 42.
    0.1500    32.6255
```

```
9.141(M)
MTB > # Exercise 9.141(M)
MTB > INVCDF 0.35;
SUBC> CHIS 19.
    0.3500    16.1089
```

EXERCISES for Section 9.8

Confidence Interval for a Variance

9.143 We are given $s^2 = 15.5$ and $n = 20$. For a 90% confidence interval, $1 - \alpha = 0.90$, so $\alpha = 0.10$. We find from Table V, with df = n - 1 = 19, $X^2_{\alpha/2} = X^2_{0.05} = 30.14$ and $X^2_{(1-\alpha/2)} = X^2_{0.95} = 10.12$. The confidence interval for σ^2 is given by

$$\frac{(n-1)s^2}{X^2_{\alpha/2}} < \sigma^2 < \frac{(n-1)s^2}{X^2_{1-\alpha/2}} \rightarrow \frac{(20-1)(15.5)}{30.14} < \sigma^2 < \frac{(20-1)(15.5)}{10.12}$$

$\Rightarrow 9.77 < \sigma^2 < 29.10.$

9.145 Since standard deviation is the square root of the variance, a confidence interval for σ can be found by taking the square root of the variance confidence limits in Exercise 9.143. Therefore, a 90% confidence interval for σ is $\sqrt{9.77} < \sigma < \sqrt{29.10} \Rightarrow 3.13 < \sigma < 5.39$.

9.147 Using a statistical calculator we find $s^2 = 6.5$ and $n = 5$. For a 99% confidence level, $1 - \alpha = 0.99$, so $\alpha = 0.01$. We find from Table V, with df = n - 1 = 4, $X^2_{\alpha/2} = X^2_{0.005} = 14.86$ and $X^2_{(1-\alpha/2)} = X^2_{0.995} = 0.21$. The confidence interval for σ^2 is given by

$$\Rightarrow \frac{(n-1)s^2}{X^2_{\alpha/2}} < \sigma^2 < \frac{(n-1)s^2}{X^2_{1-\alpha/2}} \rightarrow \frac{(5-1)(6.5)}{14.86} < \sigma^2 < \frac{(5-1)(6.5)}{0.21}$$

i.e, $1.75 < \sigma^2 < 123.81$.

9.149 We are given $s^2 = 0.52$ and $n = 24$. For a 95% confidence interval, $1 - \alpha = 0.95$, so $\alpha = 0.05$. We find from Table V, with df = n - 1 = 23, $X^2_{\alpha/2} = X^2_{0.025} = 38.08$ and $X^2_{(1-\alpha/2)} = X^2_{0.975} = 11.69$. The confidence interval for σ^2 is given by

$$\frac{(n-1)s^2}{X^2_{\alpha/2}} < \sigma^2 < \frac{(n-1)s^2}{X^2_{1-\alpha/2}} \rightarrow \frac{(24-1)(0.52)}{38.08} < \sigma^2 < \frac{(24-1)(0.52)}{11.69}$$

$\Rightarrow 0.314 < \sigma^2 < 1.023.$

9.151 We are given $s = 0.24$ and $n = 20$. For a 99% confidence level, $1 - \alpha = 0.99$, so $\alpha = 0.01$. We find from Table V, with df = n - 1 = 19, $X^2_{\alpha/2} = X^2_{0.005} = 38.58$ and $X^2_{(1-\alpha/2)} = X^2_{0.995} = 6.84$. The confidence interval for σ is given by

$$\sqrt{\frac{(n-1)s^2}{X^2_{\alpha/2}}} < \sigma < \sqrt{\frac{(n-1)s^2}{X^2_{1-\alpha/2}}} \rightarrow \sqrt{\frac{(20-1)(0.24)^2}{38.58}} < \sigma < \sqrt{\frac{(20-1)(0.24)^2}{6.84}}$$

$\Rightarrow 0.168 < \sigma < 0.400.$

9.153 Since standard deviation is the square root of the variance, a confidence interval for σ can be found by taking the square root of the variance confidence limits in Exercise 9.152. Therefore, a 95% confidence interval for σ is: $\sqrt{0.0114} < \sigma < \sqrt{0.080}$; i.e, 0.107 to 0.283.

9.155 Using a statistical calculator we find $s^2 = 3.643$ and $n = 8$. For a 95% confidence interval, $1 - \alpha = 0.95$, so $\alpha = 0.05$. We find from Table V, with df = n - 1 = 7, $X^2_{\alpha/2} = X^2_{0.025} = 16.01$ and $X^2_{(1-\alpha/2)} = X^2_{0.975} = 1.69$. The confidence interval for σ^2 is given by

$$\frac{(n-1)s^2}{X^2_{\alpha/2}} < \sigma^2 < \frac{(n-1)s^2}{X^2_{1-\alpha/2}} \rightarrow \frac{(8-1)(3.643)}{16.01} < \sigma^2 < \frac{(8-1)(3.643)}{1.69}$$

$\Rightarrow 1.59 < \sigma^2 < 15.09.$

REVIEW EXERCISES
CHAPTER 9

9.157 We are given $\bar{x} = 478.7$, $s=29.5$, and $n = 17$. For a 90% confidence level, the subscript for the t-value is (1 - 0.90)/2 = 0.05. From Table IV, using df = n - 1 = 16, we find that the t-value is $t_{\alpha/2} = t_{0.05} = 1.746$. Therefore, we find the endpoints of the confidence interval using:

$$\bar{x} \pm t_{\alpha/2}\frac{s}{\sqrt{n}} \Rightarrow 478.7 \pm 1.746\frac{29.5}{\sqrt{17}} \Rightarrow 478.7 \pm 12.49 \Rightarrow 466.2 < \mu < 491.2.$$

9.158 We are given $s = 29.5$ and $n = 17$. For a 90% confidence interval, $1 - \alpha = 0.90$, so $\alpha = 0.10$. We find from Table V, with df = n - 1 = 16, $X^2_{\alpha/2} = X^2_{0.05} = 26.30$ and $X^2_{(1-\alpha/2)} = X^2_{0.95} = 7.96$. The confidence interval for σ^2 is given by

$$\frac{(n-1)s^2}{X^2_{\alpha/2}} < \sigma^2 < \frac{(n-1)s^2}{X^2_{1-\alpha/2}} \rightarrow \frac{(17-1)(29.5)^2}{26.3} < \sigma^2 < \frac{(17-1)(29.5)^2}{7.96}$$

$\Rightarrow 529.43 < \sigma^2 < 1749.25.$

9.159 We calculate $\bar{x} = 6.3$, $s = 0.2582$, and note that $n = 10$. For a 95% confidence level, the subscript for the t-value is (1 - 0.95)/2 = 0.025. From Table IV, using df = n - 1 = 9, we find that the t-value is $t_{\alpha/2} = t_{0.025} = 2.262$. Therefore, we find the endpoints of the confidence interval using:

$$\bar{x} \pm t_{\alpha/2}\frac{s}{\sqrt{n}} \Rightarrow 6.3 \pm 2.262\frac{0.2582}{\sqrt{10}} \Rightarrow 6.3 \pm 0.18 \Rightarrow 6.12 < \mu < 6.48.$$

9.160 We calculate $s^2 = 0.0667$ and $n = 10$. For a 90% confidence interval, $1 - \alpha = 0.90$, so $\alpha = 0.10$. We find from Table V, with df = n - 1 = 9, $X^2_{\alpha/2} = X^2_{0.05} = 16.92$ and $X^2_{(1-\alpha/2)}$

$= X^2_{0.95} = 3.33$. The confidence interval for σ^2 is given by

$$\frac{(n-1)s^2}{X^2_{\alpha/2}} < \sigma^2 < \frac{(n-1)s^2}{X^2_{1-\alpha/2}} \rightarrow \frac{(10-1)(0.0667)}{16.92} < \sigma^2 < \frac{(10-1)(0.0667)}{3.33}$$

$\Rightarrow 0.0355 < \sigma^2 < 0.180 \Rightarrow 0.188 < \sigma < 0.424$.

9.161 We are given $\bar{x} = 1002.8$, s= 3.4, and n = 71. For a 99% confidence level, the subscript for the z-value is (1 - 0.99)/2 = 0.005. From Table III, we find that the z-value is $z_{\alpha/2} = z_{0.005} = 2.576$. Therefore we have $\bar{x} \pm z_{\alpha/2}\frac{\sigma}{\sqrt{n}} \Rightarrow 1002.8 \pm 2.576\frac{3.4}{\sqrt{71}}$, where we replace σ with s = 3.4, $\Rightarrow$ 1002.8 $\pm$ 1.04 $\Rightarrow 1001.76 < \mu < 1003.84$.

9.162 We are given s = 3.4 and n = 71. For a 95% confidence interval, $1 - \alpha = 0.95$, so $\alpha = 0.05$. We find from Table V, with df = n - 1 = 70, $X^2_{\alpha/2} = X^2_{0.025} = 95.02$ and $X^2_{(1-\alpha/2)} = X^2_{0.975} = 48.76$. The confidence interval for σ^2 is given by

$$\frac{(n-1)s^2}{X^2_{\alpha/2}} < \sigma^2 < \frac{(n-1)s^2}{X^2_{1-\alpha/2}} \rightarrow \frac{(71-1)(3.4)^2}{95.02} < \sigma^2 < \frac{(71-1)(3.4)^2}{48.76}$$

$\Rightarrow 8.52 < \sigma^2 < 16.60 \Rightarrow 2.92 < \sigma < 4.07$.

9.163 We wish to be 90% confident that the maximum error of the estimate is E = 0.04 and no information is given concerning the population proportion, so we use p = 0.5. The required sample size is $n = pq\left(\frac{z_{\alpha/2}}{E}\right)^2 = (0.5)(1-0.5)\left(\frac{1.645}{0.04}\right)^2 = 422.82,$ which we round up to 423.

9.164 We are given that the population standard deviation is $\sigma = 12$, and we wish to be 95% confident that the maximum error of the estimate is E = 2. The required sample size is

$$n = \left(\frac{\sigma z_{\alpha/2}}{E}\right)^2 = \left(\frac{(12)(1.960)}{2}\right)^2 = 138.30,$$ which we round up to 139.

9.165 We wish to be 99% confident that the maximum error of the estimate is E = 0.04 and no information is given concerning the population proportion, so we use p = 0.5. The required sample size is $n = pq\left(\frac{z_{\alpha/2}}{E}\right)^2 = (0.5)(1-0.5)\left(\frac{2.576}{0.04}\right)^2 = 1036.84,$ which we round up to 1037.

9.166 We wish to be 99% confident that the maximum error of the estimate is E = 0.04, and we are given that the true percentage is about 35%, so we use p = 0.35. The required sample size is $n = pq\left(\frac{z_{\alpha/2}}{E}\right)^2 = (0.35)(1-0.35)\left(\frac{2.576}{0.04}\right)^2 = 943.52,$ which we round up to 944.

9.167 We first calculate $SS(x) = \Sigma x^2 - \frac{(\Sigma x)^2}{n} = 3983.456768 - \frac{(403.875)^2}{41} = 5.0418.$ Therefore, $s^2 = \frac{SS(x)}{n-1} = \frac{5.0418}{41-1} = 0.12604.$ For a 95% confidence interval, $1 - \alpha = 0.95$, so $\alpha = 0.05$. We find from Table V, with df = n - 1 = 40, $X^2_{\alpha/2} = X^2_{0.025} = 59.34$ and $X^2_{(1-\alpha/2)} = X^2_{0.975} = 24.43$. The confidence interval for σ^2 is given by

$$\frac{(n-1)s^2}{X^2_{\alpha/2}} < \sigma^2 < \frac{(n-1)s^2}{X^2_{1-\alpha/2}} \rightarrow \frac{(41-1)(0.12604)}{59.34} < \sigma^2 < \frac{(41-1)(0.12604)}{24.43}$$

$\Rightarrow$ $0.0850 < \sigma^2 < 0.2064$.

9.168 From Exercise 9.167 we have $0.0850 < \sigma^2 < 0.2064$. Taking square roots of the endpoints, we obtain: $0.291 < \sigma < 0.454$.

9.169 We wish to be 90% confident that the maximum error of the estimate is E = 0.04, and we are given that the true proportion is somewhere between 0.55 and 0.60, so we use p = 0.55. The required sample size is

$$n = pq\left(\frac{z_{\alpha/2}}{E}\right)^2 = (0.55)(1-0.55)\left(\frac{1.645}{0.04}\right)^2 = 418.59,$$ which we round up to 419.

9.170 We are given $\bar{x} = 52.8$, s = 4.8, and n = 374. For a 99% confidence level, the subscript for the z-value is (1 - 0.99)/2 = 0.005. From Table III, we find that the z-value is $z_{\alpha/2} = z_{0.005} = 2.576$. Therefore we have $\bar{x} \pm z_{\alpha/2}\frac{\sigma}{\sqrt{n}} \Rightarrow 52.8 \pm 2.576\frac{4.8}{\sqrt{374}}$, where we replace σ with s = 4.8, $\Rightarrow$ 52.80 ± 0.64 $\Rightarrow$ $52.16 < \mu < 53.44$.

9.171 We are given n = 300 and x = 81 so the sample proportion is $\hat{p} = x/n = 81/300 = 0.27$. For a 90% confidence level, the subscript for the z-value is (1 - 0.90)/2 = 0.05. From Table III, we find that the z-value is $z_{\alpha/2} = z_{0.05} = 1.645$. Therefore, a 90% confidence interval for p is given by: $\hat{p} \pm z_{\alpha/2}\sqrt{\frac{\hat{p}(1-\hat{p})}{n}} \Rightarrow 0.270 \pm 1.645\sqrt{\frac{(0.27)(1-0.27)}{300}} \Rightarrow$ 0.270 ± 0.042. Thus, with 90% confidence, we estimate that the proportion of all recent home sales in this state that were financed with this type of mortgage is some value between 0.228 and 0.312.

9.172 We are given n = 1200 and that the sample proportion is $\hat{p} = x/n = 55\% = 0.55$. For a 95% confidence level, the subscript for the z-value is (1 - 0.95)/2 = 0.025. From Table III, we find that the z-value is $z_{\alpha/2} = z_{0.025} = 1.960$. Since we wish to find a confidence interval for the true percentage of all U.S. adults who can not locate New York on a U.S. map, we find a 95% confidence interval for q which is given by: $\hat{q} \pm z_{\alpha/2}\sqrt{\frac{\hat{q}(1-\hat{q})}{n}}$

$$\Rightarrow 0.45 \pm 1.960\sqrt{\frac{(0.45)(1 - 0.45)}{1200}}$$ $\Rightarrow 0.45 \pm 0.028$. Thus, with 95% confidence, we estimate that the true percentage of all U.S. adults who can not locate New York on a U.S. map is some value between 42.2% and 47.8%.

9.173 We note that the range of the population is about 10 - 0 = 10, so we may estimate the population standard deviation σ as R/4 = 2.5. We wish to be 99% confident that the maximum error of the estimate is E = 1. The required sample size is

$$n = \left(\frac{\sigma z_{\alpha/2}}{E}\right)^2 = \left(\frac{(2.5)(2.576)}{1}\right)^2 = 41.4736,$$ which we round up to 42.

9.174 We wish to be 95% confident that the maximum error of the estimate is E = 0.02 and no information is given concerning the population proportion, so we use p = 0.5. The required sample size is $n = pq\left(\frac{z_{\alpha/2}}{E}\right)^2 = (0.5)(1-0.5)\left(\frac{1.960}{0.02}\right)^2 = 2401.$

9.175 **a.** From Table IV, with df = 23, we find $t_{0.025} = 2.069$.
b. From Table IV, with df = 10, we find $t_{0.01} = 2.764$.

9.176 **a.** From Table V, with df = 28, we find $X^2_{0.005} = 50.99$.
b. From Table V, with df = 14, we find $X^2_{0.05} = 23.68$.

9.177 The 10th percentile is the value of X^2 for which the cumulative area to the left is 0.10. Therefore, the area under the curve to the right of X^2 is 1 - 0.10 = 0.90. From Table V, with df = 50, we find $P_{10} = X^2_{0.90} = 37.69$.

9.178 The 95th percentile is the value of t for which the cumulative area to the left is 0.95. Therefore, the area under the curve to the right of t is 1 - 0.95 = 0.05. From Table IV, with df = 29, we find $P_{95} = t_{0.05} = 1.699$.

9.179 We calculate $\bar{x} = 3503$ and s = 351.61. For a 99% confidence level, the subscript for the t-value is (1 - 0.99)/2 = 0.005. From Table IV, using df = n - 1 = 8 - 1 = 7, we find that the t-value is $t_{\alpha/2} = t_{0.005} = 3.499$. Therefore, we find the endpoints of the confidence interval: $\bar{x} \pm t_{\alpha/2}\frac{s}{\sqrt{n}} \Rightarrow 3503 \pm 3.499\frac{351.61}{\sqrt{8}} \Rightarrow 3503 \pm 435 \Rightarrow 3068 < \mu < 3938.$

MINITAB LAB ASSIGNMENTS

9.180(M)

```
MTB > # Exercise 9.180(M)
MTB > SET C1
DATA> 3675 3597 3412 3976 2831 3597 3211 3725
DATA> END
MTB > TINTERVAL 99 C1

              N      MEAN     STDEV  SE MEAN    99.0 PERCENT C.I.
C1            8   3503.00    351.61   124.31   ( 3067.89, 3938.11)
```

9.181(M)

```
MTB > # Exercise 9.181(M)
MTB > # Data are in C1 from Exercise 9.180(M)
MTB > NSCORES C1 C2
MTB > PLOT C2 C1

C2       -
         -                                                 *
         -
     1.0+
         -                                       *
         -
         -                                   *
         -
     0.0+                                  2
         -
         -                           *
         -
         -                  *
    -1.0+
         -
         -      *
         -
           --+---------+---------+---------+---------+---------+----C1
           2750      3000      3250      3500      3750      4000

MTB > CORRELATION C1 C2

Correlation of C1 and C2 = 0.967
```

We note that the normal scores plot appears reasonably straight and that the correlation coefficient of 0.967 is close to 1. Therefore, it is reasonable to assume that the data are normally distributed.

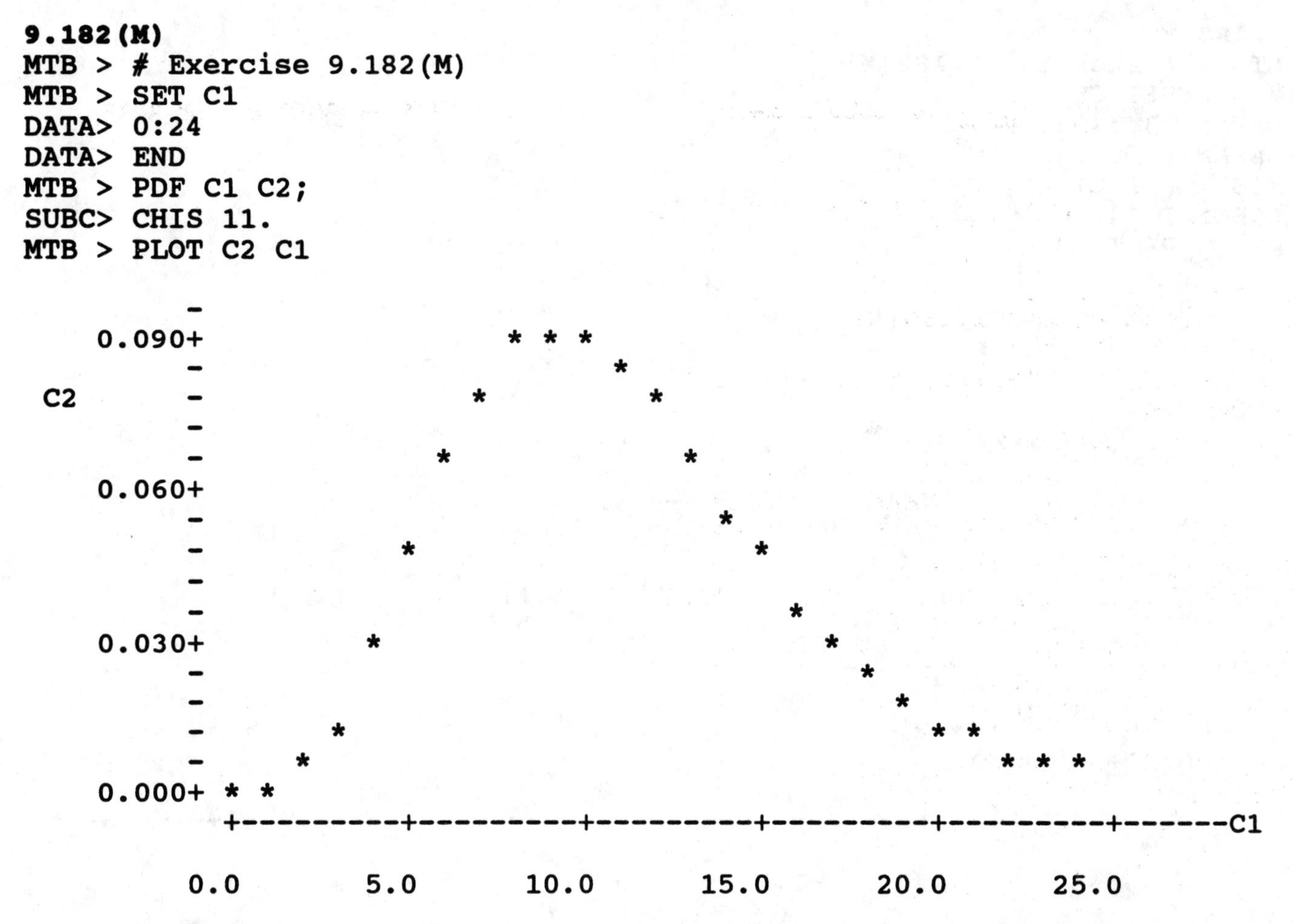

9.182(M)

```
MTB > # Exercise 9.182(M)
MTB > SET C1
DATA> 0:24
DATA> END
MTB > PDF C1 C2;
SUBC> CHIS 11.
MTB > PLOT C2 C1
```



```
MTB > # If different values for C1 are used, shape may be different.
```

9.183(M)

```
MTB > # Exercise 9.183(M)
MTB > SET C1
DATA> -3:3/0.25
DATA> END
MTB > PDF C1 C2;
SUBC> T 34.
MTB > PLOT C2 C1

      0.45+
          -
 C2       -                             * * *
          -                           *       *
          -
      0.30+                         *           *
          -
          -                       *               *
          -
          -                     *                   *
      0.15+
          -                   *                        *
          -                *                             *
          -              *                                 *
          -          * *                                     * *
      0.00+      * *                                             * *
          -
        --------+---------+---------+---------+---------+--------C1
              -2.4      -1.2       0.0       1.2       2.4

MTB > # If different values for C1 are used, shape may be different.
```

9.184(M) a.

```
MTB > # Exercise 9.184(M)
MTB > CDF 15;
SUBC> CHIS 23.
   15.0000    0.1054
MTB > # Probability = 1 - 0.1054 = 0.8946.
```

b.

```
MTB > INVCDF 0.77;
SUBC> CHIS 23.
    0.7700   27.6324
MTB > # The 77th percentile is 27.6324
```

9.185 a.

```
MTB > # Exercise 9.185(M)
MTB > CDF 1.96;
SUBC> T 43.
    1.9600    0.9718
MTB > # P(t > 1.96) = 1 - 0.9718 = 0.0282.
```

b.

```
MTB > INVCDF 0.59;
SUBC> T 43.
    0.5900    0.2289
MTB > # The 59th percentile is 0.2289.
```

9.186(M)

```
MTB > # Exercise 9.186(M)
MTB > # The data were entered into C1 using SET C1
MTB > STDEV C1
   ST.DEV. =       1.5642
MTB > ZINTERVAL 99 1.56 C1

THE ASSUMED SIGMA =1.56

             N      MEAN    STDEV  SE MEAN    99.0 PERCENT C.I.
C1          50    20.022    1.564    0.221  (  19.453,  20.591)
```

9.187(M)

```
MTB > # Exercise 9.187(M)
MTB > # The data were entered into C1 using SET C1
MTB > STDEV C1
   ST.DEV. =       9.2486
MTB > ZINTERVAL 90 9.25 C1

THE ASSUMED SIGMA =9.25

             N      MEAN    STDEV  SE MEAN    90.0 PERCENT C.I.
C1          72     67.39     9.25     1.09  (   65.59,   69.18)
```

9.188(M)

```
MTB > # Exercise 9.188(M)
MTB > ZINTERVAL 90 9 C1

THE ASSUMED SIGMA =9.00

             N      MEAN    STDEV  SE MEAN    90.0 PERCENT C.I.
C1          72     67.39     9.25     1.06  (   65.64,   69.14)
```

9.189(M)

```
MTB > # Exercise 9.189(M)
MTB > SET C1
DATA> 165 217 200 134 175 153 108 174 219 183
DATA> 123 198 261 247 176 121 202 100 207 200
DATA> END
MTB > TINTERVAL 90 C1

             N      MEAN    STDEV  SE MEAN    90.0 PERCENT C.I.
C1          20    178.15    44.53     9.96  (  160.93,  195.37)
```

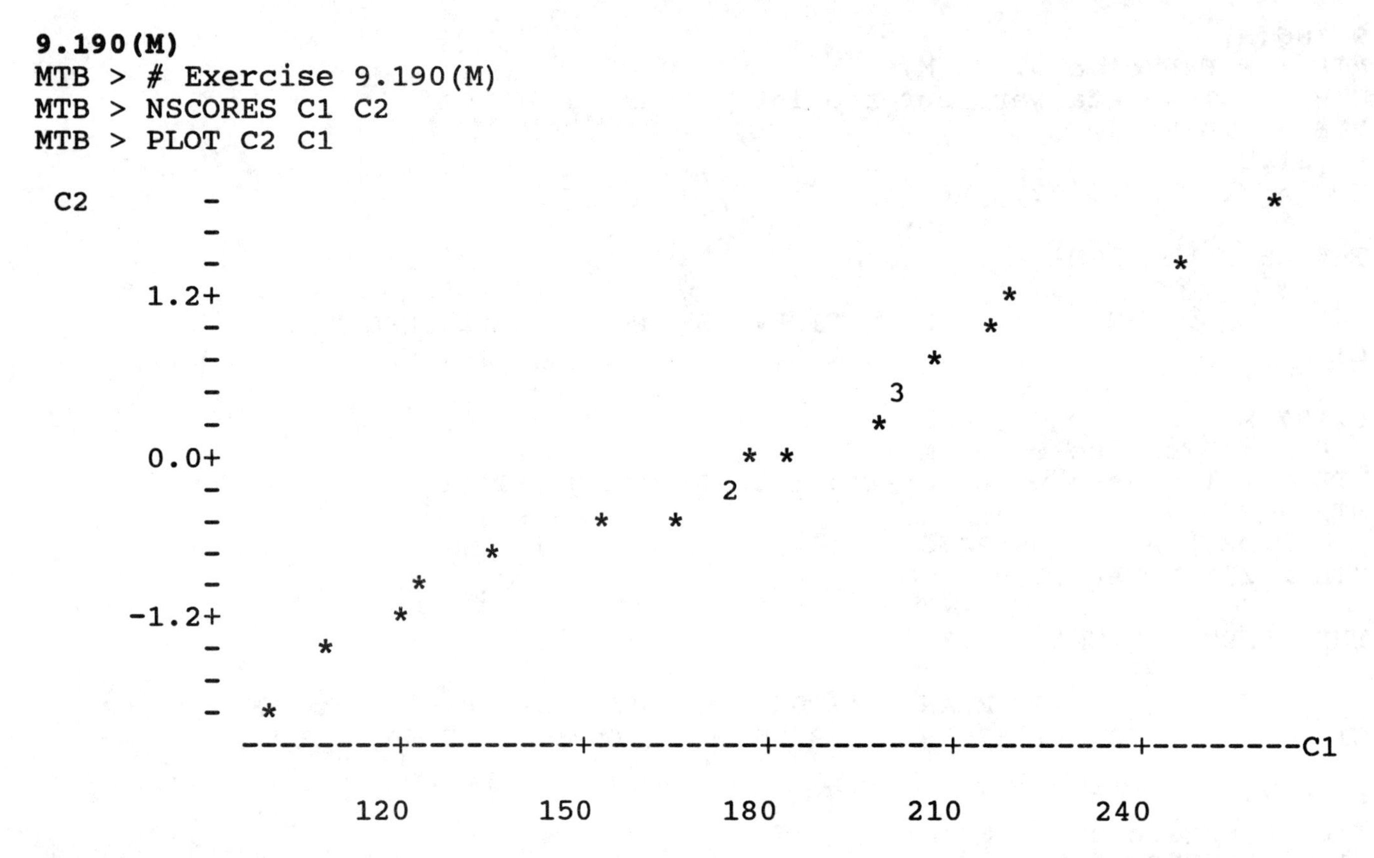


```
MTB > CORRELATION C1 C2

Correlation of C1 and C2 = 0.987
```

We note that the normal scores plot appears reasonably straight and that the correlation coefficient of 0.987 is close to 1. Therefore, it is reasonable to assume that the data are normally distributed.

CHAPTER 10
TESTS OF HYPOTHESES: SINGLE SAMPLE

EXERCISES for Section 10.1
Basic Concepts of Testing Statistical Hypotheses

10.1 **a.** The null hypothesis is falsely rejected. Therefore, a Type I error is committed.
b. The null hypothesis is falsely accepted. Therefore, a Type II error is committed.
c. Since the alternative hypothesis is falsely accepted, the null hypothesis is falsely rejected. Therefore, a Type I error is committed.
d. Since the true alternative hypothesis is not accepted, the null hypothesis is falsely accepted. Therefore, a Type II error is committed.

10.3 By decreasing the value of α, the value of β is increased, thus increasing the probability of committing a Type II error, if the null hypothesis were false.

10.5 **a.** The null hypothesis is falsely accepted. Therefore, a Type II error is committed.
b. The null hypothesis is falsely rejected. Therefore, a Type I error is committed.

10.7 **a.** The null hypothesis is falsely rejected. Therefore, a Type I error is committed.
b. The null hypothesis is falsely accepted. Therefore, a Type II error is committed.

10.9 **a.** A Type I error would be committed if we were to incorrectly conclude that the restaurant is selling underweight hamburgers.
b. A Type II error would be committed if we were to incorrectly conclude that the restaurant is not selling underweight hamburgers.
c. The restaurant would consider the Type I error to be more serious.

10.11 **a.** The manufacturer wants to demonstrate that the mean speed actually exceeds the nominal speed of 66 MHz. Therefore, the set of hypotheses which should be tested are:
H_0: $\mu = 66$
H_a: $\mu > 66$
b. A competing manufacturer would want to demonstrate that the mean speed is actually less than the nominal speed of 66 MHz. Therefore, the competing manufacturer should test the hypotheses:
H_0: $\mu = 66$
H_a: $\mu < 66$

10.13 In order to demonstrate that the modification will increase the mean top speed of this model from the present value of 950 miles-per-hour, the engineer would formulate the hypotheses:
H_0: $\mu = 950$
H_a: $\mu > 950$

EXERCISES for Section 10.2
Hypothesis Test for a Mean: Large Sample

10.15 **a.** For a one-tailed test with a significance level $\alpha = 0.01$, we have $z_\alpha = z_{0.01} = 2.326$. Since the alternative hypothesis is H_a: $\mu < 87.6$, we reject H_0 if $z < -2.326$.
b. For a two-tailed test with a significance level $\alpha = 0.01$, we have $z_{\alpha/2} = z_{0.005} = 2.576$. We reject H_0 if $z < -2.576$ or if $z > 2.576$.
c. Since the alternative hypothesis is H_a: $\mu > 87.6$, we reject H_0 if $z > 2.326$.

10.17 **a.** Since the researcher wishes to test the belief that the population mean has changed from 95, the null and alternative hypotheses are:
H_0: $\mu = 95$
H_a: $\mu \neq 95$
b. The significance level of the test is $\alpha = 0.01$. This is the probability that the researcher will falsely conclude that the population mean is no longer 95.
c. The value of the test statistic is $z = \dfrac{\overline{x} - \mu}{\sigma/\sqrt{n}} = \dfrac{90 - 95}{16/\sqrt{64}} = -2.5$, where σ is approximated by $s = 16$.
d. For a two-tailed test with a significance level $\alpha = 0.01$, we have $z_{\alpha/2} = z_{0.005} = 2.576$. We reject H_0 if $z < -2.576$ or if $z > 2.576$.
e. For $z = -2.5$ we fail to reject H_0. This means that there is insufficient evidence at the 1% significance level to show that the population mean has changed.

10.19 **Step 1: Formulate the null and alternative hypotheses.**
In order to show that the mean household income of the subscribers exceeds $80,000, we test
H_0: $\mu = 80000$
H_a: $\mu > 80000$
Step 2: Select the significance level for the test.
We are given that $\alpha = 0.05$.
Step 3: Calculate the value of the test statistic.
For the sample, $n = 539$, $\overline{x} = 80{,}634$ and $s = 7{,}874$, which is used to approximate σ.

Therefore, the value of the test statistic is: $z = \dfrac{\overline{x} - \mu}{\sigma/\sqrt{n}} = \dfrac{80634 - 80000}{7874/\sqrt{539}} = 1.87$.

Step 4: Determine the rejection region of H_0 from the tables.
For a one-tailed test with a significance level $\alpha = 0.05$, we have $z_\alpha = z_{0.05} = 1.645$. Since the alternative hypothesis is H_a: $\mu > 80000$, we reject H_0 if $z > 1.645$.
Step 5: State the conclusion of the test.
For $z = 1.87$, we reject H_0. Therefore, there is sufficient evidence at the 5% level to support the magazine's statement.

10.21 **Step 1: Formulate the null and alternative hypotheses.**
In order to show that the new battery has a longer mean life, we test
H_0: $\mu = 150$
H_a: $\mu > 150$

Step 2: **Select the significance level for the test.**
We are given that $\alpha = 0.05$.
Step 3: **Calculate the value of the test statistic.**
For the sample, $n = 35$, $\bar{x} = 159$ and $s = 27$, which is used to approximate σ. Therefore,

the value of the test statistic is: $z = \dfrac{\bar{x} - \mu}{\sigma/\sqrt{n}} = \dfrac{159 - 150}{27/\sqrt{35}} = 1.97$.

Step 4: **Determine the rejection region of H_0 from the tables.**
For a one-tailed test with a significance level $\alpha = 0.05$, we have $z_\alpha = z_{0.05} = 1.645$.
Since the alternative hypothesis is H_a: $\mu > 150$, we reject H_0 if $z > 1.645$.
Step 5: **State the conclusion of the test.**
For $z = 1.97$, we reject H_0, so there is sufficient evidence at the 5% level to show that the new battery has a longer mean life.

10.23 Step 1: **Formulate the null and alternative hypotheses.**
In order to show a significant change in the mean SAT score for this year's applicants, we test
H_0: $\mu = 573$
H_a: $\mu \neq 573$
Step 2: **Select the significance level for the test.**
We are given that $\alpha = 0.05$.
Step 3: **Calculate the value of the test statistic.**
For the sample, $n = 235$, $\bar{x} = 579$ and $s = 39$, which is used to approximate σ. Therefore,

the value of the test statistic is: $z = \dfrac{\bar{x} - \mu}{\sigma/\sqrt{n}} = \dfrac{579 - 573}{39/\sqrt{235}} = 2.36$.

Step 4: **Determine the rejection region of H_0 from the tables.**
For a two-tailed test with a significance level $\alpha = 0.05$, we have $z_{\alpha/2} = z_{0.025} = 1.96$ and we reject H_0 if $z < -1.96$ or if $z > 1.96$.
Step 5: **State the conclusion of the test.**
For $z = 2.36$, we reject H_0 and there is sufficient evidence at the 5% level to show a significant change in the mean SAT score for this year's applicants.

10.25 Step 1: **Formulate the null and alternative hypotheses.**
In order to show that the machine is not dispensing an average of 1,600 mg, we test
H_0: $\mu = 1600$
H_a: $\mu \neq 1600$
Step 2: **Select the significance level for the test.**
We are given that $\alpha = 0.05$.
Step 3: **Calculate the value of the test statistic.**
For the sample, $n = 30$, $\bar{x} = 1594$ and $s = 19$, which is used to approximate σ. Therefore,

the value of the test statistic is: $z = \dfrac{\bar{x} - \mu}{\sigma/\sqrt{n}} = \dfrac{1594 - 1600}{19/\sqrt{30}} = -1.73$.

Step 4: **Determine the rejection region of H_0 from the tables.**
For a two-tailed test with a significance level $\alpha = 0.05$, we have $z_{\alpha/2} = z_{0.025} = 1.96$ and we reject H_0 if $z < -1.96$ or $z > 1.96$.

Step 5: State the conclusion of the test.
For z = -1.73, we fail to reject H_0. Therefore, there is insufficient evidence at the 5% level to show that the machine is not dispensing an average of 1,600 mg.

10.27 Step 1: Formulate the null and alternative hypotheses.
In order to show the EPA's contention that secondary schools will have to spend more than $5,000 a year to remain in compliance with the new regulations, we test
H_0: $\mu = 5000$
H_a: $\mu > 5000$
Step 2: Select the significance level for the test.
We are given that $\alpha = 0.05$.
Step 3: Calculate the value of the test statistic.
For the sample, n = 47, $\bar{x} = 5153$ and s = 598, which is used to approximate σ.

Therefore, the value of the test statistic is: $z = \dfrac{\bar{x} - \mu}{\sigma/\sqrt{n}} = \dfrac{5153 - 5000}{598/\sqrt{47}} = 1.75.$

Step 4: Determine the rejection region of H_0 from the tables.
For a one-tailed test with a significance level $\alpha = 0.05$, we have $z_\alpha = z_{0.05} = 1.645$. Since the alternative hypothesis is H_a: $\mu > 5000$, we reject H_0 if $z > 1.645$.
Step 5: State the conclusion of the test.
For z = 1.75, we reject H_0. Therefore, there is sufficient evidence at the 5% level to support the EPA's contention.

MINITAB LAB ASSIGNMENTS

10.29

```
MTB > # Exercise 10.29(M)
MTB > # The data were entered into C1 using SET C1
MTB > STDEV C1
   ST.DEV. =        1.5642
MTB > ZTEST 19.9 1.5642 C1;
SUBC> ALTERNATIVE = +1.

TEST OF MU = 19.900 VS MU G.T. 19.900
THE ASSUMED SIGMA = 1.56

                N      MEAN     STDEV   SE MEAN         Z    P VALUE
C1             50    20.022     1.564     0.221      0.55       0.29

MTB > # Since z = 0.55 < 2.326, we fail to reject Ho.
MTB > # Insufficient evidence exists to conclude that the mean weight
MTB > # exceeds 19.9 ounces.
```

EXERCISES for Section 10.3

Hypothesis Test for a Mean: Small Sample

10.31 For a one-tailed test with $\alpha = 0.01$ and $n = 16$, we find $t_\alpha = t_{0.01} = 2.602$, where $df = n - 1 = 15$. Since the alternative hypothesis is H_a: $\mu > 76$, we reject the null hypothesis for $t > 2.602$.

10.33 For a two-tailed test with $\alpha = 0.10$ and $n = 11$, we find $t_{\alpha/2} = t_{0.05} = 1.812$, where $df = n - 1 = 10$. Therefore, we reject the null hypothesis for $t < -1.812$ or $t > 1.812$.

10.35 For a one-tailed test with $\alpha = 0.05$ and $n = 9$, we find $t_\alpha = t_{0.05} = 1.860$, where $df = n - 1 = 8$. Since the alternative hypothesis is H_a: $\mu < 172$, we reject the null hypothesis for $t < -1.860$.

10.37 **Step 1: Formulate the null and alternative hypotheses.**
We test the hypotheses
H_0: $\mu = 590$
H_a: $\mu \neq 590$
Step 2: Select the significance level for the test.
We are given that $\alpha = 0.01$.
Step 3: Calculate the value of the test statistic.
Since the population is normal and the sample size is less than 30, a t-test statistic is used, where $n = 15$, $\bar{x} = 576$ and $s = 44$. We calculate

$$t = \frac{\bar{x} - \mu_0}{s/\sqrt{n}} = \frac{576 - 590}{44/\sqrt{15}} = -1.23.$$

Step 4: Determine the rejection region of H_0 from the tables.
For a two-tailed test with a significance level $\alpha = 0.01$ and $n = 15$, we have $t_{\alpha/2} = t_{0.005} = 2.977$, where $df = n - 1 = 14$, and we reject H_0 if $t < -2.977$ or $t > 2.977$.
Step 5: State the conclusion of the test.
For $t = -1.23$, we fail to reject H_0, so there is not sufficient evidence at the 1% level of significance to show that the mean of the population differs from 590.

10.39 **Step 1: Formulate the null and alternative hypotheses.**
In order to show that the mean of the sampled population differs from 25, we test:
H_0: $\mu = 25$
H_a: $\mu \neq 25$
Step 2: Select the significance level for the test.
We are given that $\alpha = 0.01$.
Step 3: Calculate the value of the test statistic.
Since the population is normal and the sample size is less than 30, a t-test statistic is used. From the given data, using a statistical calculator, we calculate $n = 6$, $\bar{x} = 21$ and

$s = 1.789$, so we have $$t = \frac{\bar{x} - \mu_0}{s/\sqrt{n}} = \frac{21 - 25}{1.789/\sqrt{6}} = -5.48.$$

Step 4: Determine the rejection region of H_0 from the tables.
For a two-tailed test with significance level $\alpha = 0.01$, and $n = 6$, we have $t_{\alpha/2} = t_{0.005} = 4.032$, where $df = 6 - 1 = 5$ and we reject H_0 if $t < -4.032$ or $t > 4.032$.

Step 5: State the conclusion of the test.
For t = -5.48, we reject H_0, so there is sufficient evidence at the 1% level to show that the mean of the population differs from 25.

10.41 **Step 1: Formulate the null and alternative hypotheses.**
In order to show that card usage increased during the promotion period, we test:
H_0: $\mu = 8.4$
H_a: $\mu > 8.4$
Step 2: Select the significance level for the test.
We are given that $\alpha = 0.05$.
Step 3: Calculate the value of the test statistic.
Since the population is normal and the sample size is less than 30, a t-test statistic is used, where n = 28, $\bar{x} = 9.7$ and s = 2.6. We calculate

$$t = \frac{\bar{x} - \mu_0}{s/\sqrt{n}} = \frac{9.7 - 8.4}{2.6/\sqrt{28}} = 2.65.$$

Step 4: Determine the rejection region of H_0 from the tables.
For a one-tailed test with a significance level $\alpha = 0.05$ and n = 28, we have $t_\alpha = t_{0.05} = 1.703$, where df = n - 1 = 27. Since the alternative hypothesis is H_a: $\mu > 8.4$, we reject H_0 if t > 1.703.
Step 5: State the conclusion of the test.
For t = 2.65, we reject H_0, so there is sufficient evidence at the 5% level to show that card usage increased during the promotion period.

10.43 **Step 1: Formulate the null and alternative hypotheses.**
In order to show the average age of men at the time of their first marriage differs for males in the county, we test:
H_0: $\mu = 24.8$
H_a: $\mu \neq 24.8$
Step 2: Select the significance level for the test.
We are given that $\alpha = 0.05$.
Step 3: Calculate the value of the test statistic.
Since the population is normal and the sample size is less than 30, a t-test statistic is used, where n = 24, $\bar{x} = 23.5$ and s = 3.2. We calculate

$$t = \frac{\bar{x} - \mu_0}{s/\sqrt{n}} = \frac{23.5 - 24.8}{3.2/\sqrt{24}} = -1.99.$$

Step 4: Determine the rejection region of H_0 from the tables.
For a two-tailed test with a significance level $\alpha = 0.05$, and n = 24, we have $t_{\alpha/2} = t_{0.025} = 2.069$, where df = n - 1 = 23, and we reject H_0 if t < -2.069 or t > 2.069.
Step 5: State the conclusion of the test.
For t = -1.99, we fail to reject H_0, so there is not sufficient evidence at the 5% level to show that the average age of men at the time of their first marriage differs for males in the county.

10.45 **Step 1: Formulate the null and alternative hypotheses.**
In order to show that the mechanic can perform this service in less time than specified by the service manual, we test:
H_0: $\mu = 6.5$
H_a: $\mu < 6.5$
Step 2: Select the significance level for the test.
We are given that $\alpha = 0.05$.
Step 3: Calculate the value of the test statistic.
Since the population is normal and the sample size is less than 30, a t-test statistic is used. From the given data, using a statistical calculator, we calculate n = 10, $\bar{x} = 6.3$ and

s = 0.258 so we have $t = \dfrac{\bar{x} - \mu_0}{s/\sqrt{n}} = \dfrac{6.3 - 6.5}{0.258/\sqrt{10}} = -2.45.$

Step 4: Determine the rejection region of H_0 from the tables.
For a one-tailed test with a significance level $\alpha = 0.05$ and n = 10, we have $t_\alpha = t_{0.05} =$ 1.833, where df = n - 1 = 9.
Since the alternative hypothesis is H_a: $\mu < 6.5$, we reject H_0 if $t < -1.833$.
Step 5: State the conclusion of the test.
For t = -2.45, we reject H_0, so there is sufficient evidence at the 5% level to show that the mechanic can perform this service in less than 6.5 hours.

10.47 **Step 1: Formulate the null and alternative hypotheses.**
In order to show the sample mean is significantly different from 50 hours, we test:
H_0: $\mu = 50$
H_a: $\mu \neq 50$
Step 2: Select the significance level for the test.
We are given that $\alpha = 0.05$.
Step 3: Calculate the value of the test statistic.
Since the population is normal and the sample size is less than 30, a t-test statistic is used, where n = 25, $\bar{x} = 52.7$ and s = 4.2. We calculate

$$t = \frac{\bar{x} - \mu_0}{s/\sqrt{n}} = \frac{52.7 - 50}{4.2/\sqrt{25}} = 3.21.$$

Step 4: Determine the rejection region of H_0 from the tables.
For a two-tailed test with a significance level $\alpha = 0.05$ and n = 25, we have
$t_{\alpha/2} = t_{0.025} = 2.064$, where df = n - 1 = 24, and we reject H_0 if $t < -2.064$ or $t > 2.064$.
Step 5: State the conclusion of the test.
For t = 3.21, we reject H_0, so there is sufficient evidence at the 5% level to show that the sample mean is significantly different from 50 hours.

MINITAB LAB ASSIGNMENTS

10.49(M)

```
MTB > # Exercise 10.49(M)
MTB > # The data were entered into C1 using SET C1
MTB > TTEST 38 C1;
SUBC> ALTERNATIVE = 0.

TEST OF MU = 38.000 VS MU N.E. 38.000

              N      MEAN     STDEV   SE MEAN        T    P VALUE
C1           23    27.783     9.010     1.879    -5.44     0.0000

MTB > # Since t = -5.44 < -2.074, we reject Ho.
```

EXERCISES for Section 10.4

Hypothesis Test for a Proportion: Large Sample

10.51 **Step 1: Hypotheses**

H_0: p = 0.40

H_a: p < 0.40

Step 2: Significance Level

We are given that $\alpha = 0.01$.

Step 3: Calculations

For the sample, n = 500 and x = 182 so $\hat{p} = \frac{x}{n} = \frac{182}{500} = 0.364$. The value of the test statistic is:

$$z = \frac{\hat{p} - p_0}{\sqrt{\frac{p_0(1 - p_0)}{n}}} = \frac{0.364 - 0.40}{\sqrt{\frac{(0.40)\cdot(0.60)}{500}}} = -1.64.$$

Step 4: Rejection Region

For a one-tailed test with a significance level $\alpha = 0.01$, we have $z_\alpha = z_{0.01} = 2.326$. Since the alternative hypothesis is H_a: p < 0.40, we reject H_0 if z < -2.326.

Step 5: Conclusion

Since the value of the test statistic is z = -1.64, at the 1% level of significance, we fail to reject H_0. There is not sufficient evidence to show that the the population proportion is less than 0.40.

10.53 **Step 1: Hypotheses**

In order to show that the population proportion differs from 0.70, we test

H_0: p = 0.70

H_a: p ≠ 0.70

Step 2: Significance Level

We are given that $\alpha = 0.05$.

Step 3: Calculations

For the sample, n = 100 and $\hat{p}$ = 0.78. The value of the test statistic is:

$$z = \frac{\hat{p} - p_0}{\sqrt{\frac{p_0(1-p_0)}{n}}} = \frac{0.78 - 0.70}{\sqrt{\frac{(0.70)\cdot(0.30)}{100}}} = 1.75.$$

Step 4: Rejection Region
For a two-tailed test with a significance level α = 0.05, we have $z_{\alpha/2} = z_{0.025} = 1.96$, so we reject H_0 if z > 1.96 or z < -1.96.
Step 5: Conclusion
For z = 1.75, at the 5% level of significance, we fail to reject H_0 and do not conclude that the population proportion differs from 0.70.

10.55 **Step 1: Hypotheses**
In order to show that more than 50% of American adults are aware of the fact that the sun is a star, we test:
H_0: p = 0.5
H_a: p > 0.5
Step 2: Significance Level
We are given that α = 0.05.
Step 3: Calculations
For the sample, n = 1000 and $\hat{p}$ = 52.5% = 0.525. The value of the test statistic is:

$$z = \frac{\hat{p} - p_0}{\sqrt{\frac{p_0(1-p_0)}{n}}} = \frac{0.525 - 0.50}{\sqrt{\frac{(0.50)\cdot(0.50)}{1000}}} = 1.58.$$

Step 4: Rejection Region
For a one-tailed test with a significance level α = 0.05, we have $z_\alpha = z_{0.05} = 1.645$. Since the alternative hypothesis is H_a: p > 0.50, we reject H_0 if z > 1.645.
Step 5: Conclusion
For z = 1.58, we fail to reject H_0, so at the 5% level of significance, we are unable to show that more than 50% of American adults are aware of the fact that the sun is a star.

10.57 **Step 1: Hypotheses**
In order to show that less than 90% of automobile dealer customers would recommend them to a friend, we test:
H_0: p = 0.90
H_a: p < 0.90
Step 2: Significance Level
We are given that α = 0.01.
Step 3: Calculations
For the sample, n = 89 and x = 71 so $\hat{p} = \frac{x}{n} = \frac{71}{89}$. The value of the test statistic is:

$$z = \frac{\hat{p} - p_0}{\sqrt{\frac{p_0(1-p_0)}{n}}} = \frac{\frac{71}{89} - 0.90}{\sqrt{\frac{(0.90)\cdot(0.10)}{89}}} = -3.22.$$

Step 4: Rejection Region
For a one-tailed test with a significance level $\alpha = 0.01$, we have $z_{\alpha} = z_{0.01} = 2.326$. Since the alternative hypothesis is H_a: $p < 0.90$, we reject H_0 if $z < -2.326$.
Step 5: Conclusion
Since the value of the test statistic is $z = -3.22$, we reject H_0, so there is sufficient evidence, at the 1% level of significance, to show that less than 90% of the automobile dealer's customers would recommend them to a friend.

10.59 **Step 1: Hypotheses**
In order to show that the actual percentage of all Americans who take multiple vitamins regularly is different from 37%, we test:
H_0: $p=0.37$
H_a: $p \neq 0.37$
Step 2: Significance Level
We are given that $\alpha = 0.01$.
Step 3: Calculations
For the sample, n = 750 and x = 290 so $\hat{p} = \frac{x}{n} = \frac{290}{750} = 0.3867$. The value of the test statistic is:

$$z = \frac{\hat{p} - p_0}{\sqrt{\frac{p_0(1-p_0)}{n}}} = \frac{0.3867 - 0.37}{\sqrt{\frac{(0.37)\cdot(0.63)}{750}}} = 0.95.$$

Step 4: Rejection Region
For a two-tailed test with a significance level $\alpha = 0.01$, we have $z_{\alpha/2} = z_{0.005} = 2.576$, so we reject H_0 if $z > 2.576$ or $z < -2.576$.
Step 5: Conclusion
For $z = 0.95$, we fail to reject H_0, so there is not sufficient evidence, at the 1% level of significance, to show that the actual percentage of all Americans who take multiple vitamins regularly is different from 37%.

10.61 **Step 1: Hypotheses**
In order to show that the coin is biased, we test:
H_0: $p = 0.50$
H_a: $p \neq 0.50$
Step 2: Significance Level
We are given that $\alpha = 0.05$.
Step 3: Calculations
For the sample, n = 1000 and x = 535 so $\hat{p} = \frac{x}{n} = \frac{535}{1000} = 0.535$. The value of the test statistic is:

$$z = \frac{\hat{p} - p_0}{\sqrt{\frac{p_0(1-p_0)}{n}}} = \frac{0.535 - 0.50}{\sqrt{\frac{(0.50)\cdot(0.50)}{1000}}} = 2.21.$$

Step 4: Rejection Region
For a two-tailed test with a significance level $\alpha = 0.05$, we have $z_{\alpha/2} = z_{0.025} = 1.96$ so we reject H_0 if $z > 1.96$ or $z < -1.96$.

<u>Step 5</u>: **Conclusion**
Since the value of the test statistic is z = 2.21, we reject H_0, so there is sufficient evidence, at the 5% level of significance, to show that the coin is biased.

EXERCISES for Section 10.5
Hypothesis Test for a Variance

10.63 Since the sample size is 20, we use the X^2 distribution with df = n - 1 = 19.
a. The significance level is $\alpha = 0.05$ and the alternative hypothesis is: H_a: $\sigma^2 < 200$, so we find $X^2_{(1-\alpha)} = X^2_{0.95} = 10.12$. We reject the null hypothesis for $X^2 < 10.12$.
b. The significance level is $\alpha = 0.05$ and the alternative hypothesis is: H_a: $\sigma^2 > 14.9$, so we find $X^2_{\alpha} = X^2_{0.05} = 30.14$. We reject the null hypothesis for $X^2 > 30.14$.
c. For a two-tailed test and significance level $\alpha = 0.05$, we find $X^2_{(1-\alpha/2)} = X^2_{0.975} = 8.91$ and $X^2_{(\alpha/2)} = X^2_{0.025} = 32.85$. We reject the null hypothesis for $X^2 < 8.91$ or for $X^2 > 32.85$.

10.65 <u>Step 1:</u> **Hypotheses**
We test the following hypotheses at the 0.05 significance level:
H_0: $\sigma^2 = 28$
H_a: $\sigma^2 < 28$
<u>Step 2</u>: **Significance Level**
We are given that $\alpha = 0.05$.
<u>Step 3</u>: **Calculations**
For the sample, n = 21 and s = 4.8. Therefore, the value of the test statistic is:

$$X^2 = \frac{(n-1)s^2}{\sigma_0^2} = \frac{(21-1)(4.8)^2}{28} = 16.46.$$

<u>Step 4</u>: **Rejection Region**
Since $\alpha = 0.05$ and the alternative hypothesis is H_a: $\sigma^2 < 28$, we find $X^2_{(1-\alpha)} = X^2_{0.95} = 10.85$, where df = n - 1 = 20, so we reject H_0 if $X^2 < 10.85$.
<u>Step 5</u>: **Conclusion**
For $X^2 = 16.46$, we fail to reject H_0 at the 5% level of significance, and do not conclude that the variance of the population is less than 28.

10.67 <u>Step 1:</u> **Hypotheses**
We test the following hypotheses at the 0.10 significance level:
H_0: $\sigma^2 = 6^2$
H_a: $\sigma^2 \neq 6^2$
<u>Step 2</u>: **Significance Level**
We are given that $\alpha = 0.10$.
<u>Step 3</u>: **Calculations**
For the sample, n = 20 and $s^2 = 59.69$. Therefore, the value of the test statistic is:

$$X^2 = \frac{(n-1)s^2}{\sigma_0^2} = \frac{(20-1)(59.69)}{6^2} = 31.50.$$

Step 4: Rejection Region
Since $\alpha = 0.10$ and the alternative hypothesis is H_a: $\sigma \neq 6$, we find $X^2_{(1-\alpha/2)} = X^2_{0.95} = 10.12$ and $X^2_{(\alpha/2)} = X^2_{0.05} = 30.14$, where df = n - 1 = 19, so we reject H_0 if $X^2 < 10.12$ or if $X^2 > 30.14$.
Step 5: Conclusion
For $X^2 = 31.50$, we reject H_0 at the 10% level of significance, and conclude that the standard deviation of the population is different from 6.

10.69 **Step 1: Hypotheses**
In order to determine if the standard deviation of the sampled population differs from 2.4, we test:
H_0: $\sigma^2 = 2.4^2$
H_a: $\sigma^2 \neq 2.4^2$
Step 2: Significance Level
We are given that $\alpha = 0.05$.
Step 3: Calculations
For the sample, n = 6 and we calculate $s^2 = 15.5$. Therefore, the value of the test statistic is:

$$X^2 = \frac{(n-1)s^2}{\sigma_0^2} = \frac{(6-1)(15.5)}{(2.4)^2} = 13.45.$$

Step 4: Rejection Region
Since $\alpha = 0.05$ and the alternative hypothesis is H_a: $\sigma \neq 2.4$, we find $X^2_{(1-\alpha/2)} = X^2_{0.975} = 0.83$ and $X^2_{(\alpha/2)} = X^2_{0.025} = 12.83$, where df = n - 1 = 5, so we reject H_0 if $X^2 < 0.83$ or if $X^2 > 12.83$.
Step 5: Conclusion
Since the value of the test statistic is $X^2 = 13.45$, we reject H_0, and there is sufficient evidence, at the 5% level of significance, to show that the standard deviation of the sampled population differs from 2.4.

10.71 **Step 1: Hypotheses**
In order to show that the variation in the amounts dispensed, as measured by the standard deviation, exceeds 0.4 ounces, we test:
H_0: $\sigma^2 = 0.4^2$
H_a: $\sigma^2 > 0.4^2$
Step 2: Significance Level
We are given that $\alpha = 0.05$.
Step 3: Calculations
For the sample, n = 10 and we calculate $s^2 = 0.4379$. Therefore, the value of the test statistic is:

$$X^2 = \frac{(n-1)s^2}{\sigma_0^2} = \frac{(10-1)(0.4379)}{(0.4)^2} = 24.63.$$

Step 4: Rejection Region
Since $\alpha = 0.05$ and the alternative hypothesis is H_a: $\sigma > 0.4$, we find $X^2_{\alpha} = X^2_{0.05} = 16.92$, where df = n - 1 = 9, so we reject H_0 if $X^2 > 16.92$.

Step 5: Conclusion
Since the value of the test statistic is $X^2 = 24.63$, we reject H_0, so there is sufficient evidence, at the 0.05 level of significance, to show that the standard deviation exceeds 0.4 oz.

10.73 **Step 1: Hypotheses**
In order to show that the standard deviation of the calcium content of all tablets in the production run differs from 2 mg, we test:
H_0: $\sigma^2 = 2^2$
H_a: $\sigma^2 \neq 2^2$
Step 2: Significance Level
We are given that $\alpha = 0.10$.
Step 3: Calculations
For the sample, n = 71 and s = 3.4. Therefore, the value of the test statistic is:

$$X^2 = \frac{(n-1)s^2}{\sigma_0^2} = \frac{(71-1)(3.4)^2}{(2)^2} = 202.30.$$

Step 4: Rejection Region
Since $\alpha = 0.10$ and the alternative hypothesis is H_a: $\sigma \neq 2$, we find $X^2_{(1-\alpha/2)} = X^2_{0.95} = 51.74$ and $X^2_{(\alpha/2)} = X^2_{0.05} = 90.53$, where df = n - 1 = 70, so we reject H_0 if $X^2 < 51.74$ or if $X^2 > 90.53$.
Step 5: Conclusion
Since the value of the test statistic is $X^2 = 202.30$, we reject H_0, and there is sufficient evidence, at the 10% level of significance, to show that the standard deviation of the calcium content of all tablets in the production run differs from 2 mg.

10.75 **Step 1: Hypotheses**
In order to show that there sufficient evidence to conclude that the standard deviation in rates at the time of the survey exceeds one-quarter of a percentage point, we test:
H_0: $\sigma = 0.25$
H_a: $\sigma > 0.25$
Step 2: Significance Level
We are given that $\alpha = 0.05$.
Step 3: Calculations
For the sample, n = 41 and we calculate

$SS(x) = \Sigma x^2 - \frac{(\Sigma x)^2}{n} = 3983.456768 - \frac{(403.875)^2}{41} = 5.041753$ and

$s^2 = \frac{SS(x)}{n-1} = \frac{5.041753}{41-1} = 0.12604$. (See Exercise 9.167.) Therefore, the value of the test statistic is: $X^2 = \frac{(n-1)s^2}{\sigma_0^2} = \frac{(41-1)(0.12604)}{(0.25)^2} = 80.67.$

Step 4: Rejection Region
Since $\alpha = 0.05$ and the alternative hypothesis is H_a: $\sigma > 0.25$, we find $X^2_\alpha = X^2_{0.05} = 55.76$, where df = n - 1 = 40, so we reject H_0 if $X^2 > 55.76$.

Step 5: Conclusion
Since the value of the test statistic is $X^2 = 80.67$, we reject H_0, so there is sufficient evidence, at the 0.05 level of significance, to show that the standard deviation in rates exceeds one-quarter of a percentage point.

EXERCISES for Section 10.6
Using P-Values to Report Test Results

10.77 We reject the null hypothesis only if the p-value is less than or equal to α.
a. We are given $0.05 <$ p-value < 0.10, which means p-value $> \alpha = 0.05$. Therefore, we fail to reject the null hypothesis.
b. We are given $0.05 <$ p-value < 0.10, which means p-value $< \alpha = 0.10$. Therefore, we reject the null hypothesis.
c. We are given $0.025 <$ p-value < 0.05, which means p-value $> \alpha = 0.01$. Therefore, we fail to reject the null hypothesis.
d. We are given $0.005 <$ p-value < 0.01, which means p-value $< \alpha = 0.01$. Therefore, we reject the null hypothesis.
e. We are given p-value > 0.10, which means p-value $> \alpha = 0.05$. Therefore, we fail to reject the null hypothesis.
f. We are given p-value < 0.005, which means p-value $< \alpha = 0.01$. Therefore, we reject the null hypothesis.

10.79 For $z = -1.66$ and $H_a: \mu < 230$, p-value $= P(z < -1.66) = 0.0485$.

10.81 For $z = 0.98$ and $H_a: \mu \neq 89.5$, p-value $= 2P(z > 0.98) = 2(1 - 0.8365) = 0.3270$.

10.83 Since a t-test is used, where the alternative hypothesis is $H_a: \mu < 347$, the p-value is the area under the t distribution to the left of $t = -2.83$. We use Table IV with df $= n - 1 = 25 - 1 = 24$ and note that -2.83 lies to the left of $-t_{0.005} = -2.797$. Therefore, the area under the t distribution to the left of -2.83 is a value less than 0.005. Thus, p-value < 0.005.

10.85 Since a X^2-test is used, where the alternative hypothesis is $H_a: \sigma^2 > 28$, the p-value is the area under the X^2 distribution to the right of $X^2 = 23.52$. We use Table V with df $= n - 1 = 16 - 1 = 15$ and note that 23.52 lies between $X^2_{0.05} = 25.00$ and $X^2_{0.10} = 22.31$. Therefore, the area under the X^2 distribution to the right of 23.52 is a value between 0.05 and 0.10. Thus, $0.05 <$ p-value < 0.10.

10.87 **Step 1: Hypotheses**
In order to show that the new battery has a longer mean life, we test
$H_0: \mu = 150$
$H_a: \mu > 150$
Step 2: Significance Level
We are given that $\alpha = 0.05$.

Step 3: Calculations
For the sample, $n = 35$, $\bar{x} = 159$ and $s = 27$ is used to approximate σ. Therefore, the value of the test statistic is: $z = \dfrac{\bar{x} - \mu}{\sigma/\sqrt{n}} = \dfrac{159 - 150}{27/\sqrt{35}} = 1.97$.

Step 4: Calculation of p-value
Since the alternative hypothesis is H_a: $\mu > 150$, we calculate p-value = $P(z > 1.97)$ = $1 - 0.9756 = 0.0244$.
Step 5: Conclusion
Since p-value = $0.0244 < \alpha = 0.05$, we reject H_0, so there is sufficient evidence at the 5% level to show that the new battery has a longer mean life.

10.89 **Step 1: Hypotheses**
In order to show that the mechanic can perform this service in less time than specified by the service manual, we test:
H_0: $\mu = 6.5$
H_a: $\mu < 6.5$
Step 2: Significance Level
We are given that $\alpha = 0.05$.
Step 3: Calculations
Since the population is normal and the sample size is less than 30, a t-test statistic is used. From the given data, we calculate $n = 10$, $\bar{x} = 6.3$ and $s = 0.258$ so we have

$$t = \frac{\bar{x} - \mu}{s/\sqrt{n}} = \frac{6.3 - 6.5}{0.258/\sqrt{10}} = -2.45.$$

Step 4: Calculation of p-value
Since a t-test is used, where the alternative hypothesis is H_a: $\mu < 6.5$, the p-value is the area under the t distribution to the left of $t = -2.45$. We use Table IV with df = n - 1 = 10 - 1 = 9 and note that -2.45 lies between $-t_{0.025} = -2.262$ and $-t_{0.01} = -2.821$. Therefore, the area under the t distribution to the left of -2.45 is a value between 0.025 and 0.01. Thus $0.01 <$ p-value < 0.025.
Step 5: Conclusion
Since p-value $< 0.025 < \alpha = 0.05$, we reject H_0, so there is sufficient evidence at the 5% level to show that the mechanic can perform this service in less than 6.5 hours.

10.91 **Step 1: Hypotheses**
In order to show that less than ten percent of the peaches have serious blemishes, we test:
H_0: $p = 0.10$
H_a: $p < 0.10$
Step 2: Significance Level
We are given that $\alpha = 0.05$.
Step 3: Calculations
For the sample, $n = 1500$ and $x = 138$ so $\hat{p} = \dfrac{x}{n} = \dfrac{138}{1500} = 0.092$. The value of the test statistic is:

$$z = \frac{\hat{p} - p_0}{\sqrt{\dfrac{p_0(1 - p_0)}{n}}} = \frac{0.092 - 0.10}{\sqrt{\dfrac{(0.10)\cdot(0.90)}{1500}}} = -1.03.$$

<u>Step 4</u>: **Calculation of p-value**
Since the alternative hypothesis is H_a: $p < 0.10$, we calculate p-value = $P(z < -1.03)$ = 0.1515.
<u>Step 5</u>: **Conclusion**
Since p-value = 0.1515 is not less than $\alpha = 0.05$, we fail to reject H_0, so there is not sufficient evidence, at the 5% level of significance, to show that less than ten percent of the peaches have serious blemishes.

10.93 <u>Step 1</u>: **Hypotheses**
In order to show that the length of a critical component varies with a standard deviation of less than 0.15 mm., we test:
H_0: $\sigma = 0.15$
H_a: $\sigma < 0.15$
<u>Step 2</u>: **Significance Level**
We are given that $\alpha = 0.05$.
<u>Step 3</u>: **Calculations**
For the sample, n = 30 and s = 0.13. Therefore, the value of the test statistic is:

$$X^2 = \frac{(n-1)s^2}{\sigma_0^2} = \frac{(30-1)(0.13)^2}{(0.15)^2} = 21.78.$$

<u>Step 4</u>: **Calculation of p-value**
Since a X^2-test is used, where the alternative hypothesis is H_a: $\sigma < 0.15$, the p-value is the area under the X^2 distribution to the left of $X^2 = 21.78$. We use Table V with df = n - 1 = 30 - 1 = 29 and note that 21.78 lies to the right of $X^2_{0.90} = 19.77$. Therefore, the area under the X^2 distribution to the left of 21.78 is a value greater than 1 - 0.90 = 0.10, so we have p-value > 0.10.
<u>Step 5</u>: **Conclusion**
Since $\alpha = 0.05 < 0.10 <$ p-value, we fail to reject H_0, so there is not sufficient evidence, at the 5% level of significance, to show that the length of the component varies with a standard deviation of less than 0.15 mm.

MINITAB LAB ASSIGNMENTS

10.95(M)
```
MTB > # Exercise 10.95(M)
MTB > CDF -1.27;
SUBC> NORMAL 0 1.
   -1.2700     0.1020
MTB > # P-value = 0.1020.
```

10.97(M)
```
MTB > # Exercise 10.97(M)
MTB > CDF 22.14;
SUBC> CHIS 14.
   22.1400     0.9242
MTB > # P-value = 1 - 0.9242 = 0.0758.
```

REVIEW EXERCISES
Chapter 10

10.98 For a large sample one-tailed test with a significance level $\alpha = 0.01$, we have $z_\alpha = z_{0.01} = 2.326$. Since the alternative hypothesis is H_a: $\mu > 163$, we reject H_0 if $z > 2.326$.

10.99 For a large sample one-tailed test with a significance level $\alpha = 0.05$, we have $z_\alpha = z_{0.05} = 1.645$. Since the alternative hypothesis is H_a: $\mu < 23.6$, we reject H_0 if $z < -1.645$.

10.100 For a two-tailed test with $\alpha = 0.05$ and $n = 11$, we find $t_{\alpha/2} = t_{0.025} = 2.228$, where df = $n - 1 = 10$. Therefore, we reject the null hypothesis for $t < -2.228$ or $t > 2.228$.

10.101 For a one-tailed test with $\alpha = 0.01$ and $n = 23$, we find $t_\alpha = t_{0.01} = 2.508$, where df = $n - 1 = 22$. Since the alternative hypothesis is H_a : $\mu < 16.2$, we reject the null hypothesis for $t < -2.508$.

10.102 For a large sample two-tailed test with $\alpha = 0.10$, we find $z_{\alpha/2} = z_{0.05} = 1.645$. Therefore, we reject the null hypothesis for $z < -1.645$ or $z > 1.645$.

10.103 For a large sample, one-tailed hypothesis test for p and significance level $\alpha = 0.10$, we have $z_\alpha = z_{0.10} = 1.282$. Since the alternative hypothesis is H_a: $p < 1/3$, we reject H_0 if $z < -1.282$.

10.104 For a large sample, two-tailed hypothesis test for p and significance level $\alpha = 0.05$, we find $z_{\alpha/2} = z_{0.025} = 1.96$. Therefore, we reject the null hypothesis for $z < -1.96$ or $z > 1.96$.

10.105 Since the sample size is 28, we use the X^2 distribution with df = $n - 1 = 27$. For significance level $\alpha = 0.05$ and the alternative hypothesis: H_a: $\sigma^2 > 0.81$, we find $X^2_\alpha = X^2_{0.05} = 40.11$. We reject the null hypothesis for $X^2 > 40.11$.

10.106 Since the sample size is 25, we use the X^2 distribution with df = $n - 1 = 24$. For a two-tailed test and significance level $\alpha = 0.10$, we find $X^2_{(1-\alpha/2)} = X^2_{0.95} = 13.85$ and $X^2_{(\alpha/2)} = X^2_{0.05} = 36.42$. We reject the null hypothesis for $X^2 < 13.85$ or for $X^2 > 36.42$.

10.107

a. The null hypothesis is falsely accepted. Therefore, a Type II error is committed.

b. The null hypothesis is falsely rejected. Therefore, a Type I error is committed.

c. **Step 1: Hypotheses**

The researcher wants to test the following hypotheses:

H_0: $\mu = 60$

H_a: $\mu < 60$

Step 2: Significance Level

We are given that $\alpha = 0.05$.

Step 3: Calculations

For the sample, $n = 32$, $\bar{x} = 59.2$ and $s = 3.9$, which is used to approximate σ.

Therefore, the value of the test statistic is: $Z = \dfrac{\bar{x} - \mu}{\sigma/\sqrt{n}} = \dfrac{59.2 - 60}{3.9/\sqrt{32}} = -1.16$.

Step 4: Rejection Region
For a one-tailed test with a significance level $\alpha = 0.05$, we have $z_\alpha = z_{0.05} = 1.645$. Since the alternative hypothesis is H_a: $\mu < 60$, we reject H_0 if $z < -1.645$.
Step 5: Conclusion
Since $z = -1.16$, we fail to reject H_0 at level of significance 0.05 and do not find sufficient evidence that the mean of the population is less than 60.

d. For $z = -1.16$ and H_a: $\mu < 60$, p-value $= P(z < -1.16) = 0.1230$.

10.108 Step 1: Hypotheses
In order to show that less than one-third of the facelifts performed in his state are on people 50 or younger, we test:
H_0: $p = 1/3$
H_a: $p < 1/3$
Step 2: Significance Level
We are given that $\alpha = 0.05$.
Step 3: Calculations
For the sample, n = 200 and x = 59 so $\hat{p} = \frac{x}{n} = \frac{59}{200} = 0.295$. The value of the test statistic is:

$$z = \frac{\hat{p} - p_0}{\sqrt{\frac{p_0(1-p_0)}{n}}} = \frac{0.295 - \frac{1}{3}}{\sqrt{\frac{(\frac{1}{3})\cdot(\frac{2}{3})}{200}}} = -1.15.$$

Step 4: Rejection Region
For a one-tailed test with a significance level $\alpha = 0.05$, we have $z_\alpha = z_{0.05} = 1.645$. Since the alternative hypothesis is H_a: $p < 1/3$, we reject H_0 if $z < -1.645$.
Step 5: Conclusion
Since the value of the test statistic is $z = -1.15$, we fail to reject H_0, so there is not sufficient evidence, at the 5% level of significance, to show that less than one-third of the facelifts performed in his state are on people 50 or younger.

10.109 Step 1: Hypotheses
In order to show that the mean life of telephone poles can be increased to more than 50 years, we test
H_0: $\mu = 50$
H_a: $\mu > 50$
Step 2: Significance Level
We are given that $\alpha = 0.05$.
Step 3: Calculations
For the sample, n = 35, $\bar{x} = 52.1$ and s = 5.7, which is used to approximate σ. Therefore, the value of the test statistic is: $z = \frac{\bar{x} - \mu}{\sigma/\sqrt{n}} = \frac{52.1 - 50}{5.7/\sqrt{35}} = 2.18.$
Step 4: Rejection Region
For a one-tailed test with a significance level $\alpha = 0.05$, we have $z_\alpha = z_{0.05} = 1.645$. Since the alternative hypothesis is H_a: $\mu > 50$, we reject H_0 if $z > 1.645$.

Step 5: Conclusion
For $z = 2.18$, we reject H_0. Therefore, there is sufficient evidence at the 5% level to show that the mean life of telephone poles can be increased to more than 50 years.

10.110 For $z = 2.18$ and H_a: $\mu > 50$, p-value $= P(z > 2.18) = 1 - 0.9854 = 0.0146$. Since p-value $< \alpha = 0.05$, we reject H_0, as in Exercise 10.109.

10.111 a. Step 1: Hypotheses
In order to show that the average age of blood donors during the last 12 months has fallen below the previous value of 40.3 years, we test
H_0: $\mu = 40.3$
H_a: $\mu < 40.3$
Since the sample size is less than 30, a t-test statistic is used.
Step 2: Significance Level
We are given that $\alpha = 0.05$.
Step 3: Calculations
From the given data, we calculate $n = 20$, $\bar{x} = 34.45$ and $s = 10.52$ so we have

$$t = \frac{\bar{x} - \mu}{s/\sqrt{n}} = \frac{34.45 - 40.3}{10.52/\sqrt{20}} = -2.49.$$

Step 4: Rejection Region
For a one-tailed test with a significance level $\alpha = 0.05$, and $n = 20$, we have $t_\alpha = t_{0.05} = 1.729$, where $df = n - 1 = 19$. Since the alternative hypothesis is H_a: $\mu < 40.3$, we reject H_0 if $t < -1.729$.
Step 5: Conclusion
For $t = -2.49$, we reject H_0, and there is sufficient evidence at the 5% level to show that the average age of blood donors during the last 12 months has fallen below the previous value of 40.3 years.
b. In order for the test to be valid, we must assume that the sample is random from a normal population.

10.112 Since a t-test is used, where the alternative hypothesis is H_a: $\mu < 40.3$, the p-value is the area under the t distribution to the left of $t = -2.49$. We use Table IV with $df = n - 1 = 20 - 1 = 19$ and note that -2.49 lies between $-t_{0.01} = -2.539$ and $-t_{0.025} = -2.093$. Therefore, the area under the t distribution to the left of -2.49 is a value between 0.01 and 0.025. Thus, $0.01 <$ p-value < 0.025.

10.113 Step 1: Hypotheses
In order to show that the true standard deviation of the temperature exceeds 0.6 degrees, we test:
H_0: $\sigma = 0.6$
H_a: $\sigma > 0.6$
Step 2: Significance Level
We are given that $\alpha = 0.05$.

Step 3: Calculations
For the sample, n = 15 and s = 0.74. Therefore, the value of the test statistic is:

$$X^2 = \frac{(n-1)s^2}{\sigma_0^2} = \frac{(15-1)(0.74)^2}{(0.6)^2} = 21.30.$$

Step 4: Rejection Region
Since $\alpha = 0.05$ and the alternative hypothesis is H_a: $\sigma > 0.6$, we find $X^2_\alpha = X^2_{0.05} = 23.68$, where df = n - 1 = 14, so we reject H_0 if $X^2 > 23.68$.
Step 5: Conclusion
Since the value of the test statistic is $X^2 = 21.30$, we fail to reject H_0, so there is not sufficient evidence, at the 0.05 level of significance, to show that the standard deviation exceeds 0.6 degrees.

10.114 Since a X^2 test is used, where the alternative hypothesis is H_a: $\sigma > 0.6$, the p-value is the area under the X^2 distribution to the right of $X^2 = 21.30$. We use Table V with df = n - 1 = 15 - 1 = 14, and note that 21.30 lies between $X^2_{0.05} = 23.68$ and $X^2_{0.10} = 21.06$. Therefore, the area under the X^2 distribution to the right of 21.30 is a value between 0.05 and 0.10. Thus, 0.05 < p-value < 0.10.

10.115 Step 1: Hypotheses
In order to show that the average percentage of saturated fat in this brand differs from 66, we test:
H_0: $\mu = 66$
H_a: $\mu \neq 66$
Step 2: Significance Level
We are given that $\alpha = 0.01$.
Step 3: Calculations
For the sample, n = 96, $\bar{x} = 65.6$ and s = 1.4 is used to approximate σ. Therefore, the value of the test statistic is: $z = \frac{\bar{x}-\mu}{\sigma/\sqrt{n}} = \frac{65.6-66}{1.4/\sqrt{96}} = -2.80.$

Step 4: Rejection Region
For a two-tailed test with a significance level $\alpha = 0.01$, we have $z_{\alpha/2} = z_{0.005} = 2.576$, so we reject H_0 if z > 2.576 or if z < -2.576.
Step 5: Conclusion
For z = -2.80, we reject H_0, so there is sufficient evidence at the 1% level to show that the average percentage of saturated fat in this brand differs from 66.

10.116 For z = -2.80 and H_a: $\mu \neq 66$, p-value = 2P(z > | -2.80 |) = 2P(z > 2.80) = 2(1 - 0.9974) = 0.0052.

10.117 Step 1: Hypotheses
In order to show that the percentage of bullish analysts has changed from the previous week's value of 35%, we test:
H_0: p = 0.35
H_a: p $\neq$ 0.35

Step 2: Significance Level
We are given that $\alpha = 0.05$.
Step 3: Calculations
For the sample, n = 150 and x = 60 so $\hat{p} = \frac{x}{n} = \frac{60}{150} = 0.4$. The value of the test statistic is:

$$z = \frac{\hat{p} - p_0}{\sqrt{\frac{p_0(1-p_0)}{n}}} = \frac{0.4 - 0.35}{\sqrt{\frac{(0.35)\cdot(0.65)}{150}}} = 1.28.$$

Step 4: Rejection Region
For a two-tailed test with a significance level $\alpha = 0.05$, we have $z_{\alpha/2} = z_{0.025} = 1.96$ so we reject H_0 if $z > 1.96$ or $z < -1.96$.
Step 5: Conclusion
For $z = 1.28$, we fail to reject H_0, so there is not sufficient evidence, at the 5% level of significance, to show that the percentage of bullish analysts has changed from the previous week's value of 35%.

10.118 For $z = 1.28$ and H_a: $p \neq 0.35$, p-value $= 2P(z > 1.28) = 2(1 - 0.8997) = 0.2006$.

10.119 Step 1: Hypotheses
In order to show that the mean honey yield for this type of bee is greater than 53 pounds, we test:
H_0: $\mu = 53$
H_a: $\mu > 53$
Step 2: Significance Level
We are given that $\alpha = 0.01$.
Step 3: Calculations
Since the sample size is less than 30, a t-test statistic is used, where n = 18, $\bar{x} = 59.7$ and $s = 3.6$. We calculate $t = \frac{\bar{x} - \mu}{s/\sqrt{n}} = \frac{59.7 - 53}{3.6/\sqrt{18}} = 7.90$.

Step 4: Rejection Region
For a one-tailed test with a significance level $\alpha = 0.01$, and n = 18, we have $t_\alpha = t_{0.01} = 2.567$, where df = n - 1 = 17, so we reject H_0 if $t > 2.567$.
Step 5: Conclusion
For $t = 7.90$, we reject H_0, so there is sufficient evidence at the 1% level to show that the mean honey yield for this type of bee is greater than 53.

10.120 Step 1: Hypotheses
In order to determine if the variance of the population differs from 0.04, we test:
H_0: $\sigma^2 = 0.04$
H_a: $\sigma^2 \neq 0.04$
Step 2: Significance Level
We are given that $\alpha = 0.01$.

Step 3: Calculations
For the sample, n = 5 and we calculate $s^2 = 0.145$. Therefore, the value of the test statistic

is: $X^2 = \frac{(n-1)s^2}{\sigma_0^2} = \frac{(5-1)(0.145)}{0.04} = 14.50.$

Step 4: Rejection Region
Since $\alpha = 0.01$ and the alternative hypothesis is H_a: $\sigma^2 \neq 0.04$, we find $X^2_{(1-\alpha/2)} = X^2_{0.995} = 0.21$ and $X^2_{(\alpha/2)} = X^2_{0.005} = 14.86$, where df = n - 1 = 4, so we reject H_0 if $X^2 < 0.21$ or if $X^2 > 14.86$.
Step 5: Conclusion
Since the value of the test statistic is $X^2 = 14.50$, we can not reject H_0, and there is not sufficient evidence, at the 1% level of significance, to show that the population variance differs from 0.04.

10.121 Since a X^2-test is used, where the alternative hypothesis is H_a: $\sigma^2 \neq 0.04$, the p-value is double the area under the X^2 distribution to the right of $X^2 = 14.5$. We use Table V with df = n - 1 = 5 - 1 = 4 and note that 14.5 lies between $X^2_{0.005} = 14.86$ and $X^2_{0.01} = 13.28$. Therefore, the area under the X^2 distribution to the right of 14.5 is a value between 0.005 and 0.01. Thus, 0.005 < ½(p-value) < 0.01, so we have 0.01 < p-value < 0.02.

MINITAB LAB ASSIGNMENTS

10.122(M)

```
MTB > # Exercise 10.122(M)
MTB > SET C1
DATA> 14.1 14.0 14.3 15.1 14.2 14.7 15.3 14.9
DATA> 15.1 14.2 13.7 14.1 14.2 14.3 14.2
DATA> END
MTB > TTEST 15 C1;
SUBC> ALTERNATIVE = -1.

TEST OF MU = 15.000 VS MU L.T. 15.000

              N      MEAN    STDEV   SE MEAN        T    P VALUE
C1           15    14.427    0.473     0.122    -4.70     0.0002

MTB > # Since p-value = 0.0002 < 0.05, we reject Ho.
```

10.123(M)

```
MTB > # Exercise 10.123(M)
MTB > SET C1
DATA> 1300.43 795.78 1028.94 376.86 982.22 1002.23 254.33
DATA> 1729.64 103.45 1112.87 202.32 875.68 1212.21
DATA> END
MTB > TTEST 800 C1;
SUBC> ALTERNATIVE = +1.

TEST OF MU = 800.000 VS MU G.T. 800.000

              N      MEAN    STDEV   SE MEAN        T    P VALUE
C1           13   844.382  483.763   134.172     0.33       0.37
```

```
MTB > #Since p-value = 0.37 > 0.01, we fail to reject Ho.
```

10.124(M)

```
MTB > # Exercise 10.124(M)
MTB > # The data were entered into C1 using SET C1
MTB > STDEV C1
   ST.DEV. =       2.2175
MTB > ZTEST 75 2.2175 C1;
SUBC> ALTERNATIVE = 0.

TEST OF MU = 75.000 VS MU N.E. 75.000
THE ASSUMED SIGMA = 2.22

              N       MEAN     STDEV   SE MEAN         Z    P VALUE
C1           50     73.938     2.217     0.314     -3.38     0.0007

MTB > # Since p-value = 0.0007 < 0.05, we reject Ho.
```

10.125(M)

```
MTB > # Exercise 10.125(M)
MTB > CDF 1.78;
SUBC> NORMAL 0 1.
    1.7800    0.9625
MTB > # P-value = 1 - 0.9625 = 0.0375.
```

10.126(M)

```
MTB > # Exercise 10.126(M)
MTB > CDF 2.01;
SUBC> T 20.
    2.0100    0.9709
MTB > # P-value = 2(1 - 0.9709) = 0.0582.
```

10.127(M)

```
MTB > # Exercise 10.127(M)
MTB > CDF 18.53;
SUBC> CHIS 22.
   18.5300    0.3259
MTB > # P-value = 1 - 0.3259 = 0.6741
```

10.128(M)

```
MTB > # Exercise 10.128(M)
MTB > # The data were entered into C1 using SET C1
MTB > STDEV C1
   ST.DEV =        6.1122
MTB > ZTEST 80 6.1122 C1;
SUBC> ALTERNATIVE = +1.

TEST OF MU = 80.000 VS MU G.T. 80.000
THE ASSUMED SIGMA = 6.11

              N       MEAN     STDEV   SE MEAN         Z    P VALUE
C1           36     81.111     6.112     1.019      1.09       0.14

MTB > # Since p-value = 0.14 > 0.05,we fail to reject Ho.
```

CHAPTER 11

TESTS OF HYPOTHESES AND ESTIMATION: TWO SAMPLES

EXERCISES for Section 11.1

Large Sample Inferences for Two Means: Independent Samples

11.1 **Step 1: Hypotheses**

H_0: $(\mu_1 - \mu_2) = 0$

H_a: $(\mu_1 - \mu_2) > 0$

Step 2: Significance Level

We are given that $\alpha = 0.05$.

Step 3: Calculations

We approximate σ_1^2 and σ_2^2 with s_1^2 and s_2^2, respectively, and calculate:

$$z = \frac{(\bar{x}_1 - \bar{x}_2) - d_0}{\sqrt{\frac{\sigma_1^2}{n_1} + \frac{\sigma_2^2}{n_2}}} = \frac{(257 - 245) - 0}{\sqrt{\frac{(26)^2}{45} + \frac{(19)^2}{35}}} = \frac{12}{5.034} = 2.38.$$

Step 4: Rejection Region

For $\alpha = 0.05$ we reject H_0 if $z > 1.645$.

Step 5: Conclusion

Since $z = 2.38 > 1.645$, we reject H_0 and conclude, at level of significance 0.05, that $(\mu_1 - \mu_2) > 0$.

11.3 **Step 1: Hypotheses**

H_0: $(\mu_1 - \mu_2) = 10$

H_a: $(\mu_1 - \mu_2) \neq 10$

Step 2: Significance Level

We are given that $\alpha = 0.01$.

Step 3: Calculations

We approximate σ_1^2 and σ_2^2 with s_1^2 and s_2^2, respectively, and calculate:

$$z = \frac{(\bar{x}_1 - \bar{x}_2) - d_0}{\sqrt{\frac{\sigma_1^2}{n_1} + \frac{\sigma_2^2}{n_2}}} = \frac{(60.8 - 48.3) - 10}{\sqrt{\frac{(3.6)^2}{30} + \frac{(3.1)^2}{32}}} = \frac{2.5}{0.856} = 2.92.$$

Step 4: Rejection Region

For $\alpha = 0.01$ we reject H_0 if $z < -2.576$ or if $z > 2.576$.

Step 5: Conclusion

Since $z = 2.92$, we reject H_0 and conclude, at level of significance 0.01, that $(\mu_1 - \mu_2) \neq 10$.

11.5 **Step 1: Hypotheses**

H_0: $(\mu_1 - \mu_2) = 5$

H_a: $(\mu_1 - \mu_2) > 5$

Step 2: Significance Level

We are given that $\alpha = 0.05$.

Step 3: Calculations

We approximate σ_1^2 and σ_2^2 with s_1^2 and s_2^2, respectively, and calculate:

$$z = \frac{(\bar{x}_1 - \bar{x}_2) - d_0}{\sqrt{\frac{\sigma_1^2}{n_1} + \frac{\sigma_2^2}{n_2}}} = \frac{(949 - 940) - 5}{\sqrt{\frac{(23)^2}{53} + \frac{(17)^2}{68}}} = \frac{4}{3.77} = 1.06.$$

Step 4: Rejection Region
For $\alpha = 0.05$ we reject H_0 if $z > 1.645$.
Step 5: Conclusion
Since $z = 1.06$, at level of significance 0.05, we fail to reject H_0 and do not conclude $(\mu_1 - \mu_2) > 0$.

11.7 **Step 1: Hypotheses**
Let μ_1 denote the mean number of training hours required using Program A, and μ_2 the mean number of hours required using Program B. In order to determine whether the difference in the sample means is statistically significant, we test:
H_0: $(\mu_1 - \mu_2) = 0$
H_a: $(\mu_1 - \mu_2) \neq 0$
Step 2: Significance Level
We are given that $\alpha = 0.05$.
Step 3: Calculations
We approximate σ_1^2 and σ_2^2 with s_1^2 and s_2^2, respectively, and calculate:

$$z = \frac{(\bar{x}_1 - \bar{x}_2) - d_0}{\sqrt{\frac{\sigma_1^2}{n_1} + \frac{\sigma_2^2}{n_2}}} = \frac{(24.8 - 22.9) - 0}{\sqrt{\frac{(3.1)^2}{43} + \frac{(3.3)^2}{41}}} = \frac{1.9}{0.699} = 2.72.$$

Step 4: Rejection Region
For $\alpha = 0.05$ we reject H_0 if $z < -1.96$ or if $z > 1.96$.
Step 5: Conclusion
Since $z = 2.72$, at level of significance 0.05, we reject H_0 and conclude that the difference in the sample means is statistically significant.

11.9 a. **Step 1: Hypotheses**
Let μ_1 denote the mean of the sales for salespersons with a college degree, and μ_2 the mean for salespersons without a college degree. To determine whether sales are greater for salespersons with a college degree we test:
H_0: $(\mu_1 - \mu_2) = 0$
H_a: $(\mu_1 - \mu_2) > 0$
Step 2: Significance Level
We are given that $\alpha = 0.05$.
Step 3: Calculations
We approximate σ_1^2 and σ_2^2 with s_1^2 and s_2^2, respectively, and calculate:

$$z = \frac{(\bar{x}_1 - \bar{x}_2) - d_0}{\sqrt{\frac{\sigma_1^2}{n_1} + \frac{\sigma_2^2}{n_2}}} = \frac{(3542 - 3301) - 0}{\sqrt{\frac{(468)^2}{34} + \frac{(642)^2}{37}}} = \frac{241}{132.595} = 1.82.$$

Step 4: Rejection Region
For $\alpha = 0.05$ we reject H_0 if $z > 1.645$.

Step 5: Conclusion
Since z = 1.82, at level of significance 0.05, we reject H_0 and conclude that average sales are greater for salespersons with a college degree.
b. For z = 1.82 and $H_a: (\mu_1 - \mu_2) > 0$, p-value = P(z > 1.82) = 1 - 0.9656 = 0.0344.

11.11 Step 1: Hypotheses
Let μ_1 denote the mean dioxin levels for milk in cartons not lined with foil, and μ_2 the mean for the lined cartons. In order to determine whether the difference in the mean dioxin levels for the two types of containers is significant, we test:
$H_0: (\mu_1 - \mu_2) = 0$
$H_a: (\mu_1 - \mu_2) \neq 0$
Step 2: Significance Level
We are given that $\alpha = 0.01$.
Step 3: Calculations
We approximate σ_1^2 and σ_2^2 with s_1^2 and s_2^2, respectively, and calculate:

$$z = \frac{(\bar{x}_1 - \bar{x}_2) - d_0}{\sqrt{\frac{\sigma_1^2}{n_1} + \frac{\sigma_2^2}{n_2}}} = \frac{(0.030 - 0.006) - 0}{\sqrt{\frac{(0.009)^2}{50} + \frac{(0.002)^2}{50}}} = \frac{0.024}{0.001304} = 18.41.$$

Step 4: Rejection Region
For $\alpha = 0.01$ we reject H_0 if z < -2.576 or if z > 2.576.
Step 5: Conclusion
Since z = 18.41, at level of significance 0.01, we reject H_0 and conclude that the difference in the mean dioxin levels for the two types of containers is significant.

11.13 For $1 - \alpha = 0.95$, we have $z_{\alpha/2} = z_{0.025} = 1.96$. We approximate σ_1^2 and σ_2^2 with s_1^2 and s_2^2, respectively, and calculate:

$$(\bar{x}_1 - \bar{x}_2) \pm z_{\frac{\alpha}{2}}\sqrt{\frac{\sigma_1^2}{n_1} + \frac{\sigma_2^2}{n_2}} \rightarrow (4.2 - 5.9) \pm 1.96\sqrt{\frac{(1.2)^2}{32} + \frac{(1.9)^2}{36}}.$$

This simplifies to -1.7 ± 0.747 so a 95% confidence interval for the difference in the mean anxiety level for the two methods is $-2.447 < \mu_1 - \mu_2 < -0.953$.

11.15 For $1 - \alpha = 0.90$, we have $z_{\alpha/2} = z_{0.05} = 1.645$ and calculate:

$$(\bar{x}_1 - \bar{x}_2) \pm z_{\frac{\alpha}{2}}\sqrt{\frac{\sigma_1^2}{n_1} + \frac{\sigma_2^2}{n_2}} \rightarrow (28.4 - 22.3) \pm 1.645\sqrt{\frac{(2.8)^2}{48} + \frac{(2.3)^2}{45}}.$$

This simplifies to 6.1 ± 0.872. Therefore, a 90% confidence interval for the mean increase in yield of tomato plants treated with the plant food is $5.228 < \mu_1 - \mu_2 < 6.972$.

EXERCISES for 11.2 and 11.3

Small Sample Inferences for Two Means: Independent Samples

11.17 For $1 - \alpha = 0.95$, and df $= n_1 + n_2 - 2 = 26$, we have $t_{\alpha/2} = t_{0.025} = 2.056$. The variance of each population is estimated by the pooled sample variance:

$$s_p^2 = \frac{(n_1 - 1)s_1^2 + (n_2 - 1)s_2^2}{n_1 + n_2 - 2}$$

$$= \frac{(13 - 1)(2.4)^2 + (15 - 1)(2.8)^2}{13 + 15 - 2} = \frac{178.88}{26} = 6.88.$$

We calculate:

$$(\bar{x}_1 - \bar{x}_2) \pm t_{\frac{\alpha}{2}}\sqrt{s_p^2\left(\frac{1}{n_1} + \frac{1}{n_2}\right)}$$

$$\rightarrow (65.7 - 62.1) \pm 2.056\sqrt{6.88\left(\frac{1}{13} + \frac{1}{15}\right)}.$$

This simplifies to 3.6 ± 2.044 so a 95% confidence interval is $1.556 < \mu_1 - \mu_2 < 5.644$.

11.19 The degrees of freedom of the t distribution are given by:

$$df = \frac{\left(\frac{s_1^2}{n_1} + \frac{s_2^2}{n_2}\right)^2}{\frac{\left(\frac{s_1^2}{n_1}\right)^2}{n_1 - 1} + \frac{\left(\frac{s_2^2}{n_2}\right)^2}{n_2 - 1}} = \frac{\left(\frac{(9.3)^2}{8} + \frac{(21.8)^2}{21}\right)^2}{\frac{\left(\frac{(9.3)^2}{8}\right)^2}{8 - 1} + \frac{\left(\frac{(21.8)^2}{21}\right)^2}{21 - 1}} = 26.44,$$

which we round down to 26. For $1 - \alpha = 0.95$, we have $t_{\alpha/2} = t_{0.025} = 2.056$ and calculate:

$$(\bar{x}_1 - \bar{x}_2) \pm t_{\frac{\alpha}{2}}\sqrt{\frac{s_1^2}{n_1} + \frac{s_2^2}{n_2}} \rightarrow (134 - 127) \pm 2.056\sqrt{\frac{(9.3)^2}{8} + \frac{(21.8)^2}{21}}.$$

This simplifies to 7 ± 11.890 so a 95% confidence interval is $-4.890 < \mu_1 - \mu_2 < 18.890$.

11.21 For $1 - \alpha = 0.90$, and df $= n_1 + n_2 - 2 = 13 + 11 - 2 = 22$, we have $t_{\alpha/2} = t_{0.05} = 1.717$. The variance of each population is estimated by the pooled sample variance:

$$s_p^2 = \frac{(n_1 - 1)s_1^2 + (n_2 - 1)s_2^2}{n_1 + n_2 - 2}$$

$$= \frac{(13 - 1)(0.87)^2 + (11 - 1)(0.75)^2}{13 + 11 - 2} = \frac{14.7078}{22} = 0.669.$$

We calculate:

$$(\bar{x}_1 - \bar{x}_2) \pm t_{\frac{\alpha}{2}}\sqrt{s_p^2\left(\frac{1}{n_1} + \frac{1}{n_2}\right)}$$

$$\rightarrow (5.74 - 5.02) \pm 1.717\sqrt{0.669\left(\frac{1}{13} + \frac{1}{11}\right)}.$$

This simplifies to 0.72 ± 0.575 so a 90% confidence interval is $0.145 < \mu_1 - \mu_2 < 1.295$.

11.23 The degrees of freedom of the t distribution are given by:

$$df = \frac{\left(\frac{s_1^2}{n_1} + \frac{s_2^2}{n_2}\right)^2}{\frac{\left(\frac{s_1^2}{n_1}\right)^2}{n_1 - 1} + \frac{\left(\frac{s_2^2}{n_2}\right)^2}{n_2 - 1}} = \frac{\left(\frac{(0.68)^2}{10} + \frac{(0.27)^2}{15}\right)^2}{\frac{\left(\frac{(0.68)^2}{10}\right)^2}{10 - 1} + \frac{\left(\frac{(0.27)^2}{15}\right)^2}{15 - 1}} = 10.9,$$

which we round down to 10. For $1 - \alpha = 0.95$, we have $t_{\alpha/2} = t_{0.025} = 2.228$ and calculate:

$$(\bar{x}_1 - \bar{x}_2) \pm t_{\frac{\alpha}{2}}\sqrt{\frac{s_1^2}{n_1} + \frac{s_2^2}{n_2}} \rightarrow (1.77 - 0.99) \pm 2.228\sqrt{\frac{(0.68)^2}{10} + \frac{(0.27)^2}{15}}.$$

This simplifies to 0.78 ± 0.504 so a 95% confidence interval is $0.276 < \mu_1 - \mu_2 < 1.284$.

11.25 Since a t-test is used, where the alternative hypothesis involves the relation >, the p-value is the area under the t distribution to the right of t =1.77. We use Table IV with df = $n_1 + n_2 - 2 = 10 + 12 - 2 = 20$ and note that 1.77 lies between $t_{0.05} = 1.725$ and $t_{0.025} = 2.086$. Therefore, the area under the t distribution to the right of 1.77 is a value between 0.025 and 0.05. Thus, 0.025 < p-value < 0.05.

11.27 Since a t-test is used, where the alternative hypothesis involves the relation >, the p-value is the area under the t distribution to the right of t =1.25. We use Table IV with df = $n_1 + n_2 - 2 = 7 + 6 - 2 = 11$ and note that 1.25 lies to the left of $t_{0.10} = 1.363$. Therefore, the area under the t distribution to the right of 1.25 is a value larger than 0.10 so we have p-value > 0.10.

11.29 Suppose $n_1 = n_2$. Then ,

$$s_p^2 = \frac{(n_1 - 1)s_1^2 + (n_2 - 1)s_2^2}{n_1 + n_2 - 2} = \frac{(n_1 - 1)s_1^2 + (n_1 - 1)s_2^2}{n_1 + n_1 - 2} = \frac{(n_1 - 1)(s_1^2 + s_2^2)}{2(n_1 - 1)} = \frac{(s_1^2 + s_2^2)}{2}.$$

MINITAB LAB ASSIGNMENTS

11.31(M)

```
MTB > # Exercise 11.31(M)
MTB > SET C1
DATA> 40 38 38 37 36 39 41.5 38 39.5 37.5 35 40
DATA> END
MTB > SET C2
DATA> 35 37 31 39 31.5 35 32.5 34 39 36
DATA> END
MTB > # We assume the populations have the same standard deviation
MTB > TWOSAMPLE C1 C2;
SUBC> ALTERNATIVE 0;
SUBC> POOL.
```

```
TWOSAMPLE T FOR C1 VS C2
     N       MEAN       STDEV    SE MEAN
C1   12      38.29       1.83       0.53
C2   10      35.00       2.84       0.90

95 PCT CI FOR MU C1 - MU C2: (1.20, 5.38)

TTEST MU C1 = MU C2 (VS NE): T= 3.29  P=0.0037  DF=  20

POOLED STDEV =        2.34

MTB > #Since p-value = 0.0037 < 0.05, we reject Ho. There is a
MTB > #difference.
```

11.33(M)

```
MTB > # Exercise 11.33(M)
MTB > SET C1
DATA> 4.7 5.3 5.9 4.8 5.1 6.2 6.1 6.1 5.3 6.1 4.9
DATA> END
MTB > SET C2
DATA> 6.3 5.7 5.8 4.9 6.9 6.8 7.2 6.9 6.8 7.3
DATA> END
MTB > TWOSAMPLE C1 C2;
SUBC> ALTERNATIVE = 0. # This could be omitted

TWOSAMPLE T FOR C1 VS C2
     N       MEAN       STDEV    SE MEAN
C1   11      5.500      0.588       0.18
C2   10      6.460      0.771       0.24

95 PCT CI FOR MU C1 - MU C2: (-1.60, -0.32)

TTEST MU C1 = MU C2 (VS NE): T= -3.19  P=0.0058  DF=  16

MTB > #Since p-value = 0.0058 < 0.05, we reject Ho
```

EXERCISES for Section 11.4

Inferences for Two Means: Paired Samples

11.35 **Step 1: Hypotheses**

H_0: $\mu_D = 0$

H_a: $\mu_D < 0$

Step 2: Significance Level

We are given that $\alpha = 0.05$.

Step 3: Calculations

Pair	A	B	Difference D = A - B
1	25	32	-7
2	35	38	-3
3	56	65	-9
4	52	50	2
5	24	30	-6

We calculate $\Sigma D = -23$ and $\Sigma D^2 = 179$ so $\overline{D} = \frac{\Sigma D}{n} = \frac{-23}{5} = -4.6$, and

$$s_D = \sqrt{\frac{\Sigma D^2 - \frac{(\Sigma D)^2}{n}}{n-1}} = \sqrt{\frac{179 - \frac{(-23)^2}{5}}{5-1}} = 4.278.$$

$$t = \frac{\overline{D} - \mu_D}{\frac{s_D}{\sqrt{n}}} = \frac{-4.6 - 0}{\frac{4.278}{\sqrt{5}}} = -2.40.$$

Step 4: Rejection Region

For $\alpha = 0.05$ and df = n - 1 = 4, we reject H_0 if t < -2.132.

Step 5: Conclusion

Since t = -2.40, at level of significance 0.05, we reject H_0 and conclude $\mu_D < 0$.

11.37 **Step 1: Hypotheses**

H_0: $\mu_D = 7$

H_a: $\mu_D \neq 7$

Step 2: Significance Level

We are given that $\alpha = 0.01$.

Step 3: Calculations

Pair	A	B	Difference $D = A - B$
1	85	76	9
2	28	19	9
3	76	56	20
4	99	84	15
5	51	41	10
6	46	46	0

We calculate $\Sigma D = 63$ and $\Sigma D^2 = 887$ so $\bar{D} = \frac{\Sigma D}{n} = \frac{63}{6} = 10.5$, and

$$s_D = \sqrt{\frac{\Sigma D^2 - \frac{(\Sigma D)^2}{n}}{n-1}} = \sqrt{\frac{887 - \frac{(63)^2}{6}}{6-1}} = 6.7157.$$

$$t = \frac{\bar{D} - \mu_D}{\frac{s_D}{\sqrt{n}}} = \frac{10.5 - 7}{\frac{6.716}{\sqrt{6}}} = 1.28.$$

Step 4: Rejection Region
For $\alpha = 0.01$ and df = n - 1 = 6 - 1 = 5, we reject H_0 if $t < -4.032$ or if $t > 4.032$.
Step 5: Conclusion
Since t = 1.28, at level of significance 0.01, we fail to reject H_0 and cannot conclude $\mu_D \neq 7$.

11.39 **Step 1: Hypotheses**
Let μ_D denote the mean difference in length of life between platinum-tipped spark plugs and conventional spark plugs. In order to show that the platinum plugs last longer, we test:
H_0: $\mu_D = 0$
H_a: $\mu_D > 0$
Step 2: Significance Level
We are given that $\alpha = 0.05$.
Step 3: Calculations

Engine	Platinum Plug A	Conventional Plug B	Difference = A - B
1	640	470	170
2	570	370	200
3	530	460	70
4	410	490	-80
5	600	380	220
6	580	410	170

We calculate $\Sigma D = 750$ and $\Sigma D^2 = 157500$ so $\bar{D} = \frac{\Sigma D}{n} = \frac{750}{6} = 125$, and

$$S_D = \sqrt{\frac{\Sigma D^2 - \frac{(\Sigma D)^2}{n}}{n-1}} = \sqrt{\frac{157500 - \frac{(750)^2}{6}}{6-1}} = 112.916.$$

Therefore: $$t = \frac{\bar{D} - \mu_D}{\frac{s_D}{\sqrt{n}}} = \frac{125 - 0}{\frac{112.916}{\sqrt{6}}} = 2.71.$$

Step 4: Rejection Region

For $\alpha = 0.05$ and df = n - 1 = 6 - 1 = 5, we reject H_0 if $t > 2.015$.

Step 5: Conclusion

Since t = 2.71, at level of significance 0.05, we reject H_0 and conclude that the platinum plugs last longer.

11.41 Step 1: Hypotheses

Let μ_D denote the mean difference in the time required to run a mile with the old and the new shoes. In order to show that the new running shoe enables one to run faster, we test:

H_0: $\mu_D = 0$

H_a: $\mu_D > 0$

Step 2: Significance Level

We are given that $\alpha = 0.05$.

Step 3: Calculations

Runner	track shoe A	new shoe B	Difference = A - B
1	321	318	3
2	307	299	8
3	397	401	-4
4	269	260	9
5	285	285	0
6	364	363	1
7	295	289	6
8	302	296	6

We calculate $\Sigma D = 29$ and $\Sigma D^2 = 243$ so $\bar{D} = \frac{\Sigma D}{n} = \frac{29}{8} = 3.625$, and

$$S_D = \sqrt{\frac{\Sigma D^2 - \frac{(\Sigma D)^2}{n}}{n-1}} = \sqrt{\frac{243 - \frac{(29)^2}{8}}{8-1}} = 4.438.$$

Therefore: $$t = \frac{\bar{D} - \mu_D}{\frac{s_D}{\sqrt{n}}} = \frac{3.625 - 0}{\frac{4.438}{\sqrt{8}}} = 2.31.$$

Step 4: Rejection Region

For $\alpha = 0.05$ and df = n - 1 = 8 - 1 = 7, we reject H_0 if $t > 1.895$.

Step 5: Conclusion
Since t = 2.31, at level of significance 0.05, we reject H_0 and conclude that the new running shoe enables one to run faster.

11.43 **Step 1: Hypotheses**
Let μ_D denote the mean difference in the scores on the standardized test (after - before). In order to show that the program tends to increase test scores for students of this type, we test:
H_0: $\mu_D = 0$
H_a: $\mu_D > 0$
Step 2: Significance Level
We are given that $\alpha = 0.05$.
Step 3: Calculations
We are given n = 12, $\bar{D} = 9.5$, and $s_D = 4.61$. Therefore:

$$t = \frac{\bar{D} - \mu_D}{\frac{s_D}{\sqrt{n}}} = \frac{9.5 - 0}{\frac{4.61}{\sqrt{12}}} = 7.14.$$

Step 4: Rejection Region
For $\alpha = 0.05$ and df = n - 1 = 12 - 1 = 11, we reject H_0 if t > 1.796.
Step 5: Conclusion
Since t = 7.14, at level of significance 0.05, we reject H_0 and conclude that the program tends to increase test scores for students of this type.

11.45 For $1 - \alpha = 0.95$ and df = n - 1 = 10 - 1 = 9, we have $t_{\alpha/2} = t_{0.025} = 2.262$ and calculate:

$$\bar{D} \pm t_{\frac{\alpha}{2}} \frac{s_D}{\sqrt{n}} \rightarrow 4.1 \pm (2.262) \frac{3.178}{\sqrt{10}}.$$

This simplifies to 4.1 ± 2.273 so a 95% confidence interval is $1.827 < \mu_D < 6.373$.

11.47 For $1 - \alpha = 0.95$ and df = n - 1 = 12 - 1 = 11, we have $t_{\alpha/2} = t_{0.025} = 2.201$ and calculate: $\bar{D} \pm t_{\frac{\alpha}{2}} \frac{s_D}{\sqrt{n}} \rightarrow 2.5 \pm (2.201) \frac{2.576}{\sqrt{12}}.$

This simplifies to 2.5 ± 1.637 so a 95% confidence interval is $0.863 < \mu_D < 4.137$.

MINITAB LAB ASSIGNMENTS

11.49
```
MTB > # Exercise 11.49(M)
MTB > # The data were read into C1 and C2 using READ C1 C2
MTB > LET C3 = C1 - C2
MTB > TTEST 0 C3;
SUBC> ALTER = 1.

TEST OF MU = 0.000 VS MU G.T. 0.000

              N      MEAN     STDEV   SE MEAN        T    P VALUE
C3           36     1.694     3.161     0.527     3.22     0.0014

MTB > # Since p-value = 0.0014 < 0.05, rej. Ho; Herb is effective.
```

11.51(M)

```
MTB > # Exercise 11.51(M)
MTB > # Data were read into C1 and C2 using READ C1 C2
MTB > LET C3 = C2 - C1 # The increases in the median prices
MTB > TTEST 0 C3;
SUBC> ALTERNATIVE = +1.

TEST OF MU =  0.000 VS MU G.T.  0.000

              N      MEAN     STDEV   SE MEAN        T    P VALUE
C3           32    43.944    20.691     3.658    12.01     0.0000

MTB > # Since p-value = 0.0000 < 0.01, we reject Ho. Increase has
MTB > # occurred.
```

EXERCISES for Section 11.5

Large-Sample Inferences for Two Proportions

11.53 For $1 - \alpha = 0.90$, we have $z_{\alpha/2} = z_{0.05} = 1.645$. From Exercise 11.52, the sample proportions are: $\hat{p}_1 = \frac{x_1}{n_1} = \frac{320}{400} = 0.8$ *and* $\hat{p}_2 = \frac{x_2}{n_2} = \frac{210}{300} = 0.7$.

A $1 - \alpha$ confidence interval for $p_1 - p_2$ is given by: $(\hat{p}_1 - \hat{p}_2) \pm z_{\frac{\alpha}{2}}\sqrt{\frac{\hat{p}_1\hat{q}_1}{n_1} + \frac{\hat{p}_2\hat{q}_2}{n_2}}$

$(0.80 - 0.70) \pm 1.645\sqrt{\frac{(0.80)(0.20)}{400} + \frac{(0.70)(0.30)}{300}}$. This simplifies to 0.10 ± 0.055 so a 90% confidence interval for $p_1 - p_2$ is given by $0.045 < p_1 - p_2 < 0.155$.

11.55 For $1 - \alpha = 0.99$, we have $z_{\alpha/2} = z_{0.005} = 2.576$. From Exercise 11.54, the sample sizes and proportions are given by $n_1 = 1200$, $\hat{p}_1 = 0.64$, $n_2 = 1000$ and $\hat{p}_2 = 0.56$. A $1 - \alpha$ confidence interval for $p_1 - p_2$ is given by: $(\hat{p}_1 - \hat{p}_2) \pm z_{\frac{\alpha}{2}}\sqrt{\frac{\hat{p}_1\hat{q}_1}{n_1} + \frac{\hat{p}_2\hat{q}_2}{n_2}}$

$(0.64 - 0.56) \pm 2.576\sqrt{\frac{(0.64)(0.36)}{1200} + \frac{(0.56)(0.44)}{1000}}$. This simplifies to 0.08 ± 0.054 so a 99% confidence interval for $p_1 - p_2$ is given by $0.026 < p_1 - p_2 < 0.134$.

11.57 For $z = 1.59$ and H_a involving the relation $\neq$, p-value $= 2P(z > 1.59) = 2(1 - 0.9441) = 0.1118$.

11.59 **Step 1: Hypotheses**
Let p_1 denote the proportion of men and p_2 the proportion of women who regard themselves as aggressive drivers. In order to show that a larger proportion of men than women consider themselves aggressive drivers, we test:
$H_0: p_1 - p_2 = 0$
$H_a: p_1 - p_2 > 0$
Step 2: Significance Level
We are given that $\alpha = 0.05$.

Step 3: Calculations
The sample sizes and proportions are given by $n_1 = 1200$, $\hat{p}_1 = 2/3$, $n_2 = 1200$ and $\hat{p}_2 = 1/2$. This means that $x_1 = n_1 \cdot \hat{p}_1 = 800$ and $x_2 = n_2 \cdot \hat{p}_2 = 600$. The test statistic is:

$$z = \frac{(\hat{p}_1 - \hat{p}_2)}{\sqrt{\hat{p}\hat{q}\left(\frac{1}{n_1} + \frac{1}{n_2}\right)}}, \text{ where } \hat{p} = \frac{x_1 + x_2}{n_1 + n_2} = \frac{800 + 600}{1200 + 1200} = \frac{1400}{2400} = \frac{7}{12}.$$

$$Z = \frac{\left(\frac{2}{3} - \frac{1}{2}\right)}{\sqrt{\left(\frac{7}{12}\right)\left(\frac{5}{12}\right)\left(\frac{1}{1200} + \frac{1}{1200}\right)}} = \frac{0.1667}{0.02013} = 8.28.$$

Step 4: Rejection Region
For $\alpha = 0.05$ and H_a: $p_1 > p_2$, we reject H_0 if $z > 1.645$.
Step 5: Conclusion
Since $z = 8.28$, at level of significance 0.05, we reject H_0 and conclude that a larger proportion of men than women consider themselves aggressive drivers.

11.61 For $1 - \alpha = 0.95$, we have $z_{\alpha/2} = z_{0.025} = 1.96$. From Exercise 11.60, the sample proportions are: $\hat{p}_1 = \frac{x_1}{n_1} = \frac{103}{304} = 0.3388$ *and* $\hat{p}_2 = \frac{x_2}{n_2} = \frac{155}{315} = 0.4921$.

A $1 - \alpha$ confidence interval for $p_1 - p_2$ is given by: $(\hat{p}_1 - \hat{p}_2) \pm z_{\frac{\alpha}{2}}\sqrt{\frac{\hat{p}_1\hat{q}_1}{n_1} + \frac{\hat{p}_2\hat{q}_2}{n_2}}$

$$(0.3388 - 0.4921) \pm 1.96\sqrt{\frac{(0.3388)(0.6612)}{304} + \frac{(0.4921)(0.5079)}{315}}$$

This simplifies to -0.153 ± 0.077 so a 95% confidence interval for $p_1 - p_2$ is given by $-0.230 < p_1 - p_2 < -0.076$.

11.63 For $z = 0.76$ and H_a involving the relation $>$, the p-value $= P(z > 0.76) = 1 - 0.7764 = 0.2236$.

11.65 **Step 1: Hypotheses**
Let p_1 denote the proportion of alumni from the private university and p_2 the proportion of alumni from the public university who had attended at least one reunion. In order to show that the difference in the sample proportions is statistically significant, we test:
H_0: $p_1 - p_2 = 0$
H_a: $p_1 - p_2 \neq 0$
Step 2: Significance Level
We are given that $\alpha = 0.05$.
Step 3: Calculations

The sample proportions are: $\hat{p}_1 = \frac{x_1}{n_1} = \frac{653}{1046}$ *and* $\hat{p}_2 = \frac{x_2}{n_2} = \frac{791}{1327}$. The test statistic is:

$$Z = \frac{(\hat{p}_1 - \hat{p}_2)}{\sqrt{\hat{p}\hat{q}\left(\frac{1}{n_1} + \frac{1}{n_2}\right)}}, \text{ where } \hat{p} = \frac{x_1 + x_2}{n_1 + n_2} = \frac{653 + 791}{1046 + 1327} = \frac{1444}{2373}.$$

$$z = \frac{\left(\frac{653}{1046} - \frac{791}{1327}\right)}{\sqrt{\left(\frac{1444}{2373}\right)\left(\frac{929}{2373}\right)\left(\frac{1}{1046} + \frac{1}{1327}\right)}} = \frac{0.0282}{0.0202} = 1.40.$$

Step 4: Rejection Region
For $\alpha = 0.05$ and H_a: $p_1 \neq p_2$, we reject H_0 if $z < -1.96$ or if $z > 1.96$.
Step 5: Conclusion
Since $z = 1.40$, at level of significance 0.05, we fail to reject H_0 and do not conclude that the difference in the sample proportions is statistically significant.

EXERCISES for Section 11.6
F Probability Distributions

11.67 From Table VI.d, with ndf = 9 and ddf = 15, we find $F_{0.01} = 3.89$.

11.69 From Table VI.b, with ndf = 4 and ddf = 21, we find $F_{0.05} = 2.84$.

11.71 From Table VI.c, with ndf = 10 and ddf = 30, we find $F_{0.025} = 2.51$.

11.73 The 95th percentile is the value of F for which the cumulative area to the left is 0.95. Therefore, the area under the curve to the right of F is 1 - 0.95 = 0.05. From Table VI.b, with ndf = 10 and ddf = 21, we find $P_{95} = F_{0.05} = 2.32$.

11.75 The 99th percentile is the value of F for which the cumulative area to the left is 0.99. Therefore, the area under the curve to the right of F is 1 - 0.99 = 0.01. From Table VI.d, with ndf = 8 and ddf = 7, we find $P_{99} = F_{0.01} = 6.84$.

11.77 We note in Table VI.a that, with ndf = 6 and ddf = 19, $F_{0.10} = 2.11$.
Therefore, $P(F > 2.11) = 0.10$.

11.79 We note in Table VI.b that, with ndf = 2 and ddf = 17, $F_{0.05} = 3.59$.
Therefore, $P(F > 3.59) = 0.05$, so we have $P(F < 3.59) = 1 - 0.05 = 0.95$.

11.81 With ndf = 2 and ddf = 20, we note in Table VI.d that $F_{0.01} = 5.85$ and in Table VIb that $F_{0.05} = 3.49$. Therefore, $P(3.49 < F < 5.85) =$
$P(F > 3.49) - P(F > 5.85) = 0.05 - 0.01 = 0.04$.

11.83 With ndf = 5 and ddf = 29, we note in Table VI.d that $F_{0.01} = 3.73$.
This means that $P(F > 3.73) = 0.01$, so we have $c = 3.73$.

11.85 With ndf = 20 and ddf = 8, we note in Table VI.c that $F_{0.025} = 4.00$. This means that $P(F > 4.00) = 0.025$ and $P(F < 4.00) = 1 - 0.025 = 0.975$, so we have $c = 4.00$.

MINITAB LAB ASSIGNMENTS

11.87(M)

a.

```
MTB > # Exercise 11.87(M)a.
MTB > CDF 1.12;
SUBC> F 11 39.
    1.1200     0.6275
```

b.

```
MTB > # Exercise 11.87(M)b.
MTB > INVCDF 0.93;
SUBC> F 11 39.
    0.9300     1.8980
```

$(F_{.07} = 1.8980)$

11.89(M)

```
MTB > # Exercise 11.89(M)
MTB > SET C1
DATA> 0:3/0.1
DATA> END
MTB > PDF C1 C2;
SUBC> F 9 20.
MTB > PLOT C2 C1

      0.90+
          -              *
 C2       -            *  * *
          -          *
          -                   *
      0.60+                    *
          -         *             *
          -                         *
          -                          *
          -       *                    *
      0.30+                              *
          -                               * *
          -     *                             **
          -                                      * **
          -                                           * ** * **
      0.00+   * *                                              *
          -
           --+---------+---------+---------+---------+---------+----C1
           0.00      0.60      1.20      1.80      2.40      3.00

MTB > # Graph of the F distribution with ndf = 9 and ddf = 20
```

EXERCISES for Section 11.7

Inferences for Two Variances

11.91 <u>Step 1</u>: **Hypotheses**

H_0: $\sigma_1^2 = \sigma_2^2$

H_a: $\sigma_1^2 > \sigma_2^2$

Step 2: Significance Level
We are given that $\alpha = 0.05$.
Step 3: Calculations
We calculate

$$F = \frac{s_1^2}{s_2^2} = \frac{(2.89)^2}{(1.38)^2} = 4.39.$$

Step 4: Rejection Region
Since the relation in H_a is >, with ndf = $n_1 - 1 = 13 - 1 = 12$, ddf = $n_2 - 1 = 15 - 1 = 14$, and $\alpha = 0.05$, we reject H_0 if $F > F_{0.05} = 2.53$.
Step 5: Conclusion
Therefore, for F = 4.39, we reject H_0 and conclude that $\sigma_1^2 > \sigma_2^2$.

11.93 With ndf = $n_1 - 1 = 13 - 1 = 12$, and ddf = $n_2 - 1 = 15 - 1 = 14$, we note in Table VI.e that $F_{0.005} = 4.43$ and in Table VI.d that $F_{0.01} = 3.80$. Since the p-value is the area under the curve to the right of F = 4.39, we conclude $0.005 <$ p-value < 0.01.

11.95 **Step 1: Hypotheses**
H_0: $\sigma_1^2 = \sigma_2^2$
H_a: $\sigma_1^2 \neq \sigma_2^2$
Step 2: Significance Level
We are given that $\alpha = 0.05$.
Step 3: Calculations

We calculate $F = \frac{s_1^2}{s_2^2} = \frac{(23.7)^2}{(10.8)^2} = 4.82.$

Step 4: Rejection Region
Since the relation in H_a is $\neq$, with ndf = $n_1 - 1 = 7 - 1 = 6$,
ddf = $n_2 - 1 = 9 - 1 = 8$, for $\alpha = 0.05$, we reject H_0 if $F > F_{\alpha/2} = F_{0.025} = 4.65$.
Step 5: Conclusion
Therefore, for F = 4.82, we reject H_0 and conclude that $\sigma_1^2 \neq \sigma_2^2$.

11.97 **Step 1: Hypotheses**
Let σ_1^2 denote the variance in the amount of dog food when filled by the older machine and σ_2^2 the variance for the newer machine. We wish to determine whether the variability in the amount dispensed by the older machine is greater so we test:
H_0: $\sigma_1^2 = \sigma_2^2$
H_a: $\sigma_1^2 > \sigma_2^2$
Step 2: Significance Level
We are given that $\alpha = 0.05$.
Step 3: Calculations

We calculate $F = \frac{s_1^2}{s_2^2} = \frac{(2.1)^2}{(1.3)^2} = 2.61.$

Step 4: Rejection Region
Since the relation in H_a is >, with ndf = $n_1 - 1 = 16 - 1 = 15$, ddf = $n_2 - 1 = 14 - 1 = 13$, and $\alpha = 0.05$, we reject H_0 if $F > F_{0.05} = 2.53$.

Step 5: Conclusion
Therefore, we reject H_0, and there is sufficient evidence at the 5% level to conclude that the variability in the amounts dispensed by the older machine is greater.

11.99 **Step 1: Hypotheses**
Let σ_1^2 denote the variance in incomes for New York real estate agents and σ_2^2 the variance in income for Pennsylvania agents. To determine if there is a difference in the variability of incomes for agents in these two states, we test:
H_0: $\sigma_1^2 = \sigma_2^2$
H_a: $\sigma_1^2 \neq \sigma_2^2$
Step 2: Significance Level
We are given that $\alpha = 0.10$.
Step 3: Calculations

We calculate $F = \frac{s_1^2}{s_2^2} = \frac{(3153)^2}{(2786)^2} = 1.28.$

Step 4: Rejection Region
Since the relation in H_a is $\neq$, with ndf = $n_1 - 1 = 16 - 1 = 15$, ddf = $n_2 - 1 = 15 - 1 = 14$, and $\alpha = 0.10$, we reject H_0 if $F > F_{\alpha/2} = F_{0.05} = 2.46$.
Step 5: Conclusion
Therefore, since $F = 1.28$, we can not reject H_0 and, at 10% level of significance, the data do not provide sufficient evidence to indicate a difference in the variability of incomes for agents in these two states.

MINITAB LAB ASSIGNMENTS

11.101 **a.** With ndf = $n_1 - 1 = 16 - 1 = 15$, and ddf = $n_2 - 1 = 15 - 1 = 14$, we note in Table VI.a that $F_{0.10} = 2.01$. Since the relation in H_a involves $\neq$, the p-value is two times the area under the curve to the right of $F = 1.28$, so we conclude $2(0.10) <$ p-value; that is, p-value > 0.20.

b.
```
MTB > # Exercise 11.101(M)b.
MTB > CDF 1.28;
SUBC> F 15 14.
    1.2800     0.6752
MTB > # The p-value is 2(1 - 0.6752) = 0.6496.
```

11.103 a.
```
MTB > # Exercise 11.103(M)
MTB > SET C1
DATA> 25.8 26.9 26.2 25.3 26.7 26.1 26.9
DATA> END
MTB > SET C2
DATA> 16.9 17.4 16.8 16.2 17.3 16.8
DATA> END
MTB > LET K1 = (STDEV(C1))**2
MTB > LET K2 = (STDEV(C2))**2
MTB > PRINT K1 K2
```

```
K1         0.362381
K2         0.184000
MTB > # Part a: Var(1) = 0.362381, Var(2) = 0.184
```

b. **Step 1: Hypotheses**
Let σ_1^2 denote the variance for the percentage of saturated fat for a sample of seven brands of stick margarine and σ_2^2 the percentage of fat for a sample of six brands of liquid margarine. In order to determine whether the difference in the samples variances is statistically significant, we test:
H_0: $\sigma_1^2 = \sigma_2^2$
H_a: $\sigma_1^2 \neq \sigma_2^2$
Step 2: Significance Level
We are given that $\alpha = 0.05$.
Step 3: Calculations
We have $n_1 = 7$, $s_1^2 = 0.362381$, $n_2 = 6$ and $s_2^2 = 0.184$.

We calculate $F = \dfrac{s_1^2}{s_2^2} = \dfrac{0.362381}{0.184000} = 1.97$.

Step 4: Rejection Region
Since the relation in H_a is $\neq$, with ndf = $n_1 - 1 = 7 - 1 = 6$, ddf = $n_2 - 1 = 6 - 1 = 5$, and α=0.05, we reject H_0 if $F > F_{\alpha/2} = F_{0.025} = 6.98$.
Step 5: Conclusion
Therefore, we can not reject H_o and, at 5% level of significance, the difference in the samples' variances is not statistically significant. This can be done using MINITAB as follows:

```
MTB > # Part b: F ratio follows:
MTB > LET K3 = K1/K2 #K1 and K2 from Part a
MTB > PRINT K3
K3         1.96947
```

c.

```
MTB > # Part c: We find the p-value:
MTB > CDF 1.96947;
SUBC> F 6 5.
    1.9695     0.7630
MTB > # P-value = 2(1 -0.7630) = 0.474.
MTB > # Since p-value > 0.05, fail to reject Ho
MTB > # Difference is not statistically significant
```

REVIEW EXERCISES
CHAPTER 11

11.104 **Step 1: Hypotheses**
H_0: $p_1 - p_2 = 0$
H_a: $p_1 - p_2 \neq 0$
Step 2: Significance Level
We are given that $\alpha = 0.01$.

Step 3: Calculations
The sample proportions are:

$\hat{p}_1 = \frac{x_1}{n_1} = \frac{630}{1500} = 0.42$ *and* $\hat{p}_2 = \frac{x_2}{n_2} = \frac{612}{1800} = 0.34$. The test statistic is:

$$z = \frac{(\hat{p}_1 - \hat{p}_2)}{\sqrt{\hat{p}\hat{q}\left(\frac{1}{n_1} + \frac{1}{n_2}\right)}}, \text{ where } \hat{p} = \frac{x_1 + x_2}{n_1 + n_2} = \frac{630 + 612}{1500 + 1800} = \frac{1242}{3300} = \frac{207}{550}.$$

$$z = \frac{(0.42 - 0.34)}{\sqrt{\left(\frac{207}{550}\right)\left(\frac{343}{550}\right)\left(\frac{1}{1500} + \frac{1}{1800}\right)}} = \frac{0.08}{0.01694} = 4.72.$$

Step 4: Rejection Region
For $\alpha = 0.01$ we reject H_0 if $z < -2.576$ or if $z > 2.576$.
Step 5: Conclusion
Therefore, at level of significance 0.01, we reject H_0 and conclude $p_1 \neq p_2$.

11.105 a. Step 1: Hypotheses
Let p_1 denote the proportion of smokers who filed medical claims last year and p_2 the proportion of nonsmokers who filed claims. In order to determine whether smokers file a larger percentage of claims, we test:
H_0: $p_1 - p_2 = 0$
H_a: $p_1 - p_2 > 0$
Step 2: Significance Level
We are given that $\alpha = 0.05$.
Step 3: Calculations
The sample proportions are:

$\hat{p}_1 = \frac{x_1}{n_1} = \frac{125}{600} = 0.2083$ *and* $\hat{p}_2 = \frac{x_2}{n_2} = \frac{192}{1200} = 0.16$. The test statistic is:

$$z = \frac{(\hat{p}_1 - \hat{p}_2)}{\sqrt{\hat{p}\hat{q}\left(\frac{1}{n_1} + \frac{1}{n_2}\right)}}, \text{ where } \hat{p} = \frac{x_1 + x_2}{n_1 + n_2} = \frac{125 + 192}{600 + 1200} = \frac{317}{1800} = 0.1761.$$

$$z = \frac{(0.2083 - 0.16)}{\sqrt{(0.1761)(1 - 0.1761)\left(\frac{1}{600} + \frac{1}{1200}\right)}} = \frac{0.0483}{0.01905} = 2.54.$$

Step 4: Rejection Region
For $\alpha = 0.05$ we reject H_0 if $z > 1.645$.
Step 5: Conclusion
For $z = 2.54$, at level of significance 0.05, we reject H_0 and conclude that smokers file a larger percentage of claims.
b. For $z = 2.54$ and H_a involving the relation $>$, p-value $= P(z > 2.54) = 1 - 0.9945 = 0.0055$.

11.106 For $1 - \alpha = 0.90$, we have $z_{\alpha/2} = z_{0.05} = 1.645$. From Exercise 11.105, the sample proportions are: $\hat{p}_1 = \frac{x_1}{n_1} = \frac{125}{600} = 0.2083$ *and* $\hat{p}_2 = \frac{x_2}{n_2} = \frac{192}{1200} = 0.16$.

A $1 - \alpha$ confidence interval for $p_1 - p_2$ is given by: $(\hat{p}_1 - \hat{p}_2) \pm z_{\frac{\alpha}{2}}\sqrt{\frac{\hat{p}_1\hat{q}_1}{n_1} + \frac{\hat{p}_2\hat{q}_2}{n_2}}$

$$(0.2083 - 0.16) \pm 1.645\sqrt{\frac{(0.2083)(0.7917)}{600} + \frac{(0.16)(0.84)}{1200}}.$$

This simplifies to 0.048 ± 0.032, so a 90% confidence interval for $p_1 - p_2$ is given by $0.016 < p_1 - p_2 < 0.080$.

11.107 **Step 1: Hypotheses**

H_0: $\mu_D = 0$

H_a: $\mu_D < 0$

Step 2: Significance Level

We are given that $\alpha = 0.05$.

Step 3: Calculations

Pair	I	II	Difference $D = I - II$
1	195	199	-4
2	198	205	-7
3	196	196	0
4	189	197	-8
5	191	195	-4

We calculate $\Sigma D = -23$ and $\Sigma D^2 = 145$ so : $\bar{D} = \frac{\Sigma D}{n} = \frac{-23}{5} = -4.6$, and

$$s_D = \sqrt{\frac{\Sigma D^2 - \frac{(\Sigma D)^2}{n}}{n-1}} = \sqrt{\frac{145 - \frac{(-23)^2}{5}}{5-1}} = 3.130.$$

We calculate: $t = \frac{\bar{D} - \mu_D}{\frac{s_D}{\sqrt{n}}} = \frac{-4.6 - 0}{\frac{3.130}{\sqrt{5}}} = -3.29.$

Step 4: Rejection Region

For $\alpha = 0.05$ and $df = n - 1 = 4$, we reject H_0 if $t < -t_\alpha = -t_{0.05} = -2.132$.

Step 5: Conclusion

Therefore, at level of significance 0.05, we reject H_0 and conclude $\mu_D < 0$.

11.108 For $1 - \alpha = 0.95$ and $df = n - 1 = 5 - 1 = 4$, we have $t_{\alpha/2} = t_{0.025} = 2.776$ and calculate:

$$\bar{D} \pm t_{\frac{\alpha}{2}}\frac{s_D}{\sqrt{n}} \rightarrow -4.6 \pm (2.776)\frac{3.130}{\sqrt{5}}$$

This simplifies to -4.6 ± 3.886 so a 95% confidence interval is $-8.486 < \mu_D < -0.714$.

11.109 **Step 1: Hypotheses**

H_0: $\sigma_1^2 = \sigma_2^2$

H_a: $\sigma_1^2 \neq \sigma_2^2$

Step 2: Significance Level

We are given that $\alpha = 0.10$.

Step 3: Calculations

We calculate $F = \dfrac{s_1^2}{s_2^2} = \dfrac{(19.6)^2}{(16.8)^2} = 1.36.$

Step 4: Rejection Region

Since the relation in H_a is $\neq$, with ndf = $n_1 - 1 = 10 - 1 = 9$, ddf = $n_2 - 1 = 12 - 1 = 11$, for $\alpha=0.10$, we reject H_a if $F > F_{\alpha/2} = F_{0.05} = 2.90$.

Step 5: Conclusion

Since $F = 1.36$, we can not reject H_0 and, at 10% level of significance, the data do not provide sufficient evidence to conclude that the population variances differ.

11.110 **Step 1: Hypotheses**

H_0: $(\mu_1 - \mu_2) = 0$

H_a: $(\mu_1 - \mu_2) \neq 0$

Step 2: Significance Level

We are given that $\alpha = 0.05$.

Step 3: Calculations

The variance of each population is estimated by the pooled sample variance:

$$s_p^2 = \frac{(n_1 - 1)s_1^2 + (n_2 - 1)s_2^2}{n_1 + n_2 - 2}$$

$$= \frac{(10-1)(19.6)^2 + (12-1)(16.8)^2}{10 + 12 - 2} = \frac{6562.08}{20} = 328.104.$$

The value of the test statistic is given by:

$$t = \frac{(\overline{x}_1 - \overline{x}_2) - d_0}{\sqrt{s_p^2\left(\frac{1}{n_1} + \frac{1}{n_2}\right)}} = \frac{(348-335) - 0}{\sqrt{328.104\left(\frac{1}{10} + \frac{1}{12}\right)}} = \frac{13}{7.76} = 1.68.$$

Step 4: Rejection Region

For $\alpha = 0.05$ and df = $n_1 + n_2 - 2 = 20$, we find $t_{\alpha/2} = t_{0.025} = 2.086$ and, therefore, reject H_0 if $t < -2.086$ or if $t > 2.086$.

Step 5: Conclusion

For $t = 1.68$, at level of significance 0.05, we fail to reject H_0 and can not conclude that the population means differ.

11.111 For $1 - \alpha = 0.90$, and df = $n_1 + n_2 - 2 = 10 + 12 - 2 = 20$, we have $t_{\alpha/2} = t_{0.05} = 1.725$. The variance of each population is estimated by the pooled sample variance:

$$s_p^2 = \frac{(n_1 - 1)s_1^2 + (n_2 - 1)s_2^2}{n_1 + n_2 - 2}$$

$$= \frac{(10-1)(19.6)^2 + (12-1)(16.8)^2}{10 + 12 - 2} = \frac{6562.08}{20} = 328.104.$$

We calculate:

$$(\overline{x}_1 - \overline{x}_2) \pm t_{\frac{\alpha}{2}}\sqrt{s_p^2\left(\frac{1}{n_1} + \frac{1}{n_2}\right)}$$

$$\rightarrow (348 - 335) \pm 1.725\sqrt{328.104\left(\frac{1}{10} + \frac{1}{12}\right)}.$$

This simplifies to 13 ± 13.379, so a 90% confidence interval is $-0.379 < \mu_1 - \mu_2 < 26.379$.

11.112 a. Step 1: Hypotheses

Let σ_1^2 denote the variance for documentary fees charged by dealers in the home state of the attorney general and σ_2^2 the variance in fees for a neighboring state. In order to determine whether the variances differ, we test:

H_0: $\sigma_1^2 = \sigma_2^2$

H_a: $\sigma_1^2 \neq \sigma_2^2$

Step 2: Significance Level

We are given that $\alpha = 0.10$.

Step 3: Calculations

We calculate $F = \dfrac{s_1^2}{s_2^2} = \dfrac{(55.32)^2}{(31.05)^2} = 3.17$.

Step 4: Rejection Region

Since the relation in H_a is $\neq$, with ndf = $n_1 - 1 = 16 - 1 = 15$, ddf = $n_2 - 1 = 14 - 1 = 13$, for $\alpha=0.10$, we reject H_0 if $F > F_{\alpha/2} = F_{0.05} = 2.53$.

Step 5: Conclusion

For F = 3.17, we reject H_0 and, at the 10% level of significance, the data do provide sufficient evidence to conclude that the population variances differ.

b. Step 1: Hypotheses

Let μ_1 denote the mean documentary fee charged by dealers in the home state of the attorney general and μ_2 the mean documentary fee in a neighboring state. In order to determine whether a difference exists in the mean documentary fee charged by dealers in the two states, we test:

H_0: $(\mu_1 - \mu_2) = 0$

H_a: $(\mu_1 - \mu_2) \neq 0$

Step 2: Significance Level

We are given that $\alpha = 0.05$.

Step 3: Calculations

The value of the test statistic is given by:

$$t = \frac{(\bar{x}_1 - \bar{x}_2) - d_0}{\sqrt{\dfrac{s_1^2}{n_1} + \dfrac{s_2^2}{n_2}}} = \frac{(173.83 - 142.53) - 0}{\sqrt{\dfrac{(55.32)^2}{16} + \dfrac{(31.05)^2}{14}}} = \frac{31.30}{16.129} = 1.94.$$

Since the variances differ, the degrees of freedom of the t distribution are given by:

$$df = \frac{\left(\dfrac{s_1^2}{n_1} + \dfrac{s_2^2}{n_2}\right)^2}{\dfrac{\left(\dfrac{s_1^2}{n_1}\right)^2}{n_1 - 1} + \dfrac{\left(\dfrac{s_2^2}{n_2}\right)^2}{n_2 - 1}} = \frac{\left(\dfrac{(55.32)^2}{16} + \dfrac{(31.05)^2}{14}\right)^2}{\dfrac{\left(\dfrac{(55.32)^2}{16}\right)^2}{16 - 1} + \dfrac{\left(\dfrac{(31.05)^2}{14}\right)^2}{14 - 1}} = 24.14,$$

which we round down to 24.

Step 4: Rejection Region

For $\alpha = 0.05$ and H_a: $(\mu_1 - \mu_2) \neq 0$, we find $t_{\alpha/2} = t_{0.025} = 2.064$ and reject H_0 if $t < -2.064$ or if $t > 2.064$.

Step 5: Conclusion

For t = 1.94, at level of significance 0.05, we fail to reject H_0 and do not conclude that a difference exists in the mean documentary fee charged by dealers in the two states.

Review Exercises

11.113 Since a t-test is used, where the alternative hypothesis involves the relation ≠, the p-value is the area under the t distribution to the left of $t = -1.94$ plus the area under the t distribution to the right of 1.94. We use Table IV with df = 24 and note that 1.94 lies between $t_{0.05} = 1.711$ and $t_{0.025} = 2.064$. Therefore, the area under the t distribution to the right of 1.94 is a value between 0.025 and 0.05. Thus $0.025 < \frac{1}{2}(\text{p-value}) < 0.05$ so that $0.05 < \text{p-value} < 0.10$.

11.114 For $1 - \alpha = 0.99$, we have $z_{\alpha/2} = z_{0.005} = 2.576$ and calculate:

$$(\bar{x}_1 - \bar{x}_2) \pm z_{\frac{\alpha}{2}}\sqrt{\frac{\sigma_1^2}{n_1} + \frac{\sigma_2^2}{n_2}} \rightarrow (72.6 - 71.9) \pm 2.576\sqrt{\frac{(0.8)^2}{35} + \frac{(0.4)^2}{38}}.$$

This simplifies to 0.7 ± 0.386, so a 99% confidence interval for the true difference in the mean temperatures at the two sites is $0.314 < \mu_1 - \mu_2 < 1.086$.

11.115 For $1 - \alpha = 0.99$, we have $z_{\alpha/2} = z_{0.005} = 2.576$. The sample proportions are:

$$\hat{p}_1 = \frac{x_1}{n_1} = \frac{280}{400} = 0.70 \text{ and } \hat{p}_2 = \frac{x_2}{n_2} = \frac{275}{500} = 0.55.$$

A $1 - \alpha$ confidence interval for $p_1 - p_2$ is given by:

$$(\hat{p}_1 - \hat{p}_2) \pm z_{\frac{\alpha}{2}}\sqrt{\frac{\hat{p}_1\hat{q}_1}{n_1} + \frac{\hat{p}_2\hat{q}_2}{n_2}}$$

$$(0.70 - 0.55) \pm 2.576\sqrt{\frac{(0.70)(0.30)}{400} + \frac{(0.55)(0.45)}{500}}.$$ This simplifies to 0.15 ± 0.082, so a 99% confidence interval for $p_1 - p_2$ is $0.068 < p_1 - p_2 < 0.232$.

11.116 **Step 1: Hypotheses**

Let μ_1 denote the mean time the children play with the red board and μ_2 the mean time the children play with the black board. In order to demonstrate that children play with the red board for a longer period of time, we test:

H_0: $(\mu_1 - \mu_2) = 0$
H_a: $(\mu_1 - \mu_2) > 0$

Step 2: Significance Level

We are given that $\alpha = 0.05$.

Step 3: Calculations

We approximate σ_1^2 and σ_2^2 with s_1^2 and s_2^2, respectively, and calculate:

$$z = \frac{(\bar{x}_1 - \bar{x}_2) - d_0}{\sqrt{\frac{\sigma_1^2}{n_1} + \frac{\sigma_2^2}{n_2}}} = \frac{(828 - 781) - 0}{\sqrt{\frac{(96)^2}{34} + \frac{(85)^2}{35}}} = \frac{47}{21.852} = 2.15.$$

Step 4: Rejection Region

For $\alpha = 0.05$ we reject H_0 if $z > 1.645$.

Step 5: Conclusion

For $z = 2.15$, at level of significance 0.05, we reject H_0 and conclude that children would play with the red board for a longer time.

11.117 For $z = 2.15$ and H_a: $(\mu_1 - \mu_2) > 0$, p-value $= P(z > 2.15) = 1 - 0.9842 = 0.0158$.

11.118 Step 1: Hypotheses
H_0: $\sigma_1^2 = \sigma_2^2$
H_a: $\sigma_1^2 \neq \sigma_2^2$
Step 2: Significance Level
We are given that $\alpha = 0.10$.
Step 3: Calculations

We calculate $F = \dfrac{s_1^2}{s_2^2} = \dfrac{(12.6)^2}{(3.7)^2} = 11.60.$

Step 4: Rejection Region
Since the relation in H_a is $\neq$, with ndf $= n_1 - 1 = 9 - 1 = 8$, ddf $= n_2 - 1 = 15 - 1 = 14$, for $\alpha=0.10$, we reject H_0 if $F > F_{\alpha/2} = F_{0.05} = 2.70$.
Step 5: Conclusion
Since $F = 11.60 > 2.70$, we reject H_0 and, at the 10% level of significance, conclude that the population variances differ.

11.119 Step 1: Hypotheses
H_0: $(\mu_1 - \mu_2) = 20$
H_a: $(\mu_1 - \mu_2) > 20$
Step 2: Significance Level
We are given that $\alpha = 0.05$.
Step 3: Calculations
The value of the test statistic is given by:

$$t = \frac{(\bar{x}_1 - \bar{x}_2) - d_0}{\sqrt{\dfrac{s_1^2}{n_1} + \dfrac{s_2^2}{n_2}}} = \frac{(86.9 - 50.2) - 20}{\sqrt{\dfrac{(12.6)^2}{9} + \dfrac{(3.7)^2}{15}}} = \frac{16.7}{4.307} = 3.88.$$

Since the population variances differ, the degrees of freedom of the t distribution are given by:

$$df = \frac{\left(\dfrac{s_1^2}{n_1} + \dfrac{s_2^2}{n_2}\right)^2}{\dfrac{\left(\dfrac{s_1^2}{n_1}\right)^2}{n_1 - 1} + \dfrac{\left(\dfrac{s_2^2}{n_2}\right)^2}{n_2 - 1}} = \frac{\left(\dfrac{(12.6)^2}{9} + \dfrac{(3.7)^2}{15}\right)^2}{\dfrac{\left(\dfrac{(12.6)^2}{9}\right)^2}{9 - 1} + \dfrac{\left(\dfrac{(3.7)^2}{15}\right)^2}{15 - 1}} = 8.84,$$

which we round down to 8.
Step 4: Rejection Region
For $\alpha = 0.05$ and H_a: $(\mu_1 - \mu_2) > 20$, we reject H_0 if $t > t_\alpha = t_{0.05} = 1.860$.
Step 5: Conclusion
Since $t = 3.88 > 1.860$, at level of significance 0.05, we reject H_0. Therefore, we may conclude that μ_1 exceeds μ_2 by more than 20.

11.120 The degrees of freedom of the t distribution are given by:

$$df = \frac{\left(\frac{s_1^2}{n_1} + \frac{s_2^2}{n_2}\right)^2}{\frac{\left(\frac{s_1^2}{n_1}\right)^2}{n_1 - 1} + \frac{\left(\frac{s_2^2}{n_2}\right)^2}{n_2 - 1}} = \frac{\left(\frac{(12.6)^2}{9} + \frac{(3.7)^2}{15}\right)^2}{\frac{\left(\frac{(12.6)^2}{9}\right)^2}{9 - 1} + \frac{\left(\frac{(3.7)^2}{15}\right)^2}{15 - 1}} = 8.84,$$

which we round down to 8. For $1 - \alpha = 0.99$, we have $t_{\alpha/2} = t_{0.005} = 3.355$ and calculate:

$$(\bar{x}_1 - \bar{x}_2) \pm t_{\frac{\alpha}{2}}\sqrt{\frac{s_1^2}{n_1} + \frac{s_2^2}{n_2}} \rightarrow (86.9 - 50.2) \pm 3.355\sqrt{\frac{(12.6)^2}{9} + \frac{(3.7)^2}{15}}.$$

We have 36.7 ± 14.45, so a 99% confidence interval is $22.25 < \mu_1 - \mu_2 < 51.15$.

11.121 **Step 1: Hypotheses**

Let μ_D denote the mean difference between the number of airborne particles collected by treated as opposed to untreated filters. In order to show that the chemical treatment is effective, we test:

H_0: $\mu_D = 0$

H_a: $\mu_D > 0$

Step 2: Significance Level

We are given that $\alpha = 0.01$.

Step 3: Calculations

Site	A Treated	B untreated	Dif. D A - B	Site	A Treated	B untreated	Dif. D A - B
1	58	50	8	9	25	29	-4
2	27	19	8	10	36	30	6
3	38	29	9	11	45	29	16
4	45	34	11	12	39	39	0
5	27	22	5	13	56	43	13
6	16	13	3	14	43	29	14
7	87	70	17	15	47	49	-2
8	97	82	15	16	54	51	3

We calculate $\Sigma D = 122$ and $\Sigma D^2 = 1564$ so $\bar{D} = \frac{\Sigma D}{n} = \frac{122}{16} = 7.625$, and

$$s_D = \sqrt{\frac{\Sigma D^2 - \frac{(\Sigma D)^2}{n}}{n - 1}} = \sqrt{\frac{1564 - \frac{(122)^2}{16}}{16 - 1}} = 6.5.$$

Therefore:

$$t = \frac{\bar{D} - \mu_D}{\frac{s_D}{\sqrt{n}}} = \frac{7.625 - 0}{\frac{6.5}{\sqrt{16}}} = 4.69.$$

Step 4: Rejection Region

For $\alpha = 0.01$, df = 16 - 1 = 15, and H_a: $\mu_D > 0$, we reject H_0 if $t > t_\alpha = t_{0.01} = 2.602$.

Step 5: Conclusion
Since t = 4.69 > 2.602, at level of significance 0.01, we reject H_0 , and there is sufficient evidence to conclude that the chemical treatment is effective.

MINITAB LAB ASSIGNMENTS

11.122(M)

```
MTB > # Exercise 11.122(M)
MTB > READ C1 C2
DATA> 58 50
DATA> 27 19
DATA> 38 29
DATA> 45 34
DATA> 27 22
DATA> 16 13
DATA> 87 70
DATA> 97 82
DATA> 25 29
DATA> 36 30
DATA> 45 29
DATA> 39 39
DATA> 56 43
DATA> 43 29
DATA> 47 49
DATA> 54 51
DATA> END
MTB > LET C3 = C1 - C2
MTB > TTEST 0 C3;
SUBC> ALTE 1.

TEST OF MU = 0.000 VS MU G.T. 0.000

              N      MEAN     STDEV   SE MEAN        T    P VALUE
C3           16     7.625     6.500     1.625     4.69     0.0001

MTB > # Since p-value = 0.0001 < 0.01, we reject Ho; Treatment is
MTB > # effective.
```

11.123(M)

```
MTB > # Exercise 11.123(M)
MTB > # Data were entered into C1 and C2 in Exercise 11.122(M)
MTB > LET C3 = C1 - C2
MTB > TINTERVAL 99 C3

              N      MEAN     STDEV  SE MEAN   99.0 PERCENT C.I.
C3           16      7.62      6.50     1.62  (    2.84,    12.41)

MTB > # 99% confidence interval for the difference is 2.84 to 12.41.
```

11.124(M)

```
MTB > # Exercise 11.124(M)
MTB > SET C1
DATA> 37.3 35.7 39.2 38.2 32.9 36.2 38.4 39.2 41.9 30.6
DATA> END
MTB > SET C2
DATA> 31.6 33.9 32.8 31.9 32.1 32.2 33.3 30.9 32.7 33.0 32.5 32.4
DATA> END
MTB > # Part a:
MTB > LET K1 = STDEV(C1)**2
MTB > LET K2 = STDEV(C2)**2
MTB > PRINT K1 K2
K1        10.8071
K2        0.629925
MTB > # Part b:
MTB > LET K3 = K1/K2
MTB > PRINT K3
K3        17.1562
MTB > # K3 is the F ratio. The Rejection region is F > 3.59
MTB > # Since K3 = F > 3.59, we reject Ho, so the variances are
MTB > # not the same.
MTB > # Part c:
MTB > CDF 17.1562;
SUBC> F 9 11.
   17.1562    1.0000
MTB > # P-value =2(1-1.0000)=0< alpha = 0.05. Therefore reject Ho
MTB > # The variances are not the same.
```

11.125(M)

```
MTB > # Exercise 11.125(M)
MTB > # Data were entered into C1 and C2 in Exercise 11.124(M)
MTB > TWOSAMPLE t C1 C2;
SUBC> ALTERNATIVE = 0.

TWOSAMPLE T FOR C1 VS C2
      N      MEAN     STDEV   SE MEAN
C1   10     36.96      3.29       1.0
C2   12    32.442     0.794      0.23

95 PCT CI FOR MU C1 - MU C2: (2.1, 6.93)

TTEST MU C1 = MU C2 (VS NE): T= 4.24  P=0.0022  DF=  9
MTB > # P-value = 0.0022 < alpha = 0.01. Therefore reject Ho
MTB > # The two brands differ in the mean fat content.
```

11.126(M)

```
MTB > # Exercise 11.126(M)
MTB > # Data were entered into C1 and C2 in Exercise 11.124(M)
MTB > TWOSAMPLE 90 C1 C2

TWOSAMPLE T FOR C1 VS C2
      N      MEAN     STDEV   SE MEAN
C1   10     36.96      3.29       1.0
C2   12    32.442     0.794      0.23

90 PCT CI FOR MU C1 - MU C2: (2.6, 6.47)

TTEST MU C1 = MU C2 (VS NE): T= 4.24  P=0.0022  DF=  9
```

11.127(M)

```
MTB > # Exercise 11.127(M)
MTB > SET C1
DATA> 0:3/0.1
DATA> END
MTB > PDF C1 C2;
SUBC> F 13 17.
MTB > PLOT C2 C1


       0.90+
           -             ** *
     C2    -           *      *
           -
           -         *         *
       0.60+                     *
           -
           -                       *
           -        *               *
           -                          *
       0.30+                            *
           -      *                      * *
           -                                **
           -                                   * **
           -    *                                   * ** * **
       0.00+  * *                                            *
           -
             --+---------+---------+---------+---------+---------+----C1
             0.00      0.60      1.20      1.80      2.40      3.00
```

11.128(M)

```
MTB > # Exercise 11.128(M)a:
MTB > CDF 1.23;
SUBC> F 14 34.
    1.2300    0.7001
MTB > # P(F < 1.23) = 0.7001
MTB > # 11.128(M)b:
MTB > INVCDF 0.83;
SUBC> F 14 34.
    0.8300    1.4843
MTB > # The 83rd percentile is 1.4843.
```

CHAPTER 12
CHI-SQUARE TESTS FOR ANALYZING COUNT DATA

EXERCISES for Section 12.2
Hypothesis Test for a Multinomial Experiment

12.1 **Step 1: Hypotheses**
H_0: $p_1 = 0.30$, $p_2 = 0.10$, $p_3 = 0.40$, $p_4 = 0.20$
H_a: Not all of the above are true
Step 2: Significance Level
We are given that $\alpha = 0.10$.
Step 3: Calculations

Outcomes	o	p	np = e =200p	$\frac{(o-e)^2}{e}$
1	54	0.30	60	0.6
2	29	0.10	20	4.05
3	83	0.40	80	0.1125
4	34	0.20	40	0.9
Sum:	200			5.66

$X^2 = \Sigma \frac{(o-e)^2}{e} = $ **5.66.** (See column 5 of the table.)

Step 4: Rejection Region
For $\alpha = 0.10$ and df = k - 1 = 4 - 1 = 3, we find $X^2_{0.10} = 6.25$, so we reject H_0 if $X^2 > 6.25$.
Step 5: Conclusion
Since $X^2 = 5.66 < 6.25$, there is insufficient evidence at the 0.10 level to reject the null hypothesis concerning the probabilities.

12.3 For df = 5, $X^2_{0.10} = 9.24$, so for a test value of $X^2 = 7.04$, we have p-value > 0.10.

12.5 For df = 1, $X^2_{0.005} = 7.88$, so for a test value of $X^2 = 19.60$, we have p-value < 0.005.

12.7 **Step 1: Hypotheses**
H_0: $p_1 = p_2 = p_3 = 1/3$
H_a: Not all of the above are true
Step 2: Significance Level
We are given that $\alpha = 0.10$.
Step 3: Calculations

Location	o	p	np = e =800p	$\frac{(o-e)^2}{e}$
1	247	⅓	266.67	1.45
2	269	⅓	266.67	0.02
3	284	⅓	266.67	1.13
Sum:	800			2.60

$X^2 = \Sigma \frac{(o-e)^2}{e} = 2.60.$ (See column 5 of the table.)

Step 4: Rejection Region
For $\alpha = 0.10$ and df = k - 1 = 3 - 1 = 2, we find $X^2_{0.10} = 4.61$, so we reject H_0 if $X^2 > 4.61$.
Step 5: Conclusion
Since $X^2 = 2.60 < 4.61$, we fail to reject H_0 that the proportions of claims handled at the three locations are the same.

12.9 For df = 4, $X^2_{0.005} = 14.86$, so for a test value of $X^2 = 109.67$, we have p-value < 0.005.

12.11 For df = k - 1 = 5 - 1 = 4, we find $X^2_{0.10} = 7.78$ and $X^2_{0.05} = 9.49$. Therefore, for the test statistic value of $X^2 = 8.58$, $0.05 <$ p-value < 0.10.

12.13 **Step 1: Hypotheses**

H_0: $p_1 = 0.15$, $p_2 = 0.24$, $p_3 = 0.32$, $p_4 = 0.20$, $p_5 = 0.09$

H_a: Not all of the above are true

Step 2: Significance Level

We are given that $\alpha = 0.05$.

Step 3: Calculations

Course Grade	o	p	np = e = 600p	$\frac{(o-e)^2}{e}$
A	103	0.15	90	1.8778
B	153	0.24	144	0.5625
C	189	0.32	192	0.0469
D	111	0.20	120	0.6750
F	44	0.09	54	1.8519
Sum:	600			5.01

$X^2 = \Sigma \frac{(o-e)^2}{e} = 5.01.$ (See column 5 of the table.)

Step 4: Rejection Region

For $\alpha = 0.05$ and df = k - 1 = 5 - 1 = 4, we find $X^2_{0.05} = 9.49$, so we reject H_0 if $X^2 > 9.49$.

Step 5: Conclusion

Since $X^2 = 5.01 < 9.49$, we fail to reject H_0, and there is insufficient evidence, at the 0.05 level, to show that the grade distributions are not the same for the two institutions.

MINITAB LAB ASSIGNMENTS

12.15(M)

```
MTB > # Exercise 12.15(M)
MTB > CDF 8.58;
SUBC> CHIS 4.
    8.5800      0.9275
MTB > # P-value = (1 - 0.9275) = 0.0725.
```

12.17(M)

```
MTB > # Exercise 12.17(M)
MTB > CDF 5.01;
SUBC> CHIS 4.
    5.0100      0.7137
MTB > # P-value = (1 - 0.7137) = 0.2863.
```

EXERCISES for Section 12.3

Hypothesis Test for a Contingency Table

12.19 **Step 1: Hypotheses**
H_0: Row and column classifications are independent
H_a: Row and column classifications are not independent
Step 2: Significance Level
We are given that $\alpha = 0.05$.
Step 3: Calculations
Before calculating X^2, we calculate the estimated expected cell counts. We use the formula $e_{ij} = \frac{R_i C_j}{n}$, where R_i is the total count for row i, C_j is the total count for column j, and n is the total count for the table. The results are shown in both tables (in parentheses in the contingency table below).

Row, column	o	e	$\frac{(o - e)^2}{e}$
1,1	20	21.60	0.119
1,2	75	64.80	1.606
1,3	25	33.60	2.201
2,1	25	23.40	0.109
2,2	60	70.20	1.482
2,3	45	36.40	2.032
Total:	250		7.55

	1	2	3	Total
Row 1	20 (21.60)	75 (64.80)	25 (33.60)	120
Row 2	25 (23.40)	60 (70.20)	45 (36.40)	130
Total	45	135	70	250

$X^2 = \Sigma \frac{(o - e)^2}{e} = 7.55$. (See column 4 of the first table.)

Step 4: Rejection Region
For $\alpha = 0.05$ and df = (r - 1)•(c - 1) = (2 - 1)•(3 - 1) = 2, we find $X^2_{0.05} = 5.99$, so we reject H_0 if $X^2 > 5.99$.
Step 5: Conclusion
Since $X^2 = 7.55 > 5.99$, there is sufficient evidence at the 0.05 level to reject H_0 and conclude that the row and column classifications are not independent.

12.21 For df = (r - 1)•(c - 1) = (2 - 1)•(4 - 1) = 3, $X^2_{0.10} = 6.25$ and $X^2_{0.05} = 7.81$. Since the value of the test statistic is $X^2 = 7.28$, we have 0.05 < p-value < 0.10.

Hypothesis Test for a Contingency Table

12.23 **Step 1: Hypotheses**

H_0: Fire damage is independent of presence of fire extinguisher

H_a: Fire damage is dependent on presence of fire extinguisher

Step 2: Significance Level

We are given that $\alpha = 0.01$.

Step 3: Calculations

Before calculating X^2, we calculate the estimated expected cell counts. We use the formula $e_{ij} = \frac{R_i C_j}{n}$, where R_i is the total count for row i, C_j is the total count for column j, and n is the total count for the table. The results are shown in both tables (in parentheses in the contingency table below).

Row, column	o	e	$\frac{(o - e)^2}{e}$
1,1	124	101.50	4.988
1,2	110	98.39	1.370
1,3	40	53.17	3.262
1,4	16	36.94	11.870
2,1	170	192.50	2.630
2,2	175	186.61	0.722
2,3	114	100.83	1.720
2,4	91	70.06	6.259
Total:	840		32.82

	Dollar amount of damage				
Extinguisher?	< 250	250-499	500-999	≥ 1000	Total
yes	124 (101.50)	110 (98.39)	40 (53.17)	16 (36.94)	290
no	170 (192.50)	175 (186.61)	114 (100.83)	91 (70.06)	550
Total	294	285	154	107	840

$X^2 = \Sigma \frac{(o - e)^2}{e} = 32.82$. (See column 4 of the first table.)

Step 4: Rejection Region

For $\alpha = 0.01$ and df $= (r - 1)\cdot(c - 1) = (2 - 1)\cdot(4 - 1) = 3$, we find $X^2_{0.01} = 11.34$, so we reject H_0 if $X^2 > 11.34$.

Step 5: Conclusion

Since $X^2 = 32.82 > 11.34$, there is sufficient evidence at the 0.01 level to reject the null hypothesis and conclude that the amount of fire damage sustained is dependent on the presence of a fire extinguisher.

12.25

Month of injury occurrence

Sex	Jan	Feb	Mar	Apr	May	Jun	Jul	Aug	Sep	Oct	Nov	Dec	Tot
F	22 (23.3)	23 (18.7)	46 (37.8)	29 (33.6)	34 (35.5)	49 (46.9)	23 (22.1)	21 (18.7)	10 (15.6)	17 (16.4)	11 (13.7)	7 (9.5)	292
M	39 (37.7)	26 (30.3)	53 (61.2)	59 (54.4)	59 (57.5)	74 (76.1)	35 (35.9)	28 (30.3)	31 (25.4)	26 (26.6)	25 (22.3)	18 (15.5)	473
Tot	61	49	99	88	93	123	58	49	41	43	36	25	765

Step 1: Hypotheses

H_0: Month of occurrence of dental injury is independent of sex of patient

H_a: Month of occurrence of dental injury is dependent on sex of patient

Step 2: Significance Level

We are given that $\alpha = 0.05$.

Step 3: Calculations

Before calculating X^2, we calculate the estimated expected cell counts. We use the formula $e_{ij} = \frac{R_i C_j}{n}$, where R_i is the total count for row i, C_j is the total count for column j, and n is the total count for the table. The results are shown in both tables.

$$X^2 = \Sigma \frac{(o - e)^2}{e} = 11.68.$$ (See column 4 of the table.)

Step 4: Rejection Region

For $\alpha = 0.05$ and df $= (r - 1) \cdot (c - 1) = (2 - 1) \cdot (12 - 1) = 11$, we find $X^2_{0.05} = 19.68$, so we reject H_0 if $X^2 > 19.68$.

Step 5: Conclusion

Since $X^2 = 11.68 < 19.68$, there is insufficient evidence at the 0.05 level to reject H_0, so the data do not provide sufficient evidence to conclude that the month in which a dental injury occurs is dependent on the injured's sex.

Row, column	o	e	$\frac{(o - e)^2}{e}$
1,1	22	23.28	0.070
1,2	23	18.70	0.989
1,3	46	37.79	1.784
1,4	29	33.59	0.627
1,5	34	35.50	0.063
1,6	49	46.95	0.090
1,7	23	22.14	0.033
1,8	21	18.70	0.283
1,9	10	15.65	2.040
1,10	17	16.41	0.021
1,11	11	13.74	0.546
1,12	7	9.54	0.676
2,1	39	37.72	0.043
2,2	26	30.30	0.610
2,3	53	61.21	1.101
2,4	59	54.41	0.387
2,5	59	57.50	0.039
2,6	74	76.05	0.055
2,7	35	35.86	0.021
2,8	28	30.30	0.175
2,9	31	25.35	1.259
2,10	26	26.59	0.013
2,11	25	22.26	0.337
2,12	18	15.46	0.417
Total:	765		11.68

Hypothesis Test for a Contingency Table

12.27 **Step 1: Hypotheses**

H_0: Semester of project completion is independent of student's GPA.

H_a: Semester of project completion is dependent on student's GPA .

Step 2: Significance Level

We are given that $\alpha = 0.05$.

Step 3: Calculations

Before calculating X^2, we calculate the estimated expected cell counts. We use the formula $e_{ij} = \frac{R_i C_j}{n}$, where R_i is the total count for row i, C_j is the total count for column j, and n is the total count for the table. The results are shown in both tables (in parentheses in the contingency table below).

Row, column	o	e	$\frac{(o-e)^2}{e}$
1,1	10	13.15	0.755
1,2	22	27.43	1.075
1,3	36	27.43	2.678
2,1	26	24.55	0.086
2,2	47	51.22	0.348
2,3	54	51.22	0.151
3,1	22	20.30	0.142
3,2	52	42.35	2.199
3,3	31	42.35	3.042
Total:	300		10.47

	Semester project completed			
GPA	6 th	7 th	8 th	Total
≤ 2.5	10 (13.15)	22 (27.43)	36(27.43)	68
2.6 - 3.0	26 (24.55)	47 (51.22)	54 (51.22)	127
3.1 - 4.0	22 (20.30)	52 (42.35)	31 (42.35)	105
Total	58	121	121	300

$X^2 = \Sigma \frac{(o-e)^2}{e} = 10.47$. (See column 4 of the first table.)

Step 4: Rejection Region

For $\alpha = 0.05$ and df $= (r-1)\cdot(c-1) = (3-1)\cdot(3-1) = 4$, we find $X^2_{0.05} = 9.49$, so we reject H_0 if $X^2 > 9.49$.

Step 5: Conclusion

Since $X^2 = 10.47 > 9.49$, there is sufficient evidence at the 0.05 level to reject H_0, so that the semester in which a student satisfies the requirement is dependent on his/her grade point average.

12.29 **Step 1: Hypotheses**
H_0: Severity of disability and receipt of occupational therapy are independent
H_a: Severity of disability and receipt of occupational therapy are dependent
Step 2: Significance Level
We are given that $\alpha = 0.05$.
Step 3: Calculations
Before calculating X^2, we calculate the estimated expected cell counts. We use the formula $e_{ij} = \frac{R_i C_j}{n}$, where R_i is the total count for row i, C_j is the total count for column j, and n is the total count for the table. The results are shown in both tables (in parentheses in the contingency table below).

Row, column	o	e	$\frac{(o-e)^2}{e}$
1,1	26	38.30	3.950
1,2	66	61.55	0.322
1,3	40	32.15	1.917
2,1	30	17.70	8.548
2,2	24	28.45	0.696
2,3	7	14.85	4.150
Total:	193		19.58

Received O.T.?	Severity of Disability: Mild	Moderate	Severe	Total
Yes	26 (38.30)	66 (61.55)	40 (32.15)	132
no	30 (17.70)	24 (28.45)	7 (14.85)	61
Total	56	90	47	193

$X^2 = \Sigma \frac{(o-e)^2}{e} = 19.58$. (See column 4 of the first table.)

Step 4: Rejection Region
For $\alpha = 0.05$ and df $= (r - 1)\cdot(c - 1) = (2 - 1)\cdot(3 - 1) = 2$, we find $X^2_{0.05} = 5.99$, so we reject H_0 if $X^2 > 5.99$.
Step 5: Conclusion
Since $X^2 = 19.58 > 5.99$, there is sufficient evidence at the 0.05 level to reject H_0, so the data suggests that the severity of a disability and the receipt of occupational therapy are dependent.

MINITAB LAB ASSIGNMENTS

12.31(M)

```
MTB > # Exercise 12.31(M)
MTB > READ C1-C3
DATA> 26 66 40
DATA> 30 24 7
DATA> END
      2 ROWS READ
MTB > CHIS C1-C3

Expected counts are printed below observed counts

            C1        C2        C3     Total
    1       26        66        40       132
         38.30     61.55     32.15

    2       30        24         7        61
         17.70     28.45     14.85

Total       56        90        47       193

ChiSq =  3.950 +  0.321 +  1.919 +
         8.548 +  0.695 +  4.153 = 19.588
df = 2
MTB > # Since 19.588 is in the rejection region(> 5.99), reject Ho
MTB > # There is a relationship.
```

12.33(M)

```
MTB > # Exercise 12.33(M)
MTB > READ C1, C2
DATA> 189 291
DATA> 35 62
DATA> 29 30
DATA> 17 32
DATA> 13 33
DATA> 5 13
DATA> 4 12
DATA> END
      7 ROWS READ
MTB > CHIS C1-C2
```

```
Expected counts are printed below observed counts

             C1         C2     Total
    1       189        291       480
         183.22     296.78

    2        35         62        97
          37.02      59.98

    3        29         30        59
          22.52      36.48

    4        17         32        49
          18.70      30.30

    5        13         33        46
          17.56      28.44

    6         5         13        18
           6.87      11.13

    7         4         12        16
           6.11       9.89

Total       292        473       765

ChiSq =   0.183 +   0.113 +
          0.111 +   0.068 +
          1.864 +   1.151 +
          0.155 +   0.096 +
          1.183 +   0.731 +
          0.509 +   0.314 +
          0.727 +   0.449 = 7.654
df = 6

MTB > CDF 7.654;
SUBC> CHIS 6.
    7.6540     0.7356
MTB > # P-value = 1 - 0.7356 = 0.2644 > 0.05, so we fail to reject Ho.
MTB > # Or: Since 7.654 < 12.59, we fail to reject Ho.
```

12.35(M)

```
MTB > # Exercise 12.35(M)
MTB > CDF 94.680;
SUBC> CHIS 6.
   94.6800     1.0000
MTB > # P-value = 1 - 1 = 0 < 0.01, so we reject Ho
MTB > # There is a relationship between physical fitness and salary.
```

EXERCISES for Section 12.4

Hypothesis Test for Two or More Proportions

12.37 **Step 1: Hypotheses**
H_0: $p_1 = p_2 = p_3$
H_a: Not all proportions are equal
Step 2: Significance Level
We are given that $\alpha = 0.05$.
Step 3: Calculations
Before calculating X^2, we calculate the estimated expected cell counts. We use the formula $e_{ij} = \frac{R_i C_j}{n}$, where R_i is the total count for row i, C_j is the total count for column j, and n is the total count for the table. The results are shown in both tables (in parentheses in the contingency table below).

Row, column	o	e	$\frac{(o-e)^2}{e}$
1,1	192	210	1.543
1,2	168	168	0
1,3	270	252	1.286
2,1	308	290	1.117
2,2	232	232	0
2,3	330	348	0.931
Total:	1500		4.88

	Sample One	Sample Two	Sample Three	Total
x_i	192 (210)	168 (168)	270 (252)	630
$n_i - x_i$	308 (290)	232 (232)	330 (348)	870
Total = n_i	500	400	600	1500

$X^2 = \Sigma \frac{(o-e)^2}{e} = 4.88.$ (See column 4 of the first table.)

Step 4: Rejection Region
For $\alpha = 0.05$ and df = $(r - 1)\cdot(c - 1) = (2 - 1)\cdot(3 - 1) = 2$, we find $X^2_{0.05} = 5.99$, so we reject H_0 if $X^2 > 5.99$.
Step 5: Conclusion
Since $X^2 = 4.88 < 5.99$, do not reject the null hypothesis. There is insufficient evidence at the 0.05 level to conclude that the population proportions are not all equal.

12.39 **<u>Step 1</u>: Hypotheses**
H_0: $p_1 = p_2 = p_3 = p_4$
H_a: Not all probabilities are equal
<u>Step 2</u>: Significance Level
We are given that $\alpha = 0.10$.
<u>Step 3</u>: Calculations
Before calculating X^2, we calculate the estimated expected cell counts. We use the formula $e_{ij} = \frac{R_i C_j}{n}$, where R_i is the total count for row i, C_j is the total count for column j, and n is the total count for the table. The results are shown in both tables (in parentheses in the contingency table below).

Row, column	o	e	$\frac{(o - e)^2}{e}$
1,1	236	248.75	0.654
1,2	259	248.75	0.422
1,3	239	248.75	0.382
1,4	261	248.75	0.603
2,1	264	251.25	0.647
2,2	241	251.25	0.418
2,3	261	251.25	0.378
2,4	239	251.25	0.597
Total:	2000		4.10

	Coin				
	1st	2nd	3rd	4th	Total
Heads	236 (248.75)	259 (248.75)	239 (248.75)	261 (248.75)	995
Tails	264 (251.25)	241 (251.25)	261 (251.25)	239 (251.25)	1005
Total	500	500	500	500	2000

$X^2 = \Sigma \frac{(o - e)^2}{e} = 4.10$. (See column 4 of the first table.)

<u>Step 4</u>: Rejection Region
For $\alpha = 0.10$ and df $= (r - 1)\cdot(c - 1) = (2 - 1)\cdot(4 - 1) = 3$, we find $X^2_{0.10} = 6.25$, so we reject H_0 if $X^2 > 6.25$.
<u>Step 5</u>: Conclusion
Since $X^2 = 4.10 < 6.25$, there is insufficient evidence at the 0.10 level to reject H_0, so we are not able to conclude that the probability of a head is not the same for the four coins.

12.41 For df $= 1$, we find $X^2_{0.01} = 6.63$ and $X^2_{0.005} = 7.88$, so for the test statistic value $X^2 = 7.02$, we have $0.005 < \text{p-value} < 0.01$.

12.43 **Step 1: Hypotheses**

Let p_1 denote the mortality rate for women and p_2 the mortality rate for men. We test:

H_0: $p_1 = p_2$

H_a: $p_1 \neq p_2$

Step 2: Significance Level

We are given that $\alpha = 0.05$.

Step 3: Calculations

We are given $n_1 = 332$, $x_1 = 47$, $n_2 = 790$ and $x_2 = 70$. The sample proportions are:

$\hat{p}_1 = \frac{x_1}{n_1} = \frac{47}{332}$ *and* $\hat{p}_2 = \frac{x_2}{n_2} = \frac{70}{790} = \frac{7}{79}$. The test statistic is:

$$z = \frac{(\hat{p}_1 - \hat{p}_2)}{\sqrt{\hat{p}\hat{q}\left(\frac{1}{n_1} + \frac{1}{n_2}\right)}}, \text{ where } \hat{p} = \frac{x_1 + x_2}{n_1 + n_2} = \frac{47 + 70}{332 + 790} = \frac{117}{1122} = 0.1043.$$

$$z = \frac{\left(\frac{47}{332} - \frac{7}{79}\right)}{\sqrt{(.1043)(0.8957)\left(\frac{1}{332} + \frac{1}{790}\right)}} = \frac{0.05296}{0.01999} = 2.65.$$

Step 4: Rejection Region

For $\alpha = 0.05$ and H_a: $p_1 \neq p_2$, we reject H_0 if $z < -1.96$ or if $z > 1.96$.

Step 5: Conclusion

Since $z = 2.65 > 1.96$, there is sufficient evidence at the 0.05 level to reject the null hypothesis and conclude that there is a difference in the mortality rates for men and women hospitalized after myocardial infarction.

12.45 **Step 1: Hypotheses**
H_0: $p_1 = p_2 = p_3 = p_4 = p_5$
H_a: Not all proportions are equal
Step 2: Significance Level
We are given that $\alpha = 0.05$.
Step 3: Calculations
Before calculating X^2, we calculate the estimated expected cell counts. We use the formula $e_{ij} = \frac{R_i C_j}{n}$, where R_i is the total count for row i, C_j is the total count for column j, and n is the total count for the table. The results are shown in both tables (in parentheses in the contingency table below).

Row, column	o	e	$\frac{(o-e)^2}{e}$
1,1	39	40	0.025
1,2	47	40	1.225
1,3	32	40	1.600
1,4	33	40	1.225
1,5	49	40	2.025
2,1	61	60	0.017
2,2	53	60	0.817
2,3	68	60	1.067
2,4	67	60	0.817
2,5	51	60	1.350
Total:	500		10.17

Plant Location

	NY	MI	GA	CO	CA	Total
Favor	39 (40)	47 (40)	32 (40)	33 (40)	49 (40)	200
Against	61 (60)	53 (60)	68 (60)	67 (60)	51 (60)	300
Total	100	100	100	100	100	500

$X^2 = \Sigma \frac{(o-e)^2}{e} = 10.17.$ (See column 4 of the first table.)

Step 4: Rejection Region
For $\alpha = 0.05$ and df $= (r - 1) \cdot (c - 1) = (2 - 1) \cdot (5 - 1) = 4$, we find $X^2_{0.05} = 9.49$, so we reject H_0 if $X^2 > 9.49$.
Step 5: Conclusion
Since $X^2 = 10.17 > 9.49$, there is sufficient evidence at the 0.05 level to reject the null hypothesis and conclude that a difference exists in the proportions of workers in the five states who favor the proposal.

12.47 **Step 1: Hypotheses**
H_0: $p_1 = p_2 = p_3$
H_a: Not all proportions are equal
Step 2: Significance Level
We are given that $\alpha = 0.01$.
Step 3: Calculations
Before calculating X^2, we calculate the estimated expected cell counts. We use the formula $e_{ij} = \frac{R_i C_j}{n}$, where R_i is the total count for row i, C_j is the total count for column j, and n is the total count for the table. The results are shown in both tables (in parentheses in the contingency table below).

Row, column	o	e	$\frac{(o - e)^2}{e}$
1,1	115	116.55	0.021
1,2	78	99.90	4.801
1,3	140	116.55	4.718
2,1	60	58.45	0.041
2,2	72	50.10	9.573
2,3	35	58.45	9.408
Total:	500		28.56

	Pain Reliever			
	A	B	C	Total
Effective	115 (116.55)	78 (99.90)	140 (116.55)	333
Not Effective	60 (58.45)	72 (50.10)	35 (58.45)	167
Total	175	150	175	500

$X^2 = \Sigma \frac{(o - e)^2}{e} = 28.56.$ (See column 4 of the first table.)

Step 4: Rejection Region
For $\alpha = 0.01$ and df $= (r - 1)\cdot(c - 1) = (2 - 1)\cdot(3 - 1) = 2$, we find $X^2_{0.01} = 9.21$, so we reject H_0 if $X^2 > 9.21$.
Step 5: Conclusion
Since $X^2 = 28.56 > 9.21$, there is sufficient evidence at the 0.01 level to reject H_0, and conclude that the proportions are different.

MINITAB LAB ASSIGNMENTS

12.49(M)

```
MTB > # Exercise 12.49(M)
MTB > READ C1 - C6
DATA> 375 321 315 362 358 347
DATA> 125 179 185 138 142 153
DATA> END
      2 ROWS READ
MTB > CHIS C1 - C6
```

```
Expected counts are printed below observed counts

            C1        C2        C3        C4        C5        C6     Total
    1      375       321       315       362       358       347      2078
        346.33    346.33    346.33    346.33    346.33    346.33

    2      125       179       185       138       142       153       922
        153.67    153.67    153.67    153.67    153.67    153.67

Total      500       500       500       500       500       500      3000

ChiSq =  2.373 +  1.853 +  2.835 +  0.709 +  0.393 +  0.001 +
         5.348 +  4.176 +  6.389 +  1.597 +  0.886 +  0.003 = 26.563
df = 5
MTB > # Since 26.563 is in the rejection region (> 11.07), reject Ho.
MTB > # A difference in proportions exists.
```

12.51(M)

```
MTB > # Exercise 12.51(M)
MTB > READ C1 - C3
DATA> 375 321 347
DATA> 125 179 153
DATA> END
      2 ROWS READ
MTB > CHIS C1 - C3

Expected counts are printed below observed counts

            C1        C2        C3     Total
    1      375       321       347      1043
        347.67    347.67    347.67

    2      125       179       153       457
        152.33    152.33    152.33

Total      500       500       500      1500

ChiSq =  2.149 +  2.045 +  0.001 +
         4.904 +  4.668 +  0.003 = 13.771
df = 2
MTB > # Since 13.771 is in the rejection region (> 5.99), reject Ho.
MTB > # A difference in proportions exists.
```

REVIEW EXERCISES
Chapter 12

12.53 **Step 1: Hypotheses**
H_0: Age 10 aggressiveness ratings and adult delinquency are independent
H_A: Age 10 aggressiveness ratings and adult delinquency are dependent
Step 2: Significance Level
We are given that $\alpha = 0.05$.
Step 3: Calculations
Before calculating X^2, we calculate the estimated expected cell counts. We use the formula $e_{ij} = \frac{R_i C_j}{n}$, where R_i is the total count for row i, C_j is the total count for column j, and n is the total count for the table. The results are shown in both tables (in parentheses in the contingency table below).

Row, column	o	e	$\frac{(o-e)^2}{e}$
1,1	100	77.38	6.612
1,2	197	202.96	0.175
1,3	29	45.67	6.085
2,1	15	18.99	0.838
2,2	52	49.81	0.096
2,3	13	11.21	0.286
3,1	7	25.63	13.542
3,2	71	67.24	0.210
3,3	30	15.13	14.614
Total:	514		42.46

	Aggressiveness Score at Age 10 (Boys)			
	1-2	3-5	6-7	Total
0 crimes	100 (77.38)	197 (202.96)	29 (45.67)	326
1 crime	15 (18.99)	52 (49.81)	13(11.21)	80
≥ 2 crimes	7 (25.63)	71 (67.24)	30 (15.13)	108
Total	122	320	72	514

$X^2 = \Sigma \frac{(o-e)^2}{e}$ = **42.46.** (See column 4 of the first table.)

Step 4: Rejection Region
For $\alpha = 0.05$ and df = (r - 1)•(c - 1) = (3 - 1)•(3 - 1) = 4, we find $X^2_{0.05} = 9.49$, so we reject H_0 if $X^2 > 9.49$.
Step 5: Conclusion
Since $X^2 = 42.46 > 9.49$, there is sufficient evidence at the 0.05 level to reject the null hypothesis and conclude that a relationship exists between aggressiveness ratings at age 10 and adult delinquency for boys.

12.54 For df = (r - 1)•(c - 1) = (3 - 1)•(3 - 1) = 4, we find $X^2_{0.005} = 14.86$. Therefore, for the test statistic $X^2 = 42.46$, we have p-value < 0.005.

12.55 We note that $e_{23} = R_2C_3/n = (26)(27)/507 = 1.38$ and $e_{33} = R_3C_3/n = (17)(27)/507 = 0.91$. Therefore, the contingency table for this data reveals two cells with expected counts less than 5; therefore, the chi-square approximation is probably invalid.

12.56 **Step 1: Hypotheses**

H_0: $p_1 = p_2 = p_3 = p_4 = p_5 = p_6 = p_7$

H_a: Not all proportions are equal

Step 2: Significance Level

We are given that $\alpha = 0.05$.

Step 3: Calculations

Before calculating X^2, we calculate the estimated expected cell counts. We use the formula $e_{ij} = \frac{R_i C_j}{n}$, where R_i is the total count for row i, C_j is the total count for column j, and n is the total count for the table. The results are shown in both tables (in parentheses in the contingency table below).

Row, column	o	e	$\frac{(o - e)^2}{e}$
1,1	41	30.44	3.663
1,2	59	46.93	3.104
1,3	66	59.62	0.683
1,4	95	97.04	0.043
1,5	36	46.30	2.291
1,6	22	32.35	3.311
1,7	7	13.32	2.999
2,1	7	17.56	6.350
2,2	15	27.07	5.382
2,3	28	34.38	1.184
2,4	58	55.96	0.074
2,5	37	26.70	3.973
2,6	29	18.65	5.744
2,7	14	7.68	5.201
Total:	514		44.00

Aggressiveness Score at Age 10 (Boys)

	1	2	3	4	5	6	7	Total
No Record	41 (30.44)	59 (46.93)	66 (59.62)	95 (97.04)	36 (46.30)	22 (32.35)	7 (13.32)	326
Record	7 (17.56)	15 (27.07)	28 (34.38)	58 (55.96)	37 (26.70)	29 (18.65)	14 (7.68)	188
Total	48	74	94	153	73	51	21	514

$X^2 = \Sigma \frac{(o - e)^2}{e} = 44.00.$ (See column 4 of the first table.)

Step 4: Rejection Region

For $\alpha = 0.05$ and df $= (r - 1)\cdot(c - 1) = (2 - 1)\cdot(7 - 1) = 6$, we find $X^2_{0.05} = 12.59$, so we reject H_0 if $X^2 > 12.59$.

Step 5: Conclusion

Since $X^2 = 44.00 > 12.59$, there is sufficient evidence at the 0.05 level to reject the null hypothesis and conclude that the sample proportions are significantly different.

12.57 **Step 1: Hypotheses**
H_0: $p_1 = p_2 = p_3$
H_a: Not all proportions are equal
Step 2: Significance Level
We are given that $\alpha = 0.05$.
Step 3: Calculations
Before calculating X^2, we calculate the estimated expected cell counts. We use the formula $e_{ij} = \frac{R_i C_j}{n}$, where R_i is the total count for row i, C_j is the total count for column j, and n is the total count for the table. The results are shown in both tables (in parentheses in the contingency table below).

Row, column	o	e	$\frac{(o-e)^2}{e}$
1,1	241	234.29	0.192
1,2	158	162.90	0.147
1,3	65	66.81	0.049
2,1	15	21.71	2.074
2,2	20	15.10	1.590
2,3	8	6.19	0.529
Total:	507		4.58

Aggressiveness Score at Age 10 (Girls)

	1-3	4	5-7	Total
No Record	241 (234.29)	158 (162.90)	65 (66.81)	464
Record	15 (21.71)	20 (15.10)	8 (6.19)	43
Total	256	178	73	507

$X^2 = \Sigma \frac{(o - e)^2}{e} = 4.58$. (See column 4 of the first table.)

Step 4: Rejection Region
For $\alpha = 0.05$ and df $= (r - 1)\cdot(c - 1) = (2 - 1)\cdot(3 - 1) = 2$, we find $X^2_{0.05} = 5.99$, so we reject H_0 if $X^2 > 5.99$.
Step 5: Conclusion
Since $X^2 = 4.58 < 5.99$, there is insufficient evidence at the 0.05 level to reject the null hypothesis. We are unable to conclude that the sample proportions for the three aggressiveness categories differ significantly.

12.58 For df $= (r - 1)\cdot(c - 1) = (2 - 1)\cdot(7 - 1) = 6$, we find $X^2_{0.005} = 18.55$. Therefore, for the test statistic $X^2 = 44.00$, we have p-value < 0.005.

12.59 For df $= (r - 1)\cdot(c - 1) = (2 - 1)\cdot(3 - 1) = 2$, we find $X^2_{0.10} = 4.61$. Therefore, for the test statistic $X^2 = 4.58$, we have p-value > 0.10.

12.60 **Step 1: Hypotheses**
H_0: Egg color preference is independent of size purchased
H_a: Egg color preference is dependent on size purchased
Step 2: Significance Level
We are given that $\alpha = 0.05$.

Step 3: Calculations
Before calculating X^2, we calculate the estimated expected cell counts. We use the formula $e_{ij} = \frac{R_i C_j}{n}$, where R_i is the total count for row i, C_j is the total count for column j, and n is the total count for the table. The results are shown in both tables (in parentheses in the contingency table below).

Row, column	o	e	$\frac{(o-e)^2}{e}$
1,1	123	128.24	0.214
1,2	217	223.16	0.170
1,3	286	268.10	1.195
1,4	114	120.49	0.350
2,1	208	202.76	0.135
2,2	359	352.84	0.108
2,3	406	423.90	0.756
2,4	197	190.51	0.221
Total:	1910		3.15

	Egg size				
	small	medium	large	x-large	Total
brown	123 (128.24)	217 (223.16)	286 (268.10)	114 (120.49)	740
white	208 (202.76)	359 (352.84)	406 (423.90)	197 ((190.51)	1170
Total	331	576	692	311	1910

$X^2 = \Sigma \frac{(o-e)^2}{e} = 3.15.$ (See column 4 of the first table.)

Step 4: Rejection Region
For $\alpha = 0.05$ and df $= (r - 1)\cdot(c - 1) = (2 - 1)\cdot(4 - 1) = 3$, we find $X^2_{0.05} = 7.81$, so we reject H_0 if $X^2 > 7.81$.
Step 5: Conclusion
Since $X^2 = 3.15 < 7.81$, there is insufficient evidence at the 0.05 level to reject the null hypothesis. We are not able to conclude that egg color preference is dependent on the size purchased.

12.61 For df $= (r - 1)\cdot(c - 1) = (2 - 1)\cdot(4 - 1) = 3$, we find $X^2_{0.10} = 6.25$. Therefore, for the test statistic $X^2 = 3.15$, we have p-value > 0.10.

12.62 **Step 1: Hypotheses**
H_0: $p_1 = p_2 = p_3 = p_4 = p_5 = 0.20$
H_a: Not all of the above are true
Step 2: Significance Level
We are given that $\alpha = 0.01$.

Step 3: Calculations

Color of Sector	o	p	np = e =1000p	$\frac{(o-e)^2}{e}$
red	228	0.20	200	3.92
white	207	0.20	200	0.245
blue	164	0.20	200	6.48
green	188	0.20	200	0.72
black	213	0.20	200	0.845
Total:	1000			12.21

$X^2 = \Sigma\frac{(o-e)^2}{e} = 12.21.$ (See column 4 of the table.)

Step 4: Rejection Region
For $\alpha = 0.01$ and df = k - 1 = 5 - 1 = 4, we find $X^2_{0.01} = 13.28$, so we reject H_0 if $X^2 > 13.28$.
Step 5: Conclusion
Since $X^2 = 12.21 < 13.28$, there is insufficient evidence at the 0.01 level to reject H_0, so the data do not indicate that the wheel is out of balance.

12.63 For df = k - 1 = 5 - 1 = 4, we note $X^2_{0.025} = 11.14$ and $X^2_{0.01} = 13.28$. Since $X^2 = 12.21$, we have $0.01 <$ p-value < 0.025.

12.64 **Step 1: Hypotheses**
H_0: No relationship exists between income level and type of plan
H_a: There is a relationship between income level and type of plan
Step 2: Significance Level
We are given that $\alpha = 0.05$.
Step 3: Calculations
Before calculating X^2, we calculate the estimated expected cell counts. We use the formula $e_{ij} = \frac{R_i C_j}{n}$, where R_i is the total count for row i, C_j is the total count for column j, and n is the total count for the table. The results are shown in both tables (in parentheses in the contingency table below).

Row, column	o	e	$\frac{(o-e)^2}{e}$
1,1	21	39.19	8.443
1,2	34	32.57	0.063
1,3	39	34.60	0.560
1,4	49	36.64	4.169
2,1	131	138.38	0.394
2,2	127	115.02	1.248
2,3	128	122.21	0.274
2,4	119	129.40	0.836
3,1	79	53.43	12.237
3,2	31	44.41	4.049
3,3	37	47.19	2.200
3,4	48	49.96	0.077
Total:	843		34.55

	Type of mortgage selected				
	A	B	C	D	Total
Low income lev.	21 (39.19)	34 (32.57)	39 (34.60)	49 (36.64)	143
Average income lev.	131 (138.38)	127 (115.02)	128 (122.21)	119 (129.40)	505
High income lev.	79 (53.43)	31 (44.41)	37 (47.19)	48 (49.96)	195
Total	231	192	204	216	843

$X^2 = \Sigma \frac{(o-e)^2}{e} = 34.55.$ (See column 4 of the first table.)

Step 4: Rejection Region
For $\alpha = 0.05$ and df = $(r - 1)\cdot(c - 1) = (3 - 1)\cdot(4 - 1) = 6$, we find $X^2_{0.05}=12.59$, so we reject H_0 if $X^2 > 12.59$.
Step 5: Conclusion
Since $X^2 = 34.55 > 12.59$, there is sufficient evidence at the 0.05 level to reject the null hypothesis and conclude that a relationship exists between income level and the type of plan chosen.

12.65 For df = $(r - 1)\cdot(c - 1) = (3 - 1)\cdot(4 - 1) = 6$, we note that $X^2_{0.005} = 18.55$. Since the test statistic has a value of $X^2 = 34.55$, we have p-value < 0.005.

12.66 **Step 1: Hypotheses**
H_0: $p_1 = p_2 = p_3 = p_4$
H_a: Not all proportions are equal
Step 2: Significance Level
We are given that $\alpha = 0.01$.
Step 3: Calculations
Before calculating X^2, we calculate the estimated expected cell counts. We use the formula $e_{ij} = \frac{R_i C_j}{n}$, where R_i is the total count for row i, C_j is the total count for column j, and n is the total count for the table. The results are shown in both tables (in parentheses in the contingency table below).

Row, column	o	e	$\frac{(o-e)^2}{e}$
1,1	220	187.50	5.633
1,2	180	150.00	6.000
1,3	260	300.00	5.333
1,4	90	112.50	4.500
2,1	280	312.50	3.380
2,2	220	250.00	3.600
2,3	540	500.00	3.200
2,4	210	187.50	2.700
Total:	2000		34.35

	Sample One	Sample Two	Sample Three	Sample Four	Total
	220 (187.50)	180 (150.00)	260 (300.00)	90 (112.50)	750
	280 (312.50)	220 (250.00)	540 (500.00)	210 (187.50)	1250
Total	500	400	800	300	2000

$X^2 = \Sigma \frac{(o-e)^2}{e} = \mathbf{34.35}$. (See column 4 of the first table.)

Step 4: Rejection Region
For $\alpha = 0.01$ and df = $(r - 1) \cdot (c - 1) = (2 - 1) \cdot (4 - 1) = 3$, we find $X^2_{0.01} = 11.34$, so we reject H_0 if $X^2 > 11.34$.
Step 5: Conclusion
Since $X^2 = 34.35 > 11.34$, there is sufficient evidence at the 0.01 level to reject the null hypothesis and conclude that the population proportions are not all the same.

12.67 **Step 1: Hypotheses**
H_0: $p_1 = p_2 = p_3$
H_a: Not all proportions are equal
Step 2: Significance Level
We are given that $\alpha = 0.05$.
Step 3: Calculations
Before calculating X^2, we calculate the estimated expected cell counts. We use the formula $e_{ij} = \frac{R_i C_j}{n}$, where R_i is the total count for row i, C_j is the total count for column j, and n is the total count for the table. The results are shown in both tables (in parentheses in the contingency table below).

Row, column	o	e	$\frac{(o-e)^2}{e}$
1,1	231	273	6.462
1,2	329	273	11.487
1,3	259	273	0.718
2,1	769	727	2.426
2,2	671	727	4.314
2,3	741	727	0.270
Total:	3000		25.68

	Country 1	Country 2	Country 3	Total
Failed	231 (273)	329 (273)	259 (273)	819
Passed	769 (727)	671 (727)	741 (727)	2181
Total	1000	1000	1000	3000

$X^2 = \Sigma \frac{(o-e)^2}{e} = \mathbf{25.68}$. (See column 4 of the first table.)

Step 4: Rejection Region
For $\alpha = 0.05$ and df = $(r - 1) \cdot (c - 1) = (2 - 1) \cdot (3 - 1) = 2$, we find $X^2_{0.05} = 5.99$, so we reject H_0 if $X^2 > 5.99$.
Step 5: Conclusion
Since $X^2 = 25.68 > 5.99$, there is sufficient evidence at the 0.05 level to reject H_0, which indicates a difference in compliance rates.

12.68 For df = (r - 1)·(c - 1) = (2 - 1)·(3 - 1) = 2, we find $X^2_{0.005}$ = 10.60. Therefore, since X^2 = 25.68, we have p-value < 0.005.

12.69 **Step 1: Hypotheses**
H_0: p_1 = 0.24, p_2 = 0.38, p_3 = 0.21, p_4 = 0.17
H_a: Not all of the above are true
Step 2: Significance Level
We are given that α = 0.05.
Step 3: Calculations

Exit	o	p	np = e =10000p	$\frac{(o-e)^2}{e}$
1	2250	0.24	2400	9.3750
2	3980	0.38	3800	8.5263
3	2040	0.21	2100	1.7143
4	1730	0.17	1700	0.5294
Total:	10000			20.15

$$X^2 = \Sigma \frac{(o-e)^2}{e} = 20.15.$$ (See column 5 of the table.)

Step 4: Rejection Region
For α = 0.05 and df = k - 1 = 4 - 1 = 3, we find $X^2_{0.05}$= 7.81, so we reject H_0 if X^2 > 7.81.
Step 5: Conclusion
Since X^2 = 20.15 > 7.81, there is sufficient evidence at the 0.05 level to reject H_0, so these results indicate that a change has occurred in the pattern of exiting.

12.70 For df = k - 1 = 4 - 1 = 3, $X^2_{.005}$ = 12.84. Since the test value is X^2 = 20.15, we have p-value < 0.005.

MINITAB LAB ASSIGNMENTS

12.71(M)

```
MTB > # Exercise 12.71(M)
MTB > READ C1 - C3
DATA> 697 603 571
DATA> 503 597 629
DATA> END
      2 ROWS READ
MTB > CHIS C1 - C3
```

```
Expected counts are printed below observed counts

              C1         C2         C3      Total
    1        697        603        571       1871
          623.67     623.67     623.67

    2        503        597        629       1729
          576.33     576.33     576.33

Total       1200       1200       1200       3600

ChiSq =   8.623 +   0.685 +   4.448 +
          9.331 +   0.741 +   4.813 = 28.640
df = 2
MTB > # Since 28.64 is in the rejection region (> 9.21), reject Ho.
MTB > # We conclude, at level of significance 0.01, that the three
MTB > # proportions are not equal.
```

12.72(M)

```
MTB > # Exercise 12.72(M)
MTB > CDF 28.640;
SUBC> CHIS 2.
   28.6400     1.0000
MTB > # P-value = (1 - 1.0000) = 0.
```

12.73(M)

```
MTB > # Exercise 12.73(M)
MTB > READ C1 - C4
DATA> 27 108 79 35
DATA> 18 99 71 48
DATA> END
      2 ROWS READ
MTB > CHIS C1 - C4

Expected counts are printed below observed counts

              C1         C2         C3         C4      Total
    1         27        108         79         35        249
           23.10     106.27      77.01      42.61

    2         18         99         71         48        236
           21.90     100.73      72.99      40.39

Total         45        207        150         83        485

ChiSq =   0.657 +   0.028 +   0.051 +   1.360 +
          0.694 +   0.030 +   0.054 +   1.435 = 4.309
df = 3
MTB > # Since 4.309 is not in the rejection region (> 7.81), fail to
MTB > # reject Ho.
MTB > # Insufficient evidence for dependence at the 0.05 level.
```

12.74(M)

```
MTB > # Exercise 12.74(M)
MTB > CDF 4.309;
SUBC> CHIS 3.
   4.3090     0.7700
MTB > # P-value = 1 - 0.7700 = 0.2300.
```

12.75(M)

```
MTB > # Exercise 12.75(M)
MTB > READ C1 - C3
DATA> 60 47 32
DATA> 340 353 368
DATA> END
      2 ROWS READ
MTB > CHIS C1 - C3

Expected counts are printed below observed counts

              C1        C2        C3     Total
    1         60        47        32       139
           46.33     46.33     46.33

    2        340       353       368      1061
          353.67    353.67    353.67

Total        400       400       400      1200

ChiSq =  4.031 +  0.010 +  4.434 +
         0.528 +  0.001 +  0.581 = 9.585
df = 2

MTB > # Since 9.585 is in the rejection region (> 5.99), reject Ho.
MTB > # We conclude, at level of significance 0.05, that the three
MTB > # proportions are not equal.
```

12.76(M)

```
MTB > # Exercise 12.76(M)
MTB > CDF 9.585;
SUBC> CHIS 2.
   9.5850     0.9917
MTB > # P-VALUE = 1 - 0.9917 = 0.0083.
```

CHAPTER 13
ANALYSIS OF VARIANCE

EXERCISES for Sections 13.1 & 13.2
The Analysis of Variance Technique
The Analysis of Variance Table and Computing Formulas

13.1 We are given for sample 1: $n_1 = 21$ and $s_1^2 = 46$, for sample 2: $n_2 = 16$ and $s_2^2 = 38$. We calculate:

$$s_p^2 = \frac{(n_1 - 1)s_1^2 + (n_2 - 1)s_2^2}{(n_1 - 1) + (n_2 - 1)} = \frac{(21-1)(46) + (16-1)(38)}{(21-1) + (16-1)} = \frac{1490}{35} = 42.5714.$$

13.3 We are given for sample 1: $n_1 = 8$ and $s_1^2 = 36$, for sample 2: $n_2 = 5$ and $s_2^2 = 40$, for sample 3: $n_3 = 10$ and $s_3^2 = 35$, for sample 4: $n_4 = 6$ and $s_4^2 = 42$. We calculate:

$$s_p^2 = \frac{(n_1 - 1)s_1^2 + \cdots + (n_4 - 1)s_4^2}{(n_1 - 1) + \cdots + (n_4 - 1)}$$
$$= \frac{(8-1)(36) + (5-1)(40) + (10-1)(35) + (6-1)(42)}{(8-1) + (5-1) + (10-1) + (6-1)} = \frac{937}{25} = 37.48.$$

13.5 **a.** Using the given statistics, we calculate:

$$\bar{x} = \frac{n_1\bar{x}_1 + \cdots + n_3\bar{x}_3}{n_1 + \cdots + n_3} = \frac{6\cdot 10 + 6\cdot 22 + 8\cdot 5}{6 + 6 + 8} = \frac{232}{20} = 11.6.$$

b. We use the grand mean and the three sample means to calculate the treatment sum of squares:

$$SSTr = n_1(\bar{x}_1 - \bar{x})^2 + \cdots + n_3(\bar{x}_3 - \bar{x})^2$$
$$= (6)(10 - 11.6)^2 + (6)(22 - 11.6)^2 + (8)(5 - 11.6)^2 = 1012.8.$$

c. The pooled sample variance is:

$$s_p^2 = \frac{(n_1 - 1)s_1^2 + \cdots + (n_3 - 1)s_3^2}{(n_1 - 1) + \cdots + (n_3 - 1)}$$
$$= \frac{(6-1)(20) + (6-1)(16) + (8-1)(15)}{(6-1) + (6-1) + (8-1)} = \frac{285}{17} = 16.7647.$$

d. Using the results from parts (b) and (c), we construct the ANOVA table. We have:

Total df = n - 1 = 20 - 1 = 19
Treatment df = k - 1 = 3 - 1 = 2
Error df = n - k = 20 - 3 = 17

$$MSTr = \frac{SSTr}{k-1} = \frac{1012.8}{2} = 506.4$$

MSE = s_p^2 = 16.7647
SSE = (n - k)s_p^2 = (17)(16.7647) = 285.0

$$F = \frac{MSTr}{MSE} = \frac{506.4}{16.7647} = 30.21.$$

The ANOVA table follows:

Source of Variation	df	SS	MS	F
Treatments	2	1012.8	506.4	30.21
Error	17	285.0	16.7647	
Total	19	1297.8		

13.7 **a.** We use the grand mean $\bar{x}$ and the three sample means to calculate SSTr, the treatment sum of squares:

$$SSTr = n_1(\bar{x}_1 - \bar{x})^2 + \cdots + n_3(\bar{x}_3 - \bar{x})^2$$
$$= (5)(20 - 14)^2 + (3)(12 - 14)^2 + (4)(8 - 14)^2 = 336.$$

b. We calculate SSTr using the shortcut Formula 13-2:
$T_1 = 19 + 23 + 20 + 18 + 20 = 100$; $T_2 = 10 + 11 + 15 = 36$; $T_3 = 11 + 6 + 9 + 6 = 32$; $T = T_1 + T_2 + T_3 = 168$. Therefore,

$$SSTr = \left(\frac{T_1^2}{n_1} + \frac{T_2^2}{n_2} + \frac{T_3^2}{n_3}\right) - \frac{T^2}{n} = \left(\frac{100^2}{5} + \frac{36^2}{3} + \frac{32^2}{4}\right) - \frac{168^2}{12} = 2688 - 2352 = 336.$$

c. We calculate the pooled sample variance:

$$s_p^2 = \frac{(n_1 - 1)s_1^2 + \cdots + (n_3 - 1)s_3^2}{(n_1 - 1) + \cdots + (n_3 - 1)} = \frac{(5-1)(3.5) + (3-1)(7) + (4-1)(6)}{(5-1) + (3-1) + (4-1)} = \frac{46}{9}.$$

Therefore SSE = $(df_{error})s_p^2 = (n - k)(s_p^2) = (12 - 3)(46/9) = 46$.
d. In order to use the shortcut Formula 13-3 to calculate SSE, we first note that $\Sigma x^2 = 2734$ and that, from b., T = 168. Next, we calculate:

$$SST = \sum x^2 - \frac{T^2}{n} = 2734 - \frac{168^2}{12} = 2734 - 2352 = 382.$$

Therefore SSE = SST - SSTr = 382 - 336 = 46.
e. We construct the ANOVA table:

Source of Variation	df	SS	MS	F
Treatments	2	336	168.0	32.87
Error	9	46	5.111	
Total	11	382		

13.9 Totals for the samples are computed as follows: $T_1 = 8 + 9 + 6 + 11 + 8 = 42$; $T_2 = 13 + 16 + 16 + 19 = 64$; $T_3 = 11 + 12 + 19 + 18 = 60$; $T = T_1 + T_2 + T_3 = 166$.

Therefore, $$SSTr = \left(\frac{T_1^2}{n_1} + \frac{T_2^2}{n_2} + \frac{T_3^2}{n_3}\right) - \frac{T^2}{n} = \left(\frac{42^2}{5} + \frac{64^2}{4} + \frac{60^2}{4}\right) - \frac{166^2}{13} = 157.108.$$

$\Sigma x^2 = 8^2 + 9^2 + \ldots + 18^2 = 2358.$

$$SST = \sum x^2 - \frac{T^2}{n} = 2358 - \frac{166^2}{13} = 238.308.$$

SSE = SST - SSTr = 238.308 - 157.108 = 81.200.
MSTr = SSTr/df = 157.108/2 = 78.55.
MSE = SSE/df = 81.200/10 = 8.12.
F = MSTr/MSE = 9.67.
We construct the ANOVA table:

Source of Variation	df	SS	MS	F
Treatments	2	157.108	78.55	9.67
Error	10	81.200	8.12	
Total	12	238.308		

13.11 Totals for the samples are computed as follows: $T_1 = 33 + \ldots + 30 = 146$; $T_2 = 28 + \ldots + 20 = 91$; $T_3 = 39 + \ldots + 35 = 186$; $T_4 = 21 + \ldots + 21 = 108$; $T_5 = 19 + \ldots + 27 = 133$; $T = T_1 + T_2 + T_3 + T_4 + T_5 = 664$. Therefore,

$$SSTr = \left(\frac{T_1^2}{n_1} + \frac{T_2^2}{n_2} + \frac{T_3^2}{n_3} + \frac{T_4^2}{n_4} + \frac{T_5^2}{n_5}\right) - \frac{T^2}{n}$$
$$= \left(\frac{(146)^2}{5} + \frac{(91)^2}{4} + \frac{(186)^2}{5} + \frac{(108)^2}{5} + \frac{133^2}{5}\right) - \frac{(664)^2}{24} = 752.583.$$

$\Sigma x^2 = 33^2 + 30^2 + \ldots + 27^2 = 19484.$

$$SST = \sum x^2 - \frac{T^2}{n} = 19484 - \frac{(664)^2}{24} = 1113.333.$$

SSE = SST - SSTr = 1113.333 - 752.583 = 360.750.
MSTr = SSTr/df = 752.583/4 = 188.15.
MSE = SSE/df = 360.75/19 = 18.99.
F = MSTr/MSE = 188.15/18.99 = 9.91.
We construct the ANOVA table:

Source of Variation	df	SS	MS	F
Treatments	4	752.583	188.15	9.91
Error	19	360.750	18.99	
Total	23	1113.333		

13.13 **a.** For Treatments, we have df = (SSTr)/MSTr) = (21.7)/(3.1) = 7, which means that for Total, df = 28 + 7 = 35. We next calculate SSE = (df)(MSE) = (28)(1.7) = 47.6. Therefore, SST = SSTr + SSE = 21.7 + 47.6 = 69.3. Finally, F = (MSTr)/(MSE) = (3.1)/(1.7) = 1.82. The ANOVA table follows:

Source of Variation	df	SS	MS	F
Treatments	7	21.7	3.1	1.82
Error	28	47.6	1.7	
Total	35	69.3		

b. Since df = 7 for Treatments, there are 7 + 1 = 8 treatments.
c. Since df = 35 for Total, the total number of sample observations is 35 + 1 = 36.

EXERCISES for Sections 13.3 & 13.4
Applications of the One-Way Analysis of Variance
Estimation of Means

13.15 **a.** We construct dotplots for the samples as follows:

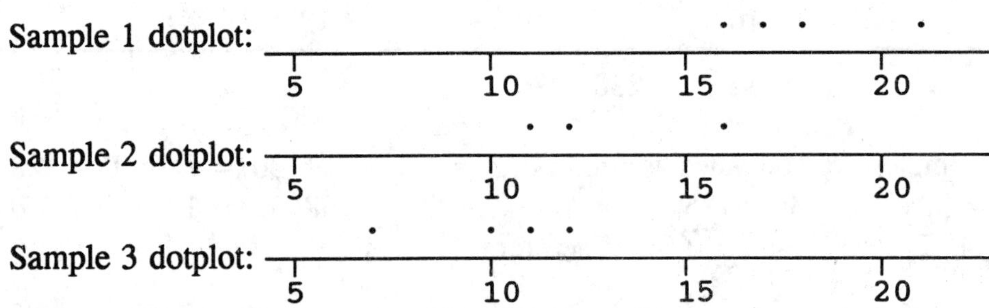

The sample means appear to differ significantly.
b. One-way Analysis of Variance
Step 1: Hypotheses
H_0: $\mu_1 = \mu_2 = \mu_3$
H_a: not all the means are equal
Step 2: Significance Level
We are given that $\alpha = 0.05$.
Step 3: Calculations
$T = T_1 + T_2 + T_3 = 90 + 39 + 40 = 169$,
$n = n_1 + n_2 + n_3 = 5 + 3 + 4 = 12$, $\Sigma x^2 = 17^2 + 21^2 + \ldots + 11^2 = 2569$.

$$SST = \sum x^2 - \frac{T^2}{n} = 2569 - \frac{(169)^2}{12} = 188.9167.$$

$$SSTr = \left(\frac{T_1^2}{n_1} + \frac{T_2^2}{n_2} + \frac{T_3^2}{n_3}\right) - \frac{T^2}{n}$$

$$= \left(\frac{(90)^2}{5} + \frac{(39)^2}{3} + \frac{(40)^2}{4}\right) - \frac{(169)^2}{12} = 146.9167$$

SSE = SST - SSTr = 188.9167 - 146.9167 = 42.
The ANOVA Table follows:

Source of Variation	df	SS	MS	F
Treatments	2	146.92	73.46	15.74
Error	9	42.00	4.667	
Total	11	188.92		

Step 4: Rejection Region
For ndf = k - 1 = 3 - 1 = 2, ddf = n - k = 12 - 3 = 9, and $\alpha = 0.05$, we find $F_\alpha = F_{0.05} = 4.26$. Therefore we reject H_0 if F > 4.26.
Step 5: Conclusion
The value of F in the ANOVA table is F = 15.74, which exceeds 4.26. Therefore, at level of significance $\alpha = 0.05$, we reject H_0 and conclude that the means are different.

13.17 From Exercise 13.15, we have $\bar{x}_1 = \frac{T_1}{n_1} = \frac{90}{5} = 18$, $\bar{x}_2 = \frac{T_2}{n_2} = \frac{39}{3} = 13$, and

$$s_p = \sqrt{MSE} = \sqrt{4.667} = 2.16.$$

a. A 95% confidence interval for μ_1 is $\bar{x}_1 \pm t_{\frac{\alpha}{2}} \frac{s_p}{\sqrt{n_1}} \rightarrow 18 \pm 2.262 \frac{2.16}{\sqrt{5}}$,

where df = n - k = 12 - 3 = 9 and $t_{\alpha/2} = t_{0.025}$.
This simplifies to 18 ± 2.19 or $15.81 < \mu_1 < 20.19$.
b. A 95% confidence interval for $\mu_1 - \mu_2$ is

$$(\bar{x}_1 - \bar{x}_2) \pm t_{\frac{\alpha}{2}} \sqrt{s_p^2\left(\frac{1}{n_1} + \frac{1}{n_2}\right)} \rightarrow (18 - 13) \pm 2.262\sqrt{4.667\left(\frac{1}{5} + \frac{1}{3}\right)}.$$

This simplifies to 5 ± 3.57 or $1.43 < \mu_1 - \mu_2 < 8.57$.

13.19 For ndf = 2 and ddf = 9, we find $F_{0.005} = 10.11$. Therefore, for the test statistic F = 15.74, we have p-value < 0.005.

13.21 **Step 1: Hypotheses**
H_0: $\mu_1 = \mu_2$
H_a: $\mu_1 \neq \mu_2$
Step 2: Significance Level
We are given that $\alpha = 0.05$.
Step 3: Calculations
$T = T_1 + T_2 = 459.5 + 350 = 809.5$,
$n = n_1 + n_2 = 12 + 10 = 22$, $\Sigma x^2 = 40^2 + 38^2 + \ldots + 36^2 = 29954.25$.

$$SST = \sum x^2 - \frac{T^2}{n} = 29954.25 - \frac{(809.5)^2}{22} = 168.33.$$

$$SSTr = \left(\frac{T_1^2}{n_1} + \frac{T_2^2}{n_2}\right) - \frac{T^2}{n} = \left(\frac{(459.5)^2}{12} + \frac{(350)^2}{10}\right) - \frac{(809.5)^2}{22} = 59.10.$$

SSE = SST - SSTr = 168.33 - 59.10 = 109.23.
The ANOVA Table follows:

Source of Variation	df	SS	MS	F
Treatments	1	59.10	59.10	10.82
Error	20	109.23	5.462	
Total	21	168.33		

Step 4: Rejection Region
For ndf = k - 1 = 2 - 1 = 1, ddf = n - k = 22 - 2 = 20, and α = 0.05, we find F_α = $F_{0.05}$ = 4.35. Therefore we reject H_0 if F > 4.35.
Step 5: Conclusion
The value of F in the ANOVA table is F = 10.82, which exceeds 4.35. Therefore, at level of significance α = 0.05, we reject H_0 and conclude a difference exists in the mean hourly rates for the two states..

13.23 From Exercise 13.21, we have $\bar{x}_1 = \frac{T_1}{n_1} = \frac{459.50}{12} = 38.29,\quad \bar{x}_2 = \frac{T_2}{n_2} = \frac{350}{10} = 35,$

s_p^2 = MSE = 5.462, and df = 22 - 2 = 20 so $t_{0.005}$ = 2.845.
A 99% confidence interval for $\mu_1 - \mu_2$ is

$$(\bar{x}_1 - \bar{x}_2) \pm t_{\frac{\alpha}{2}}\sqrt{s_p^2\left(\frac{1}{n_1} + \frac{1}{n_2}\right)} \rightarrow (38.29 - 35) \pm 2.845\sqrt{5.462\left(\frac{1}{12} + \frac{1}{10}\right)}.$$

This simplifies to 3.29 ± 2.85 or \$0.44 < $\mu_1 - \mu_2$ < \$6.14.

13.25 For ndf = 2 and ddf = 9, we find $F_{0.005}$ = 10.11. Therefore, for the test statistic F = 44.67, we have p-value < 0.005.

13.27 From Exercise 13.24, we have $\bar{x}_1 = \frac{T_1}{n_1} = \frac{6.4}{4} = 1.6$, and $\bar{x}_3 = \frac{T_3}{n_3} = \frac{16.3}{4} = 4.075,$

and s_p^2 = MSE = 0.1375. Since df = n - k = 12 - 3 = 9, we have $t_{\alpha/2} = t_{0.005}$ = 3.250.
A 99% confidence interval for $\mu_3 - \mu_1$ is

$$(\bar{x}_3 - \bar{x}_1) \pm t_{\frac{\alpha}{2}}\sqrt{s_p^2\left(\frac{1}{n_3} + \frac{1}{n_1}\right)} \rightarrow (4.075 - 1.6) \pm 3.250\sqrt{0.1375\left(\frac{1}{4} + \frac{1}{4}\right)},$$ which

simplifies to 2.475 ± 0.852 or 1.623 < $\mu_3 - \mu_1$ < 3.327.

13.29 From Exercise 13.28, we have $\bar{x}_3 = \dfrac{T_3}{n_3} = \dfrac{25.2}{6} = 4.2$, and

$$s_p = \sqrt{MSE} = \sqrt{0.571} = 0.756.$$

Since df = n - k = 24 - 4 = 20, we have $t_{0.025} = 2.086$. A 95% confidence interval for μ_3

is $\bar{x}_3 \pm t_{\frac{\alpha}{2}} \dfrac{s_p}{\sqrt{n_3}} \rightarrow 4.2 \pm 2.086 \dfrac{0.756}{\sqrt{6}}$, which simplifies to

4.2 ± 0.644 or $3.556 < \mu_3 < 4.844$.

13.31 One-way Analysis of Variance

Step 1: Hypotheses

H_0: $\mu_1 = \mu_2 = \mu_3$

H_a: not all the means are equal

Step 2: Significance Level

We are given that $\alpha = 0.05$.

Step 3: Calculations

$T = T_1 + T_2 + T_3 = 78 + 88 + 80 = 246$,

$n = n_1 + n_2 + n_3 = 8 + 8 + 7 = 23$, $\Sigma x^2 = 9^2 + 8^2 + \ldots + 13^2 = 2692$.

$$SST = \sum x^2 - \frac{T^2}{n} = 2692 - \frac{(246)^2}{23} = 60.87$$

$$SSTr = \left(\frac{T_1^2}{n_1} + \frac{T_2^2}{n_2} + \frac{T_3^2}{n_3}\right) - \frac{T^2}{n}$$

$$= \left(\frac{(78)^2}{8} + \frac{(88)^2}{8} + \frac{(80)^2}{7}\right) - \frac{(246)^2}{23} = 11.66$$

SSE = SST - SSTr = 60.87 - 11.66 = 49.21.

The ANOVA Table follows:

Source of Variation	df	SS	MS	F
Treatments	2	11.66	5.83	2.37
Error	20	49.21	2.46	
Total	22	60.87		

Step 4: Rejection Region

For ndf = k - 1 = 3 - 1 = 2, ddf = n - k = 23 - 3 = 20, and $\alpha = 0.05$, we find $F_\alpha = F_{0.05} = 3.49$. Therefore we reject H_0 if $F > 3.49$.

Step 5: Conclusion

The value of F in the ANOVA table is F = 2.37, which is less than 3.49. Therefore, at level of significance $\alpha = 0.05$, we fail to reject H_0 and do not find sufficient evidence to indicate a difference among mean ascorbic acid levels for Cortland apples grown in the three regions.

13.33 From Exercise 13.31, we have $\bar{x}_1 = \frac{T_1}{n_1} = \frac{78}{8} = 9.75$, and $\bar{x}_3 = \frac{T_3}{n_3} = \frac{80}{7} = 11.43$,

and s_p^2 = MSE = 2.46. Since df = n - k = 23 - 3 = 20, we have $t_{0.05}$ = 1.725. A 90% confidence interval for $\mu_1 - \mu_3$ is

$$(\bar{x}_1 - \bar{x}_3) \pm t_{\frac{\alpha}{2}}\sqrt{s_p^2\left(\frac{1}{n_1} + \frac{1}{n_3}\right)} \rightarrow (9.75 - 11.43) \pm 1.725\sqrt{2.46\left(\frac{1}{8} + \frac{1}{7}\right)},$$

which simplifies to -1.68 ± 1.40 or $-3.08 < \mu_1 - \mu_3 < -0.28$.

13.35 For ndf = 3 and ddf = 19, we find $F_{0.005}$ = 5.92. Therefore, for the test statistic F = 12.68, we have p-value < 0.005.

13.37 We disregard Lake 2 and test:

Step 1: Hypotheses

H_0: $\mu_1 = \mu_3 = \mu_4$

H_a: not all the means are equal

Step 2: Significance Level

We are given that α = 0.05.

Step 3: Calculations

$T = T_1 + T_3 + T_4 = 10 + 10.7 + 16.2 = 36.9$,

$n = n_1 + n_3 + n_4 = 6 + 5 + 6 = 17$, $\Sigma x^2 = 1.7^2 + 1.4^2 + \ldots + 3.5^2 = 89.67$.

$$SST = \sum x^2 - \frac{T^2}{n} = 89.67 - \frac{(36.9)^2}{17} = 9.575$$

$$SSTr = \left(\frac{T_1^2}{n_1} + \frac{T_3^2}{n_3} + \frac{T_4^2}{n_4}\right) - \frac{T^2}{n} = \left(\frac{(10)^2}{6} + \frac{(10.7)^2}{5} + \frac{(16.2)^2}{6}\right) - \frac{(36.9)^2}{17} = 3.210$$

SSE = SST - SSTr = 9.575 - 3.210 = 6.365.

The ANOVA Table follows:

Source of Variation	df	SS	MS	F
Treatments	2	3.210	1.605	3.53
Error	14	6.365	0.455	
Total	16	9.575		

Step 4: Rejection Region

For ndf = k - 1 = 3 - 1 = 2, ddf = n - k = 17 - 3 = 14, and α = 0.05, we find F_α = $F_{0.05}$ = 3.74. Therefore we reject H_0 if F > 3.74.

Step 5: Conclusion

The value of F in the ANOVA table is F = 3.53, which is less than 3.74. Therefore, at level of significance α = 0.05, we fail to reject H_0 and do not conclude that there is a difference in the mean DDT levels for the three lakes.

13.39 From the MINITAB display of Exercise 13.38, we have $\bar{x}_1 = 23958$, $\bar{x}_2 = 25512$, $s_p^2 =$ MSE $= 874482$ with df $= 27$, so $t_{0.025} = 2.052$. A 95% confidence interval for $\mu_1 - \mu_2$ is

$$(\bar{x}_1 - \bar{x}_2) \pm t_{\frac{\alpha}{2}}\sqrt{s_p^2\left(\frac{1}{n_1} + \frac{1}{n_2}\right)} \rightarrow (23958 - 25512) \pm 2.052\sqrt{874482\left(\frac{1}{10} + \frac{1}{10}\right)},$$

This simplifies to -1554 ± 858 or $-2412 < \mu_1 - \mu_2 < -696$.

MINITAB LAB ASSIGNMENTS

13.41(M) From the MINITAB display of Exercise 13.40(M), we have $\bar{x}_2 = 87.6$, $\bar{x}_4 = 79.6$ and s_p^2 = MSE = 12.9 with df = n - k = 36, so we have $t_{0.025} = 1.96$. A 95% confidence interval for $\mu_2 - \mu_4$ is

$$(\bar{x}_2 - \bar{x}_4) \pm t_{\frac{\alpha}{2}}\sqrt{s_p^2\left(\frac{1}{n_2} + \frac{1}{n_4}\right)} \rightarrow (87.6 - 79.6) \pm 1.96\sqrt{12.9\left(\frac{1}{10} + \frac{1}{10}\right)}.$$

This simplifies to 8.00 ± 3.15 or $4.85 < \mu_2 - \mu_4 < 11.15$.

13.43(M)

```
MTB > # Exercise 13.43(M)
MTB > SET C1
DATA> 1.7 1.3 0.8 1.5 0.9 1.1
DATA> SET C2
DATA> 5.3 4.2 3.9 5.7 3.8 5.2
DATA> SET C3
DATA> 5.1 4.1 5.2 3.3 2.6 4.9
DATA> SET C4
DATA> 1.2 2.9 1.5 2.3 2.1 1.8
DATA> END
MTB > AOVO C1 - C4

ANALYSIS OF VARIANCE
SOURCE      DF        SS        MS         F        p
FACTOR       3    51.123    17.041     29.82    0.000
ERROR       20    11.430     0.571
TOTAL       23    62.553
                                  INDIVIDUAL 95 PCT CI'S FOR MEAN
                                  BASED ON POOLED STDEV
 LEVEL       N      MEAN     STDEV  -------+---------+---------+---------
C1           6    1.2167    0.3488   (---*---)
C2           6    4.6833    0.8134                         (---*----)
C3           6    4.2000    1.0658                      (---*---)
C4           6    1.9667    0.6055        (---*---)
                                    -------+---------+---------+---------
POOLED STDEV =    0.7560                  1.5       3.0       4.5
```

13.45(M)

```
MTB > # Exercise 13.45(M)
MTB > SET C1
DATA> 1.7 1.4 1.9 1.1 2.1 1.8
DATA> SET C2
DATA> 0.3 0.7 0.5 0.1 1.1 0.9
DATA> SET C3
DATA> 2.7 1.9 2.0 1.5 2.6
DATA> SET C4
DATA> 1.2 3.1 1.9 3.7 2.8 3.5
DATA> END
MTB > AOVO C1 - C4

ANALYSIS OF VARIANCE
SOURCE      DF        SS        MS        F        p
FACTOR       3    14.149     4.716    12.68    0.000
ERROR       19     7.065     0.372
TOTAL       22    21.215
                                   INDIVIDUAL 95 PCT CI'S FOR MEAN
                                   BASED ON POOLED STDEV
LEVEL        N      MEAN     STDEV  ---------+---------+---------+------
C1           6    1.6667    0.3615            (-----*----)
C2           6    0.6000    0.3742  (----*----)
C3           5    2.1400    0.5030                 (----*-----)
C4           6    2.7000    0.9695                       (----*----)
                                    ---------+---------+---------+------
POOLED STDEV =    0.6098                   1.0       2.0       3.0
```

REVIEW EXERCISES
CHAPTER 13

13.46 One-way Analysis of Variance

Step 1: Hypotheses

H_0: $\mu_1 = \mu_2 = \mu_3$

H_a: not all the means are equal

Step 2: Significance Level

We are given that $\alpha = 0.01$.

Step 3: Calculations

$T = T_1 + T_2 + T_3 = 6541 + 6762 + 6812 = 20115$,

$n = n_1 + n_2 + n_3 = 7 + 7 + 7 = 21$,

$\Sigma x^2 = 899^2 + 929^2 + \ldots + 989^2 = 19287525$.

$$SST = \sum x^2 - \frac{T^2}{n} = 19287525 - \frac{(20115)^2}{21} = 20228.57$$

$$SSTr = \left(\frac{T_1^2}{n_1} + \frac{T_2^2}{n_2} + \frac{T_3^2}{n_3}\right) - \frac{T^2}{n} = \left(\frac{(6541)^2}{7} + \frac{(6762)^2}{7} + \frac{(6812)^2}{7}\right) - \frac{(20115)^2}{21} = 5942$$

SSE = SST - SSTr = 20228.57 - 5942 = 14286.57.

The ANOVA Table follows:

Source of Variation	df	SS	MS	F
Treatments	2	5942	2971	3.74
Error	18	14286.57	793.7	
Total	20	20228.57		

Step 4: Rejection Region

For ndf = k - 1 = 3 - 1 = 2, ddf = n - k = 21 - 3 = 18, and α = 0.01, we find F_α = $F_{0.01}$ = 6.01. Therefore we reject H_0 if F > 6.01.

Step 5: Conclusion

The value of F in the ANOVA table is F = 3.74, which is less than 6.01. Therefore, at level of significance α = 0.01, we fail to reject H_0 and do not conclude that there is a difference in the mean selling price for the three stores.

13.47 For ndf = 2 and ddf = 18, we find $F_{0.05}$ = 3.55 and $F_{0.025}$ = 4.56. Therefore, for the test statistic F = 3.74, we have 0.025 < p-value < 0.05.

13.48 From Exercise 13.46, we have $\bar{x}_1 = \frac{T_1}{n_1} = \frac{6541}{7} = 934.43$, and

$s_p = \sqrt{MSE} = \sqrt{793.7} = 28.173.$ Since df = n - k = 21 - 3 = 18, we have $t_{0.05}$ = 1.734.

A 90% confidence interval for μ_1 is $\bar{x}_1 \pm t_{\frac{\alpha}{2}} \frac{s_p}{\sqrt{n_1}} \rightarrow 934.43 \pm 1.734 \frac{28.173}{\sqrt{7}}$,

which simplifies to 934.43 ± 18.46 or 915.97 < μ_1 < 952.89.

13.49 From Exercise 13.46, we have $\bar{x}_1 = \frac{T_1}{n_1} = \frac{6541}{7} = 934.43$, and

$\bar{x}_3 = \frac{T_3}{n_3} = \frac{6812}{7} = 973.14$, and s_p^2 = MSE = 793.7. Since df = n - k = 21 - 3 = 18,

we have $t_{0.05}$ = 1.734. A 90% confidence interval for $\mu_1 - \mu_3$ is

$$(\bar{x}_1 - \bar{x}_3) \pm t_{\frac{\alpha}{2}} \sqrt{s_p^2\left(\frac{1}{n_1} + \frac{1}{n_3}\right)} \rightarrow (934.43 - 973.14) \pm 1.734 \sqrt{793.7\left(\frac{1}{7} + \frac{1}{7}\right)},$$

which simplifies to -38.71 ± 26.11 or -64.82 < $\mu_1 - \mu_3$ < -12.60.

13.50 We are given for sample 1: $n_1 = 7$ and $s_1^2 = 20$, for sample 2: $n_2 = 10$ and $s_2^2 = 16$, for sample 3: $n_3 = 9$ and $s_3^2 = 22$. We calculate

$$s_p^2 = \frac{(n_1 - 1)s_1^2 + \cdots + (n_3 - 1)s_3^2}{(n_1 - 1) + \cdots + (n_3 - 1)} = \frac{(7-1)(20) + (10-1)(16) + (9-1)(22)}{(7-1) + (10-1) + (9-1)}$$

$$= \frac{440}{23} = 19.13.$$

13.51 If $n_1 = n_2 = \cdots\cdots = n_k = n$, then

$$MSE = s_p^2 = \frac{(n_1 - 1)s_1^2 + (n_2 - 1)s_2^2 + \cdots + (n_k - 1)s_k^2}{(n_1 - 1) + (n_2 - 1) + \cdots + (n_k - 1)}$$

$$= \frac{(n - 1)s_1^2 + (n - 1)s_2^2 + \cdots + (n - 1)s_k^2}{(n - 1) + (n - 1) + \cdots + (n - 1)}$$

$$= \frac{(n - 1)(s_1^2 + s_2^2 + \cdots + s_k^2)}{k(n - 1)}$$

$$= \frac{s_1^2 + s_2^2 + \cdots + s_k^2}{k}$$

13.52 **a.** Using the given statistics, we calculate:

$$\bar{x} = \frac{n_1\bar{x}_1 + \cdots + n_4\bar{x}_4}{n_1 + \cdots + n_4} = \frac{9 \cdot 42 + 10 \cdot 39 + 5 \cdot 54 + 6 \cdot 50}{9 + 10 + 5 + 6} = \frac{1338}{30} = 44.6.$$

We use the grand mean and the three sample means to calculate the treatment sum of squares:

$$SSTr = n_1(\bar{x}_1 - \bar{x})^2 + \cdots + n_4(\bar{x}_4 - \bar{x})^2$$

$$= (9)(42 - 44.6)^2 + (10)(39 - 44.6)^2 + (5)(54 - 44.6)^2 + (6)(50 - 44.6)^2 = 991.2.$$

b. The pooled sample variance is:

$$s_p^2 = \frac{(n_1 - 1)s_1^2 + \cdots + (n_4 - 1)s_4^2}{(n_1 - 1) + \cdots + (n_4 - 1)}$$

$$= \frac{(9-1)(18) + (10-1)(25) + (5-1)(22) + (6-1)(20)}{(9-1) + (10-1) + (5-1) + (6-1)} = \frac{557}{26} = 21.423.$$

c. Using the results from parts (a) and (b), we construct the ANOVA table:

Total df = n - 1 = 30 - 1 = 29
Treatment df = k - 1 = 4 - 1 = 3
Error df = n - k = 30 - 4 = 26

$$MSTr = \frac{SSTr}{k-1} = \frac{991.2}{3} = 330.4$$

MSE = s_p^2 = 21.423
SSE = (n - k)s_p^2 = (26)(21.423) = 557

$$F = \frac{MSTr}{MSE} = \frac{330.4}{21.423} = 15.42.$$

The ANOVA table follows:

Source of Variation	df	SS	MS	F
Treatments	3	991.2	330.4	15.42
Error	26	557	21.423	
Total	29	1548.2		

d. Step 1: Hypotheses
H_0: $\mu_1 = \mu_2 = \mu_3 = \mu_4$
H_a: not all the means are equal
Step 2: Significance Level
We are given that $\alpha = 0.05$.
Step 3: Calculations
See above.
Step 4: Rejection Region
For ndf = k - 1 = 4 - 1 = 3, ddf = n - k = 30 - 4 = 26, and $\alpha = 0.05$, we find $F_\alpha = F_{0.05} = 2.98$. Therefore we reject H_0 if F > 2.98.
Step 5: Conclusion
The value of F in the ANOVA table is F = 15.42, which is larger than 2.98. Therefore, at level of significance $\alpha = 0.05$, we reject H_0 and conclude the means are different.

13.53 From Exercise 13.52, we have $\bar{x}_2 = 39$. Since df = n - k = 30 - 4 = 26, we have $t_{0.05} = 1.706$. $s_p = \sqrt{MSE} = \sqrt{21.423} = 4.6285$. A 90% confidence interval for μ_2 is

$$\bar{x}_2 \pm t_{\frac{\alpha}{2}} \frac{s_p}{\sqrt{n_2}} \rightarrow 39 \pm 1.706 \frac{4.6285}{\sqrt{10}},$$

which simplifies to 39 ± 2.497 or $36.503 < \mu_2 < 41.497$.

13.54 From Exercise 13.52, we have $\bar{x}_3 = 54$, $\bar{x}_1 = 42$ and $s_p^2 = MSE = 21.423$. Since df = n - k = 30 - 4 = 26, we have $t_{0.05} = 1.706$. A 90% confidence interval for $\mu_3 - \mu_1$ is

$$(\bar{x}_3 - \bar{x}_1) \pm t_{\frac{\alpha}{2}} \sqrt{s_p^2\left(\frac{1}{n_3} + \frac{1}{n_1}\right)} \rightarrow (54 - 42) \pm 1.706 \sqrt{21.423\left(\frac{1}{5} + \frac{1}{9}\right)},$$

which simplifies to 12 ± 4.404 or $7.596 < \mu_3 - \mu_1 < 16.404$.

13.55 **a.** We construct dot plots for the samples as follows:

```
                          .  .  .  .  .
Sample 1 dotplot:  -+---------+---------+---------+-------
                    4         8        12        16

                                .
                                .  .       .
Sample 2 dotplot:  -+---------+---------+---------+-------
                    4         8        12        16

                                .  .  .  .
Sample 3 dotplot:  -+---------+---------+---------+-------
                    4         8        12        16

                                         .
                                   .  .  .     .
Sample 4 dotplot:  -+---------+---------+---------+-------
                    4         8        12        16
```

The dotplots suggest that the sample means differ significantly.

b. **Step 1: Hypotheses**

H_0: $\mu_1 = \mu_2 = \mu_3 = \mu_4$

H_a: not all the means are equal

Step 2: Significance Level

We are given that $\alpha = 0.05$.

Step 3: Calculations

$T = T_1 + T_2 + T_3 + T_4 = 35 + 41 + 42 + 62 = 180$,

$n = n_1 + n_2 + n_3 + n_4 = 5 + 4 + 4 + 5 = 18$,

$\Sigma x^2 = 6^2 + 7^2 + \ldots + 15^2 = 1916$.

$$SST = \sum x^2 - \frac{T^2}{n} = 1916 - \frac{(180)^2}{18} = 116.$$

$$SSTr = \left(\frac{T_1^2}{n_1} + \frac{T_2^2}{n_2} + \frac{T_3^2}{n_3} + \frac{T_4^2}{n_4}\right) - \frac{T^2}{n}$$

$$= \left(\frac{(35)^2}{5} + \frac{(41)^2}{4} + \frac{(42)^2}{4} + \frac{(62)^2}{5}\right) - \frac{(180)^2}{18} = 75.05.$$

SSE = SST - SSTr = 116 - 75.05 = 40.95.

The ANOVA Table follows:

Source of Variation	df	SS	MS	F
Treatments	3	75.05	25.017	8.55
Error	14	40.95	2.925	
Total	17	116		

Step 4: Rejection Region

For ndf = k - 1 = 4 - 1 = 3, ddf = n - k = 18 - 4 = 14, and $\alpha = 0.05$, we find $F_\alpha = F_{0.05} = 3.34$. Therefore we reject H_0 if $F > 3.34$.

Step 5: Conclusion
The value of F in the ANOVA table is F = 8.55, which is greater than 3.34. Therefore, at level of significance α = 0.05, we reject H_0 and conclude that the means of the four populations differ.

13.56 Using information given in the partial ANOVA table, we compute:
ndf = (SSTr) / (MSTr) = 493/70.43 =7,
Total df = ndf + ddf = 7 + 22 = 29,
SSE = SST - SSTr = 721 - 493 = 228,
MSE = SSE/ddf = 228/22 = 10.364,
F = MSTr/MSE = 70.43/10.364 = 6.80.
We may now complete the ANOVA table:

Source of Variation	df	SS	MS	F
Treatments	7	493	70.43	6.80
Error	22	228	10.364	
Total	29	721		

For ndf = 7, ddf = 22, and α = 0.01, we find $F_\alpha = F_{0.01}$ = 3.59. Therefore we reject H_0 if F > 3.59. The value of F in the ANOVA table is F = 6.80, which is greater than 3.59. Therefore, at level of significance α = 0.01, we reject H_0 and conclude that the means of the eight populations differ.

13.57 One-way Analysis of Variance
Step 1: Hypotheses
H_0: $\mu_1 = \mu_2 = \mu_3$
H_a: not all the means are equal
Step 2: Significance Level
We are given that α = 0.05.
Step 3: Calculations
$T = T_1 + T_2 + T_3 = 489 + 527 + 551 = 1567$,
$n = n_1 + n_2 + n_3 = 8 + 8 + 8 = 24$,
$\Sigma x^2 = 63^2 + 64^2 + \ldots + 68^2 = 102699$.

$$SST = \sum x^2 - \frac{T^2}{n} = 102699 - \frac{(1567)^2}{24} = 386.958$$

$$SSTr = \left(\frac{T_1^2}{n_1} + \frac{T_2^2}{n_2} + \frac{T_3^2}{n_3}\right) - \frac{T^2}{n}$$

$$= \left(\frac{(489)^2}{8} + \frac{(527)^2}{8} + \frac{(551)^2}{8}\right) - \frac{(1567)^2}{24} = 244.333$$

SSE = SST - SSTr = 386.958 - 244.333 = 142.625.

The ANOVA Table follows:

Source of Variation	df	SS	MS	F
Treatments	2	244.333	122.17	17.99
Error	21	142.625	6.792	
Total	23	386.958		

Step 4: Rejection Region
For ndf = k - 1 = 3 - 1 = 2, ddf = n - k = 24 - 3 = 21, and $\alpha = 0.05$, we find $F_\alpha = F_{0.05} = 3.47$. Therefore we reject H_0 if $F > 3.47$.
Step 5: Conclusion
The value of F in the ANOVA table is F = 17.99, and is greater than 3.47. Therefore, at level of significance $\alpha = 0.05$, we reject H_0 and conclude that there is a difference in the mean selling price for a quart of milk in the three geographical regions.

13.58 For ndf = 2 and ddf = 21, we find $F_{0.005} = 6.89$. Therefore, for the test statistic F = 17.99, we have p-value < 0.005.

13.59 From Exercise 13.57, we have $\bar{x}_2 = \frac{T_2}{n_2} = \frac{527}{8} = 65.875$, and $s_p = \sqrt{MSE} = \sqrt{6.792}$.

Since df = n - k = 24 - 3 = 21, we have $t_{0.025} = 2.080$. A 95% confidence interval for μ_2

is $\bar{x}_2 \pm t_{\frac{\alpha}{2}} \frac{s_p}{\sqrt{n_2}} \rightarrow 65.875 \pm 2.080 \frac{\sqrt{6.792}}{\sqrt{8}}$,

which simplifies to 65.875 ± 1.917 or $63.958 < \mu_2 < 67.792$.

13.60 From Exercise 13.46, we have $\bar{x}_1 = \frac{T_1}{n_1} = \frac{489}{8} = 61.125$, and

$\bar{x}_3 = \frac{T_3}{n_3} = \frac{551}{8} = 68.875$, and $s_p^2 = MSE = 6.792$. Since df = n - k = 24 - 3 = 21,

we have $t_{0.025} = 2.080$. A 95% confidence interval for $\mu_3 - \mu_1$ is

$$(\bar{x}_3 - \bar{x}_1) \pm t_{\frac{\alpha}{2}} \sqrt{s_p^2\left(\frac{1}{n_3} + \frac{1}{n_1}\right)} \rightarrow (68.875 - 61.125) \pm 2.080 \sqrt{6.792\left(\frac{1}{8} + \frac{1}{8}\right)},$$

which simplifies to 7.75 ± 2.710 or $5.04 < \mu_3 - \mu_1 < 10.46$.

13.61 Step 1: Hypotheses

H_0: $\mu_1 = \mu_2 = \mu_3 = \mu_4$

H_a: not all the means are equal

Step 2: Significance Level

We are given that $\alpha = 0.05$.

Step 3: Calculations

$T = T_1 + T_2 + T_3 + T_4 = 108 + 152 + 140 + 120 = 520$,

$n = n_1 + n_2 + n_3 + n_4 = 5 + 6 + 5 + 5 = 21$,

$\Sigma x^2 = 24^2 + 22^2 + \ldots + 23^2 = 13192$.

$$SST = \sum x^2 - \frac{T^2}{n} = 13192 - \frac{(520)^2}{21} = 315.81.$$

$$SSTr = \left(\frac{T_1^2}{n_1} + \frac{T_2^2}{n_2} + \frac{T_3^2}{n_3} + \frac{T_4^2}{n_4}\right) - \frac{T^2}{n}$$

$$= \left(\frac{(108)^2}{5} + \frac{(152)^2}{6} + \frac{(140)^2}{5} + \frac{(120)^2}{5}\right) - \frac{(520)^2}{21} = 107.28.$$

SSE = SST - SSTr = 315.81 - 107.28 = 208.53.

The ANOVA Table follows:

Source of Variation	df	SS	MS	F
Treatments	3	107.28	35.76	2.92
Error	17	208.53	12.266	
Total	20	315.81		

Step 4: Rejection Region

For ndf = k - 1 = 4 - 1 = 3, ddf = n - k = 21 - 4 = 17, and $\alpha = 0.05$, we find $F_\alpha = F_{0.05} = 3.20$. Therefore we reject H_0 if $F > 3.20$.

Step 5: Conclusion

The value of F in the ANOVA table is F = 2.92, which is less than 3.20. Therefore, at level of significance $\alpha = 0.05$, we fail to reject H_0 and can not conclude that there is a difference in the mean ratings for the four word processing programs.

13.62 For ndf = 3 and ddf = 17, we find $F_{0.05} = 3.20$ and $F_{0.10} = 2.44$. Therefore, for the test statistic F = 2.92, we have $0.05 < \text{p-value} < 0.10$.

13.63 From Exercise 13.61, we have $\bar{x}_3 = \frac{T_3}{n_3} = \frac{140}{5} = 28$, and $\bar{x}_1 = \frac{T_1}{n_1} = \frac{108}{5} = 21.6$,

and $s_p^2 = MSE = 12.266$. Since df = n - k = 21 - 4 = 17, we have $t_{0.05} = 1.740$. A 90% confidence interval for $\mu_3 - \mu_1$ is:

$$(\bar{x}_3 - \bar{x}_1) \pm t_{\frac{\alpha}{2}}\sqrt{s_p^2\left(\frac{1}{n_3} + \frac{1}{n_1}\right)} \rightarrow (28 - 21.6) \pm 1.740\sqrt{12.266\left(\frac{1}{5} + \frac{1}{5}\right)},$$

which simplifies to 6.4 ± 3.85 or $2.55 < \mu_3 - \mu_1 < 10.25$.

13.64 One-way Analysis of Variance

Step 1: Hypotheses

H_0: $\mu_1 = \mu_2 = \mu_3$

H_a: not all the means are equal

Step 2: Significance Level

We are given that $\alpha = 0.05$.

Step 3: Calculations

$T = T_1 + T_2 + T_3 = 36.1 + 32.2 + 40.4 = 108.7$,

$n = n_1 + n_2 + n_3 = 5 + 5 + 5 = 15$,

$\Sigma x^2 = 7.3^2 + 6.9^2 + \ldots + 8.3^2 = 799.55$.

$$SST = \sum x^2 - \frac{T^2}{n} = 799.55 - \frac{(108.7)^2}{15} = 11.8373$$

$$SSTr = \left(\frac{T_1^2}{n_1} + \frac{T_2^2}{n_2} + \frac{T_3^2}{n_3}\right) - \frac{T^2}{n}$$

$$= \left(\frac{(36.1)^2}{5} + \frac{(32.2)^2}{5} + \frac{(40.4)^2}{5}\right) - \frac{(108.7)^2}{15} = 6.7293$$

SSE = SST - SSTr = 11.8373 - 6.7293 = 5.1080.

The ANOVA Table follows:

Source of Variation	df	SS	MS	F
Treatments	2	6.7293	3.365	7.90
Error	12	5.1080	0.426	
Total	14	11.8373		

Step 4: Rejection Region

For ndf = k - 1 = 3 - 1 = 2, ddf = n - k = 15 - 3 = 12, and $\alpha = 0.05$, we find $F_\alpha = F_{0.05} = 3.89$. Therefore we reject H_0 if $F > 3.89$.

Step 5: Conclusion

The value of F in the ANOVA table is F = 7.90, and is greater than 3.89. Therefore, at level of significance $\alpha = 0.05$, we reject H_0 and conclude that there is a difference in the mean length of life among the three brands of batteries.

MINITAB LAB ASSIGNMENTS

13.65(M)

```
MTB > # Exercise 13.65(M)
MTB > SET C1
DATA> 40 38 38 37 36 39 41.5 38 39.5 37.5 35 40
DATA> SET C2
DATA> 35 37 31 39 31.5 35 32.5 34 39 36
DATA> SET C3
DATA> 40 42 40 39.5 37.5 45 43 42.5 39.5 41 44
DATA> END
MTB > AOVO C1 - C3

ANALYSIS OF VARIANCE
SOURCE      DF        SS         MS         F        p
FACTOR       2    206.10     103.05     19.39    0.000
ERROR       30    159.41       5.31
TOTAL       32    365.52
                                     INDIVIDUAL 95 PCT CI'S FOR MEAN
                                     BASED ON POOLED STDEV
 LEVEL       N      MEAN      STDEV  ------+---------+---------+---------+
C1          12    38.292      1.827               (----*-----)
C2          10    35.000      2.838  (-----*-----)
C3          11    41.273      2.240                          (-----*-----)

                                     ------+---------+---------+---------+
POOLED STDEV =    2.305                  35.0      37.5      40.0      42.5
MTB > # a. Since p-value = 0.000 < 0.01, we reject Ho.  We conclude that
MTB > # a difference exists in the mean hourly rates for the three
MTB > # states.
MTB > # b. From the visual display, it appears that the mean for C2
MTB > # differs from the means for C1 and C3.  The mean for C1 may
MTB > # differ from the mean for C3.
```

13.66(M) From the MINITAB output we find $\bar{x}_1 = 38.292$, $\bar{x}_3 = 41.273$ and $s_p^2 = MSE = 5.31$. Since ddf = 30, we have $t_{0.025} = 1.96$. A 95% confidence interval for $\mu_1 - \mu_3$ is

$$(\bar{x}_1 - \bar{x}_3) \pm t_{\frac{\alpha}{2}}\sqrt{s_p^2\left(\frac{1}{n_1} + \frac{1}{n_3}\right)} \Rightarrow (38.292 - 41.273) \pm 1.96\sqrt{5.31\left(\frac{1}{12} + \frac{1}{11}\right)},$$

which simplifies to -2.98 ± 1.89 or $-4.87 < \mu_1 - \mu_3 < -1.09$.

13.67(M)

```
MTB > # Exercise 13.67(M)
MTB > # The data were entered into columns C1, C2, and C3.
MTB > AOVO C1-C3

ANALYSIS OF VARIANCE
SOURCE     DF        SS         MS         F        p
FACTOR      2    2556.8     1278.4     29.90    0.000
ERROR      45    1924.2       42.8
TOTAL      47    4481.0
                                      INDIVIDUAL 95 PCT CI'S FOR MEAN
                                      BASED ON POOLED STDEV
 LEVEL      N      MEAN      STDEV   -----+---------+---------+---------+-
C1         16    79.187      7.626                (----*----)
C2         16    70.000      6.673   (----*----)
C3         16    87.875      5.058                           (----*---)
                                     -----+---------+---------+---------+-
POOLED STDEV =     6.539                 70.0      77.0      84.0      91.0
MTB > # Since p-value = 0.000 < 0.05, we reject Ho. This indicates a
MTB > # difference in the mean quality ratings for the three
MTB > # sources of pulp.
```

13.68(M)

```
MTB > # Exercise 13.68(M)
MTB > # The data are in C1 and C3 from Exercise 13.67(M)
MTB > AOVO C1 C3

ANALYSIS OF VARIANCE
SOURCE     DF        SS         MS         F        p
FACTOR      1     603.8      603.8     14.42    0.001
ERROR      30    1256.2       41.9
TOTAL      31    1860.0
                                      INDIVIDUAL 95 PCT CI'S FOR MEAN
                                      BASED ON POOLED STDEV
 LEVEL      N      MEAN      STDEV   ---------+---------+---------+-------
C1         16    79.187      7.626    (-----*------)
C3         16    87.875      5.058                   (------*-----)
                                     ---------+---------+---------+-------
POOLED STDEV =     6.471                     80.0      85.0      90.0
MTB > # Since p-value = 0.001 < 0.05, we reject Ho. This indicates a
MTB > # difference in the mean quality ratings for sources A and C.
```

CHAPTER 14
LINEAR REGRESSION ANALYSIS

EXERCISES for Sections 14.1 & 14.2
The Simple Linear Model and Related Assumptions
Fitting the Model by the Method of Least Squares

14.1 A deterministic model assumes that the phenomenum of interest can be predicted precisely, while a probabilistic model allows for variation in the predictions.

14.3 A probabilistic model produces different values of y for the same x because it contains a random error component. A deterministic model has no random component and thus gives the same y value for a given value of x.

14.5 The equation for the model is $\hat{y} = 50 + 45x$, where x denotes the number of hours required for a repair and y denotes its cost in dollars. The model is deterministic since there is no random component.

14.7 **a.** The scatter diagram is shown below.
b. From the data we calculate $\Sigma x = 23$, $\Sigma y = 23$, $\Sigma x^2 = 143$, $\Sigma y^2 = 171$, $\Sigma xy = 61$. We may now find SS(x) and SS(xy):

$$SS(x) = \Sigma x^2 - \frac{(\Sigma x)^2}{n} = 143 - \frac{(23)^2}{5} = 37.2, \text{ and}$$

$$SS(xy) = \Sigma xy - \frac{(\Sigma x)(\Sigma y)}{n} = 61 - \frac{(23)(23)}{5} = -44.8.$$

Therefore, the slope of the least squares line is: $\beta_1 = \frac{SS(xy)}{SS(x)} = \frac{-44.8}{37.2} = -1.2043$, and the y-intercept is $\beta_0 = \bar{y} - \beta_1 \bar{x} = \frac{23}{5} - (-1.2043)\frac{23}{5} = 10.1398$. The least squares line is: $\hat{y} = 10.1398 - 1.2043x$ and appears to fit the points quite well.

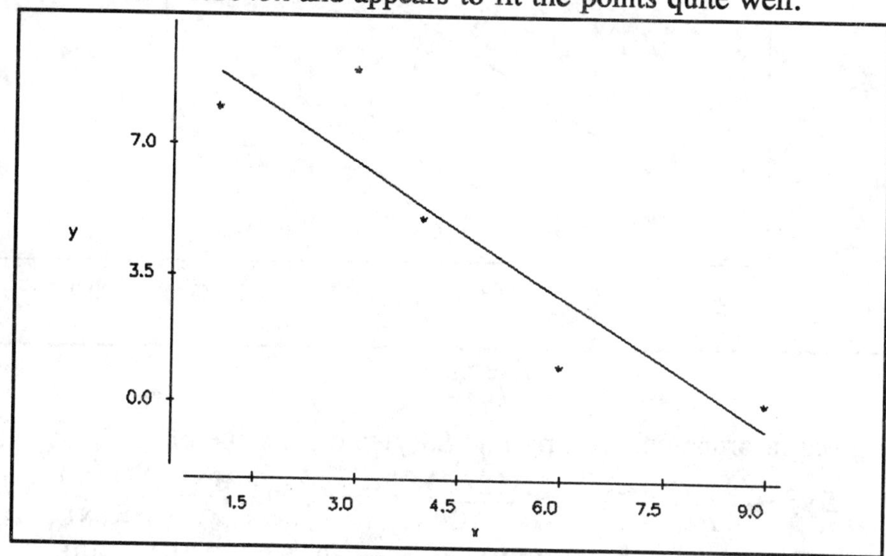

Exercise 14.7

14.9 **a.** The scatter diagram is shown below.

b. For the data we calculate $\Sigma x = 25$, $\Sigma y = 24$, $\Sigma x^2 = 193$, $\Sigma y^2 = 162$, $\Sigma xy = 174$. We find SS(x) and SS(xy):

$$SS(x) = \Sigma x^2 - \frac{(\Sigma x)^2}{n} = 193 - \frac{(25)^2}{5} = 68, \text{ and}$$

$$SS(xy) = \Sigma xy - \frac{(\Sigma x)(\Sigma y)}{n} = 174 - \frac{(25)(24)}{5} = 54.$$

Therefore, the slope of the least squares line is: $\beta_1 = \frac{SS(xy)}{SS(x)} = \frac{54}{68} = 0.79412$, and the y-intercept is $\beta_0 = \bar{y} - \beta_1 \bar{x} = \frac{24}{5} - (0.79412)\frac{25}{5} = 0.8294$. The least squares line is: $\hat{y} = 0.8294 + 0.7941x$. The line appears to fit the points well.

c. In order to calculate SSE, we need SS(y):

$$SS(y) = \Sigma y^2 - \frac{(\Sigma y)^2}{n} = 162 - \frac{(24)^2}{5} = 46.8. \text{ Therefore,}$$

$$SSE = SS(y) - \beta_1 SS(xy) = 46.8 - (0.79412)(54) = 3.918.$$

d. $s^2 = \frac{SSE}{n-2} = \frac{3.918}{5-2} = 1.306.$

e. $s = \sqrt{s^2} = \sqrt{1.306} = 1.143.$

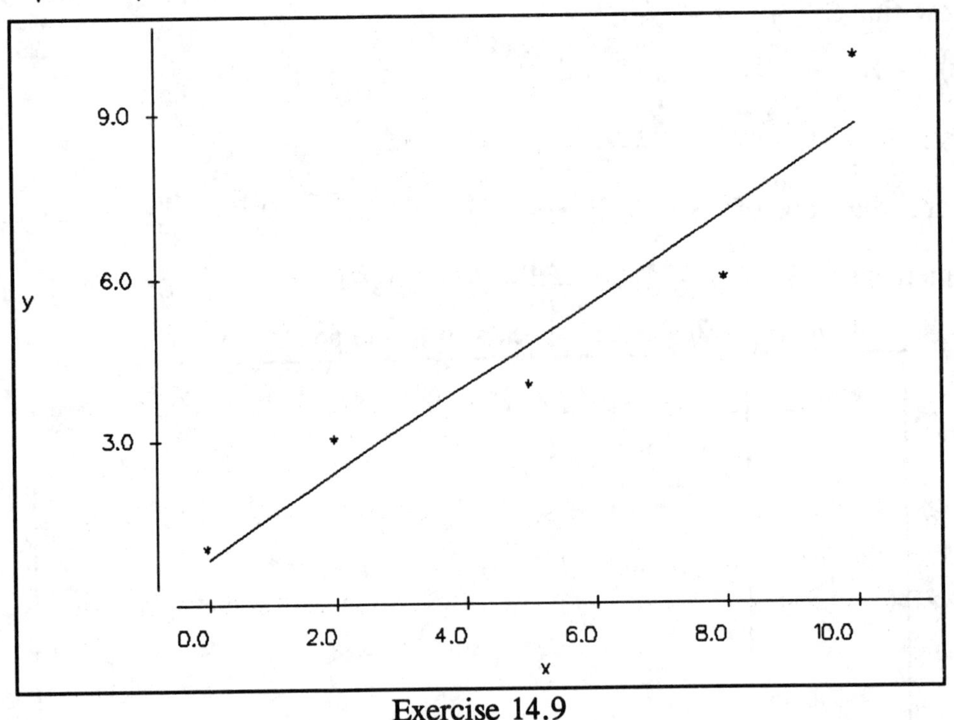

Exercise 14.9

14.11 Using the given information, we first find SS(y) and the slope of the least squares line:

$$SS(y) = \Sigma y^2 - \frac{(\Sigma y)^2}{n} = 1750 - \frac{(280)^2}{100} = 966 \text{ and } \beta_1 = \frac{SS(xy)}{SS(x)} = \frac{-155}{310} = -0.5.$$

Therefore: $SSE = SS(y) - \beta_1 SS(xy) = 966 - (-0.5)(-155) = 888.5$

and $s^2 = \frac{SSE}{n-2} = \frac{888.5}{100-2} = 9.066.$

The Simple Linear Model and Related Assumptions
Fitting the Model by the Method of Least Squares

14.13 **a.** The scatter diagram with the least squares line follows:

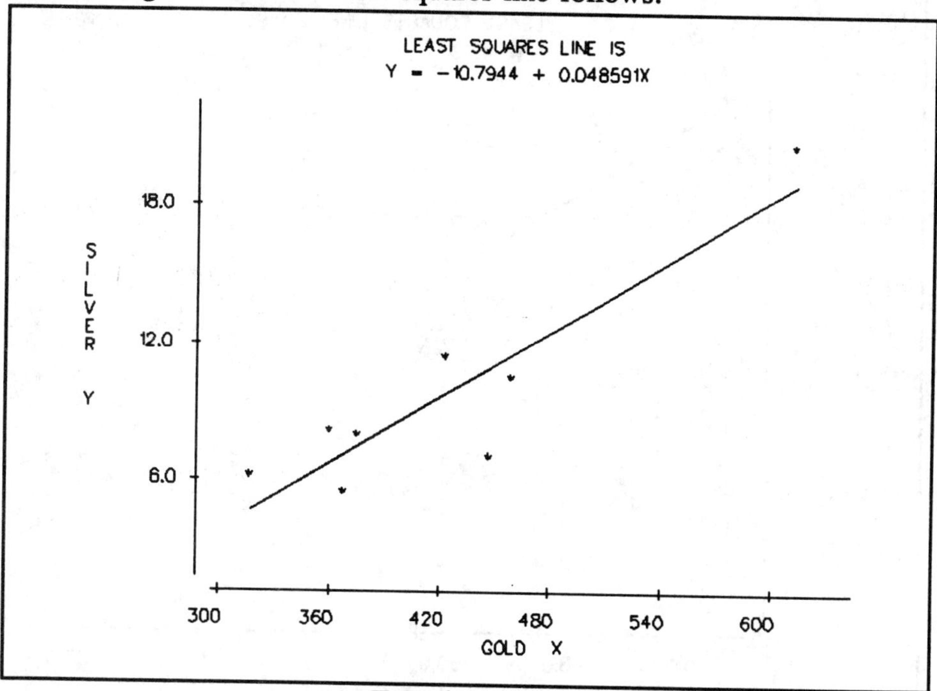

Exercise 14.13

b. For the data we are given $\Sigma x = 3368$; $\Sigma y = 77.3$; $\Sigma x^2 = 1476094$; $\Sigma y^2 = 913.3636$; $\Sigma xy = 35369.65$. We calculate:

$$SS(x) = \Sigma x^2 - \frac{(\Sigma x)^2}{n} = 1476094 - \frac{(3368)^2}{8} = 58166,$$

$$SS(y) = \Sigma y^2 - \frac{(\Sigma y)^2}{n} = 913.3636 - \frac{(77.3)^2}{8} = 166.4524, \text{ and}$$

$$SS(xy) = \Sigma xy - \frac{(\Sigma x)(\Sigma y)}{n} = 35369.65 - \frac{(3368)(77.3)}{8} = 2826.35.$$

The slope of the least squares line is: $\hat{\beta}_1 = \frac{SS(xy)}{SS(x)} = \frac{2826.35}{58166} = 0.048591$, and the

y-intercept is $\hat{\beta}_0 = \bar{y} - \hat{\beta}_1\bar{x} = \frac{77.3}{8} - 0.048591\frac{3368}{8} = -10.794.$

Therefore, the least squares line is $\hat{y} = -10.794 + 0.0486x$.

c. The graph of the least squares line is shown on the scatter diagram above.

14.15 **a.** The scatter diagram with the least squares line follows:

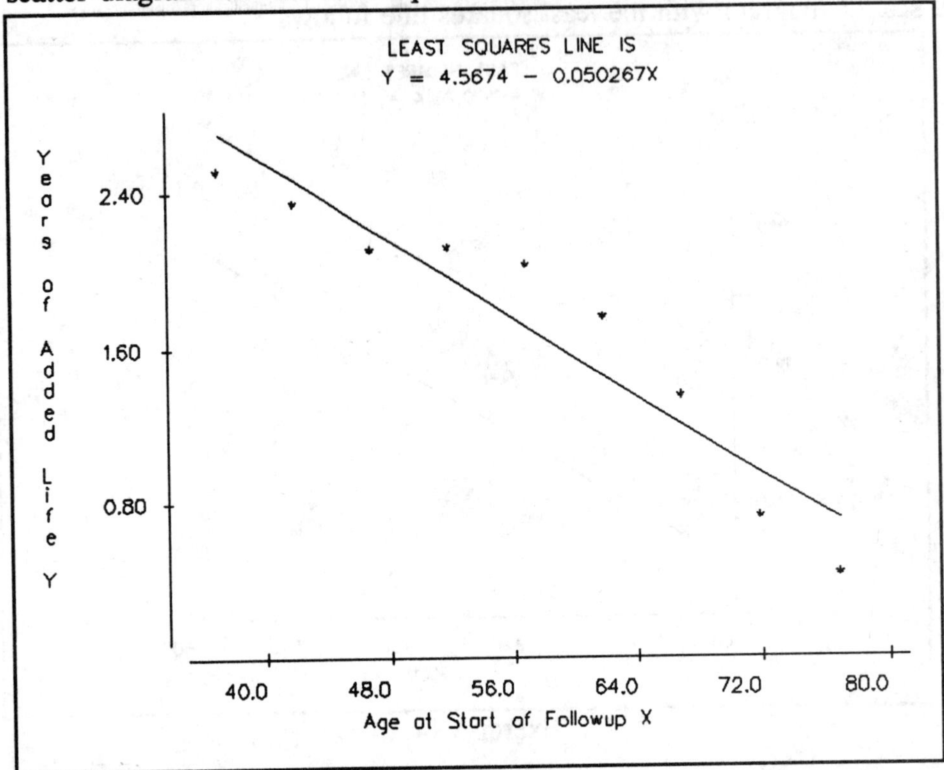

Exercise 14.15

b. For the data we are given $\Sigma x=513$; $\Sigma y=15.32$; $\Sigma x^2=30741$; $\Sigma y^2=30.298$; $\Sigma xy=797.84$. We calculate:

$$SS(x) = \Sigma x^2 - \frac{(\Sigma x)^2}{n} = 30741 - \frac{(513)^2}{9} = 1500,$$

$$SS(y) = \Sigma y^2 - \frac{(\Sigma y)^2}{n} = 30.298 - \frac{(15.32)^2}{9} = 4.21996, \text{ and}$$

$$SS(xy) = \Sigma xy - \frac{(\Sigma x)(\Sigma y)}{n} = 797.84 - \frac{(513)(15.32)}{9} = -75.4.$$

The slope of the least squares line is: $\beta_1 = \frac{SS(xy)}{SS(x)} = \frac{-75.4}{1500} = -0.050267$, and

the y-intercept is $\beta_0 = \bar{y} - \beta_1\bar{x} = \frac{15.32}{9} - (-0.050267)\frac{513}{9} = 4.5674.$

Therefore, the least squares line is $\hat{y} = 4.5674 - 0.050267x$, which we show on the scatter diagram.

c. We use the above results and calculate

$$SSE = SS(y) - \beta_1\, SS(xy) = 4.21996 - (-0.050267)(-75.4) = 0.4298.$$

$$s^2 = \frac{SSE}{n-2} = \frac{0.4298}{9-2} = 0.0614.$$

EXERCISES for Sections 14.3 & 14.4

Inferences About the Slope to Assess the Usefulness of the Model

The Coefficients of Correlation and Determination to Measure the Usefulness of the Model

14.17 **a.** In Exercises 14.7 and 14.8 we calculated $\Sigma x = 23$, $\Sigma y = 23$, $\Sigma x^2 = 143$, $\Sigma y^2 = 171$, $\Sigma xy = 61$ and then found $SS(x) = 37.2$, $SS(y) = 65.2$ and $SS(xy) = -44.8$. Therefore, the correlation coefficient is: $r = \dfrac{SS(xy)}{\sqrt{SS(x)\,SS(y)}} = \dfrac{-44.8}{\sqrt{(37.2)(65.2)}} = -0.90967.$

b. Therefore, the coefficient of determination is $r^2 = (-0.90967)^2 = 0.827$.

c. The least squares line accounts for 82.7% of the total variation in the y values.

14.19 For 90% confidence and df = n - 2 = 3 we find $t_{\alpha/2} = t_{0.05} = 2.353$. The confidence interval is given by $\beta_1 \pm t_{\frac{\alpha}{2}} \dfrac{s}{\sqrt{SS(x)}} \rightarrow 1.127 \pm 2.353 \dfrac{0.961}{\sqrt{26.8}}$ which simplifies to 1.127 ± 0.437. Therefore, with 90% confidence, we estimate the slope of the regression line is between 0.690 and 1.564.

14.21 **Step 1: Hypotheses**

In order to determine whether x and y are negatively correlated, we formulate the hypotheses in terms of the slope β_1:

H_0: $\beta_1 = 0$

H_a: $\beta_1 < 0$

Step 2: Significance Level

We are given that $\alpha = 0.01$.

Step 3: Calculations

The slope of the least squares line is: $\beta_1 = \dfrac{SS(xy)}{SS(x)} = \dfrac{-1004.1}{1464.1} = -0.68581.$

Next we calculate s:

$SSE = SS(y) - \beta_1\, SS(xy) = 790.1 - (-0.68581)(-1004.1) = 101.478.$

$s^2 = \dfrac{SSE}{n-2} = \dfrac{101.478}{10-2} = 12.685$ so $s = \sqrt{s^2} = \sqrt{12.685} = 3.562.$

The test statistic is $t = \dfrac{\beta_1\sqrt{SS(x)}}{s} = \dfrac{-0.68581\sqrt{1464.1}}{3.562} = -7.37.$

Step 4: Rejection Region

For df = n - 2 = 8 and $\alpha = 0.01$, we find $t_\alpha = t_{0.01} = 2.896$. Therefore, we reject H_0 if $t < -2.896$.

Step 5: Conclusion

Since $t = -7.37$, we reject H_0. This indicates that x and y are negatively correlated.

14.23 We are given n = 100, $\Sigma x = 387$, $\Sigma y = 588$, $\Sigma x^2 = 1981$, $\Sigma y^2 = 5023$, and $\Sigma xy = 2530$. Therefore:

$$SS(x) = \Sigma x^2 - \frac{(\Sigma x)^2}{n} = 1981 - \frac{(387)^2}{100} = 483.31$$

$$SS(y) = \Sigma y^2 - \frac{(\Sigma y)^2}{n} = 5023 - \frac{(588)^2}{100} = 1565.56$$

$$SS(xy) = \Sigma xy - \frac{(\Sigma x)(\Sigma y)}{n} = 2530 - \frac{(387)(588)}{100} = 254.44.$$

The correlation coefficient is:

$$r = \frac{SS(xy)}{\sqrt{SS(x)\,SS(y)}} = \frac{254.44}{\sqrt{(483.31)(1565.56)}} = 0.2925.$$

Therefore, the coefficient of determination is $r^2 = 0.086$. The least squares line accounts for 8.6% of the total variation in the y values.

14.25 **a.** We are given: $\Sigma x=37$; $\Sigma y=42$; $\Sigma x^2=219$; $\Sigma y^2=244$; $\Sigma xy=165$. Therefore:

$$SS(x) = \Sigma x^2 - \frac{(\Sigma x)^2}{n} = 219 - \frac{(37)^2}{8} = 47.875,$$

$$SS(y) = \Sigma y^2 - \frac{(\Sigma y)^2}{n} = 244 - \frac{(42)^2}{8} = 23.5, \text{ and}$$

$$SS(xy) = \Sigma xy - \frac{(\Sigma x)(\Sigma y)}{n} = 165 - \frac{(37)(42)}{8} = -29.25.$$

The correlation coefficient is:

$$r = \frac{SS(xy)}{\sqrt{SS(x)\,SS(y)}} = \frac{-29.25}{\sqrt{(47.875)(23.5)}} = -0.872.$$

b. The negative value of r indicates that the subtest scores tend to decrease as the time after injury increases.
c. The coefficient of determination is $r^2 = 0.760$.
d. The least squares line accounts for 76% of the total variation in the subtest scores.

14.27 Since a t-test is used, where the alternative hypothesis is H_a: $\beta_1 \neq 0$, the p-value is the area under the t distribution to the left of t = -4.36 plus the area under the t distribution to the right of 4.36. We use Table IV with df = n - 2 = 6. We note that 4.36 lies to the right of $t_{0.005} = 3.707$. Therefore, the area under the t distribution to the right of 4.36 is less than 0.005. Thus ½(p-value) < 0.005 so that p-value < 0.01.

14.29 **a.** For the data we are given $\Sigma x = 3368$; $\Sigma y = 77.3$; $\Sigma x^2 = 1476094$; $\Sigma y^2 = 913.3636$; and $\Sigma xy = 35369.65$. We calculate:

$$SS(x) = \Sigma x^2 - \frac{(\Sigma x)^2}{n} = 1476094 - \frac{(3368)^2}{8} = 58166,$$

$$SS(y) = \Sigma y^2 - \frac{(\Sigma y)^2}{n} = 913.3636 - \frac{(77.3)^2}{8} = 166.4524, \text{ and}$$

$$SS(xy) = \Sigma xy - \frac{(\Sigma x)(\Sigma y)}{n} = 35369.65 - \frac{(3368)(77.3)}{8} = 2826.35.$$

The correlation coefficient is:

$$r = \frac{SS(xy)}{\sqrt{SS(x)\,SS(y)}} = \frac{2826.35}{\sqrt{(58166)(166.4524)}} = 0.9083.$$

The coefficient of determination is $r^2 = 0.825$, so the least squares line accounts for 82.5% of the total variation in the price of silver.

b. Step 1: Hypotheses

In order to determine whether knowledge of the gold price contributes information for predicting the price of silver, we formulate hypotheses in terms of the slope β_1:

H_0: $\beta_1 = 0$

H_a: $\beta_1 \neq 0$

Step 2: Significance Level

We are given that $\alpha = 0.01$.

Step 3: Calculations

The slope of the least squares line is: $\beta_1 = \dfrac{SS(xy)}{SS(x)} = \dfrac{2826.35}{58166} = 0.04859.$

Next we calculate s:

$$SSE = SS(y) - \beta_1 SS(xy) = 166.4524 - (0.04859)(2826.35) = 29.1201.$$

$$s^2 = \frac{SSE}{n-2} = \frac{29.1201}{8-2} = 4.853 \text{ so } s = \sqrt{s^2} = \sqrt{4.853} = 2.203.$$

The test statistic is $t = \dfrac{\beta_1\sqrt{SS(x)}}{s} = \dfrac{0.04859\sqrt{58166}}{2.203} = 5.32.$

Step 4: Rejection Region

For df = n - 2 = 6 and $\alpha = 0.01$, we find $t_{\alpha/2} = t_{0.005} = 3.707$. Therefore, we reject H_0 if $t < -3.707$ or if $t > 3.707$.

Step 5: Conclusion

Since t = 5.32, we reject H_0. This indicates, at level of significance $\alpha = 0.01$, that knowledge of the gold price contributes information for predicting the price of silver.

14.31 For 95% confidence, and df = n - 2 = 6 we find $t_{\alpha/2} = t_{0.025} = 2.447$. The confidence interval is given by $\beta_1 \pm t_{\frac{\alpha}{2}} \dfrac{s}{\sqrt{SS(x)}} \rightarrow 0.0486 \pm 2.447 \dfrac{2.203}{\sqrt{58166}}$ which simplifies to 0.0486 ± 0.0224. Therefore, with 95% confidence, we estimate the slope of the regression line is between 0.0262 and 0.0710.

14.33 Since a t-test is used, where the alternative hypothesis is H_a: $\beta_1 \neq 0$, the p-value is the area under the t distribution to the left of t = -4.90 plus the area under the t distribution to the right of 4.90. We use Table IV with df = n - 2 = 12. We note that 4.90 lies to the right of $t_{0.005} = 3.055$. Therefore, the area under the t distribution to the right of 4.90 is less than 0.005. Thus ½(p-value) < 0.005 so that p-value < 0.01.

EXERCISES for Section 14.5

Using the Model to Estimate and Predict

14.35 **a.** A scatter diagram for the data points follows:

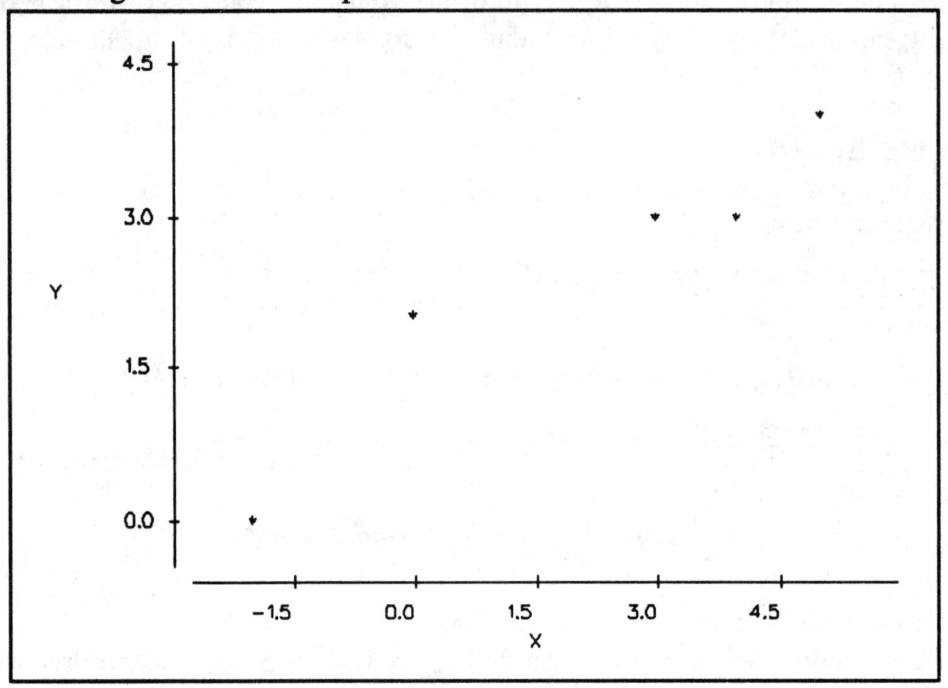

Exercise 14.35

b. We calculate: $\Sigma x = 10$; $\Sigma y = 12$; $\Sigma x^2 = 54$; $\Sigma y^2 = 38$; $\Sigma xy = 41$. Therefore:

$$SS(x) = \Sigma x^2 - \frac{(\Sigma x)^2}{n} = 54 - \frac{(10)^2}{5} = 34,$$

$$SS(y) = \Sigma y^2 - \frac{(\Sigma y)^2}{n} = 38 - \frac{(12)^2}{5} = 9.2, \text{ and}$$

$$SS(xy) = \Sigma xy - \frac{(\Sigma x)(\Sigma y)}{n} = 41 - \frac{(10)(12)}{5} = 17.$$

The slope of the least squares line is: $\beta_1 = \frac{SS(xy)}{SS(x)} = \frac{17}{34} = 0.5$, and the y-intercept is $\beta_0 = \bar{y} - \beta_1\bar{x} = \frac{12}{5} - (0.5)\frac{10}{5} = 1.4$.

The equation of the least squares line is: $\hat{y} = 1.4 + 0.5x$.

c. The correlation coefficient is: $r = \frac{SS(xy)}{\sqrt{SS(x)\,SS(y)}} = \frac{17}{\sqrt{(34)(9.2)}} = 0.9612$, so the coefficient of determination is $r^2 = 0.924$. Therefore, the least squares line accounts for 92.4% of the total variation in the y-values.

d. **<u>Step 1</u>: Hypotheses**

H_0: $\beta_1 = 0$

H_a: $\beta_1 \neq 0$

<u>Step 2</u>: Significance Level

We are given that $\alpha = 0.05$.

Step 3: Calculations
We first calculate s: $SSE = SS(y) - \hat{\beta}_1\, SS(xy) = 9.2 - (0.5)\,(17) = 0.7.$

$$s^2 = \frac{SSE}{n-2} = \frac{0.7}{5-2} = 0.2333 \text{ so } s = \sqrt{s^2} = \sqrt{0.2333} = 0.483.$$

The test statistic is

$$t = \frac{\hat{\beta}_1\sqrt{SS(x)}}{s} = \frac{(0.5)\sqrt{34}}{0.483} = 6.04.$$

Step 4: Rejection Region
For df = n - 2 = 3 and $\alpha = 0.05$, we find $t_{\alpha/2} = t_{0.025} = 3.182$. Therefore, we reject H_0 if $t < -3.182$ or $t > 3.182$.
Step 5: Conclusion
Since t = 6.04, we reject H_0. This indicates that knowledge of x contributes information for predicting y.
e. The confidence interval estimate for the mean of y at x_0 is given by

$$\hat{y} \pm t_{\frac{\alpha}{2}}\, s\sqrt{\frac{1}{n} + \frac{(x_0 - \bar{x})^2}{SS(x)}},$$ where for df = n - 2 = 3 and 95% confidence, we have

$t_{\alpha/2} = t_{0.025} = 3.182$. When $x_0 = 3$, we have $\hat{y} = 1.4 + 0.5x_0 = 1.4 + (0.5)\cdot(3) = 2.9$.
From above, we have s = 0.483, $\bar{x} = 10/5 = 2$, and SS(x) = 34. We calculate

$$2.9 \pm (3.182)\cdot(0.483)\sqrt{\frac{1}{5} + \frac{(3-2)^2}{34}},$$ or 2.9 ± 0.736. Therefore, with 95%

confidence, we estimate the mean of y for x = 3 is between 2.164 and 3.636.
f. The prediction interval for predicting y when $x = x_0$ is given by

$$\hat{y} \pm t_{\frac{\alpha}{2}}\, s\sqrt{1 + \frac{1}{n} + \frac{(x_0 - \bar{x})^2}{SS(x)}},$$ where the variables are as above. We calculate

$$2.9 \pm (3.182)\cdot(0.483)\sqrt{1 + \frac{1}{5} + \frac{(3-2)^2}{34}},$$ or 2.9 ± 1.704. Therefore, with 95%

probability, we predict that when x = 3, y is between 1.196 and 4.604.
g. The width of the confidence interval is 1.472 and the width of the prediction interval is 3.408.

14.37 **a.** We are given that n = 20, $\bar{x} = 20$, SS(x) = 22.2, SSE = 44.8 and the least squares line is $\hat{y} = 21.7 + 10.8x$. The confidence interval estimate for the mean of y at x_0 is given by

$$\hat{y} \pm t_{\frac{\alpha}{2}}\, s\sqrt{\frac{1}{n} + \frac{(x_0 - \bar{x})^2}{SS(x)}},$$ where for df = n - 2 = 18 and 90% confidence, we have

$t_{\alpha/2} = t_{0.05} = 1.734$. When $x_0 = 25$, we have $\hat{y} = 21.7 + 10.8x_0 = 21.7 + (10.8)\cdot(25) =$

291.7. Also, since SSE = 44.8, $s^2 = \frac{SSE}{n-2} = \frac{44.8}{20-2} = 2.48889$ so

$s = \sqrt{s^2} = \sqrt{2.48889} = 1.578.$ We calculate

$291.7 \pm (1.734)\cdot(1.578)\sqrt{\frac{1}{20}+\frac{(25-20)^2}{22.2}}$, or 291.7 ± 2.967. Therefore, with 90% confidence, we estimate the mean of y for x = 25 is between 288.733 and 294.667.

b. If x_0 had been 21, the confidence interval would have been narrower since 21 is closer to $\bar{x} = 20$.

c. The smallest width would occur for $x_0 = \bar{x} = 20$.

14.39 We are given for the data, $\Sigma x = 54.69$, $\Sigma y = 1133$, $\Sigma x^2 = 214.9087$, $\Sigma y^2 = 92815$, and $\Sigma xy = 4395.19$. From Exercise 14.32, the estimated standard deviation of the model is $s = 5.583$, and the least squares line is $\hat{y} = 175.94 - 24.32x$. Also $n = 14$, $\bar{x} = 54.69/14 = 3.9064$, and $SS(x) = \Sigma x^2 - \frac{(\Sigma x)^2}{n} = 214.9087 - \frac{(54.69)^2}{14} = 1.2661$.

The confidence interval estimate for the mean of y at x_0 is given by

$\hat{y} \pm t_{\frac{\alpha}{2}} s\sqrt{\frac{1}{n}+\frac{(x_0-\bar{x})^2}{SS(x)}}$, where for df = n - 2 = 12 and 95% confidence, we have $t_{\alpha/2} = t_{0.025} = 2.179$. When $x_0 = 4.08$, we have $\hat{y} = 175.94 - 24.32x_0 = 175.94 - (24.32)\cdot(4.08) = 76.71$. We calculate

$76.71 \pm (2.179)\cdot(5.583)\sqrt{\frac{1}{14}+\frac{(4.08-3.9064)^2}{1.2661}}$, or 76.71 ± 3.75.

Therefore, with 95% confidence, we estimate the mean of y for x = 4.08 is between 72.96 and 80.46.

14.41 We are given for the data, $\Sigma x = 3368$, $\Sigma y = 77.3$, $\Sigma x^2 = 1476094$, $\Sigma y^2 = 913.3636$, and $\Sigma xy = 35369.65$. From Exercise 14.29, the estimated standard deviation of the model is $s = 2.203$, and, from Exercise 14.13, the least squares line is $\hat{y} = -10.79 + 0.0486x$. Also $n = 8$, $\bar{x} = 3368/8 = 421$, and

$$SS(x) = \Sigma x^2 - \frac{(\Sigma x)^2}{n} = 1476094 - \frac{(3368)^2}{8} = 58166.$$

The prediction interval for predicting y at x_0 is given by $\hat{y} \pm t_{\frac{\alpha}{2}} s\sqrt{1+\frac{1}{n}+\frac{(x_0-\bar{x})^2}{SS(x)}}$,

where for df = n - 2 = 6 and 95% probability, we have $t_{\alpha/2} = t_{0.025} = 2.447$. When $x_0 = 400$, we have $\hat{y} = -10.79 + 0.0486x_0 = -10.79 + (0.0486)\cdot(400) = 8.65$. We

calculate $8.65 \pm (2.447)\cdot(2.203)\sqrt{1+\frac{1}{8}+\frac{(400-421)^2}{58166}}$, or 8.65 ± 5.74.

Therefore, with 95% probability, we predict that, when the price of gold is $x_0 = 400$, the price of silver is between 2.91 and 14.39.

14.43 **a.** The scatter diagram follows:

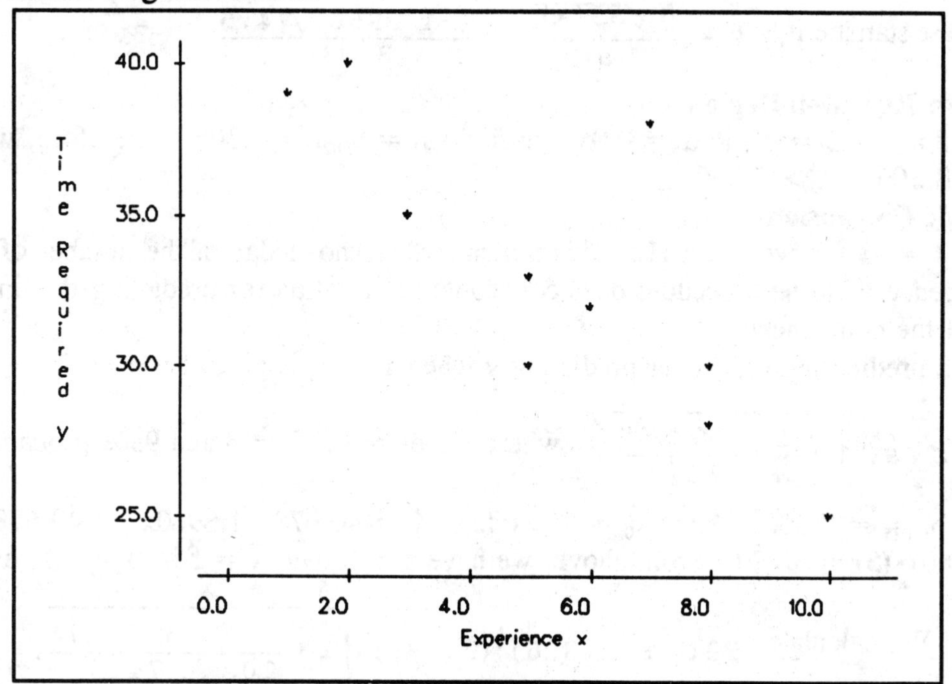

Exercise 14.43

b. We calculate: $\Sigma x = 55$; $\Sigma y = 330$; $\Sigma x^2 = 377$; $\Sigma y^2 = 11112$; $\Sigma xy = 1711$. Therefore:

$$SS(x) = \Sigma x^2 - \frac{(\Sigma x)^2}{n} = 377 - \frac{(55)^2}{10} = 74.5,$$

$$SS(y) = \Sigma y^2 - \frac{(\Sigma y)^2}{n} = 11112 - \frac{(330)^2}{10} = 222, \text{ and}$$

$$SS(xy) = \Sigma xy - \frac{(\Sigma x)(\Sigma y)}{n} = 1711 - \frac{(55)(330)}{10} = -104.$$

The slope of the least squares line is: $\hat{\beta}_1 = \frac{SS(xy)}{SS(x)} = \frac{-104}{74.5} = -1.3960$, and the y-intercept is $\hat{\beta}_0 = \bar{y} - \hat{\beta}_1\bar{x} = \frac{330}{10} - (-1.3960)\frac{55}{10} = 40.678.$

The equation of the least squares line is: $\hat{y} = 40.678 - 1.3960x$.

c. The correlation coefficient is: $r = \frac{SS(xy)}{\sqrt{SS(x)\,SS(y)}} = \frac{-104}{\sqrt{(74.5)(222)}} = -0.809,$ so the coefficient of determination is $r^2 = 0.654$. Therefore, the least squares line accounts for 65.4% of the total variation in the y-values.

d. Step 1: Hypotheses

H_0: $\beta_1 = 0$

H_a: $\beta_1 \neq 0$

Step 2: Significance Level

We are given that $\alpha = 0.05$.

Step 3: Calculations

We first calculate s: $SSE = SS(y) - \hat{\beta}_1\, SS(xy) = 222 - (-1.3960)(-104) = 76.816.$

$$s^2 = \frac{SSE}{n-2} = \frac{76.816}{10-2} = 9.602 \text{ so } \quad s = \sqrt{s^2} = \sqrt{9.602} = 3.099.$$

The test statistic is $t = \frac{\hat{\beta}_1\sqrt{SS(x)}}{s} = \frac{(-1.3960)\sqrt{74.5}}{3.099} = -3.89$.

Step 4: Rejection Region
For df = n - 2 = 8 and $\alpha = 0.05$, we find $t_{\alpha/2} = t_{0.025} = 2.306$. Therefore, we reject H_0 if $t < -2.306$ or $t > 2.306$.
Step 5: Conclusion
Since t = -3.89, we reject H_0. This indicates that knowledge of the number of days of experience with the procedure does contribute information for predicting the time required to install the component.
e. The prediction interval for predicting y when $x = x_0$ is given by

$\hat{y} \pm t_{\frac{\alpha}{2}} s\sqrt{1 + \frac{1}{n} + \frac{(x_0 - \bar{x})^2}{SS(x)}}$, where for df = n - 2 = 8 and 95% probability, we have

$t_{\alpha/2} = t_{0.025} = 2.306$. When $x_0 = 5$, we have $\hat{y} = 40.678 - 1.3960x_0 = 40.678 - (1.3960)\cdot(5) = 33.70$. From above, we have s = 3.099, $\bar{x} = 55/10 = 5.5$, and SS(x) =

74.5. We calculate $33.7 \pm (2.306)\cdot(3.099)\sqrt{1 + \frac{1}{10} + \frac{(5-5.5)^2}{74.5}}$,

or 33.70 ± 7.51. Therefore, with 95% probability, we predict that a randomly selected worker with 5 days experience will take between 26.19 and 41.21 seconds to install the component.

EXERCISES for Section 14.6

Conducting a Regression Analysis with MINITAB

14.45(M) **a.** From the MINITAB output, we see that the equation of the least squares line is $\hat{y} = 4.57 - 0.0503x$.
b. The coefficient of determination is 89.8% = 0.898. This indicates that the least squares line accounts for 89.8% of the total variation in the y values.
c. The value of the test statistic is the t-ratio = t = -7.86, with a p-value of 0.000 which is less than 0.05. Therefore, the model is useful.
d. For x = 50, we are 95% confident that the long run average years of added life is between 1.8318 and 2.2763.

14.47(M) **a.** From the MINITAB output, we see that the equation of the least squares line is $\hat{y} = -10.8 + 0.0486x$.
b. The coefficient of determination is 82.5% = 0.825, which indicates that the fit is quite good.
c. The value of the test statistic is the t-ratio = t = 5.32, with a p-value of 0.002 which is less than 0.05. Therefore, the model is useful.
d. For a randomly selected year when x = 400, the probability is 0.95 that y, the price for an ounce of silver, is between 2.904 and 14.380.
e. This interval is too wide to be of practical value.

14.49(M) a. From the MINITAB output, the coefficient of determination of the model is r^2 = 82.5% = 0.825. This indicates that the least squares line accounts for 82.5% of the total variation in the y values.

b. We note from the analysis of variance that SS(y) = SST = 166.45 and SSE = 29.12. Therefore, the coefficient of determination is

$$r^2 = 1 - \frac{SSE}{SS(y)} = 1 - \frac{29.12}{166.45} = 0.825.$$

This is the same value as found in part (a).

14.51(M)

```
MTB > # The data were entered into C1 and C2
MTB > # Part a
MTB > NAME C1 'FEE' C2 'ERROR'
MTB > PLOT C2 C1
```

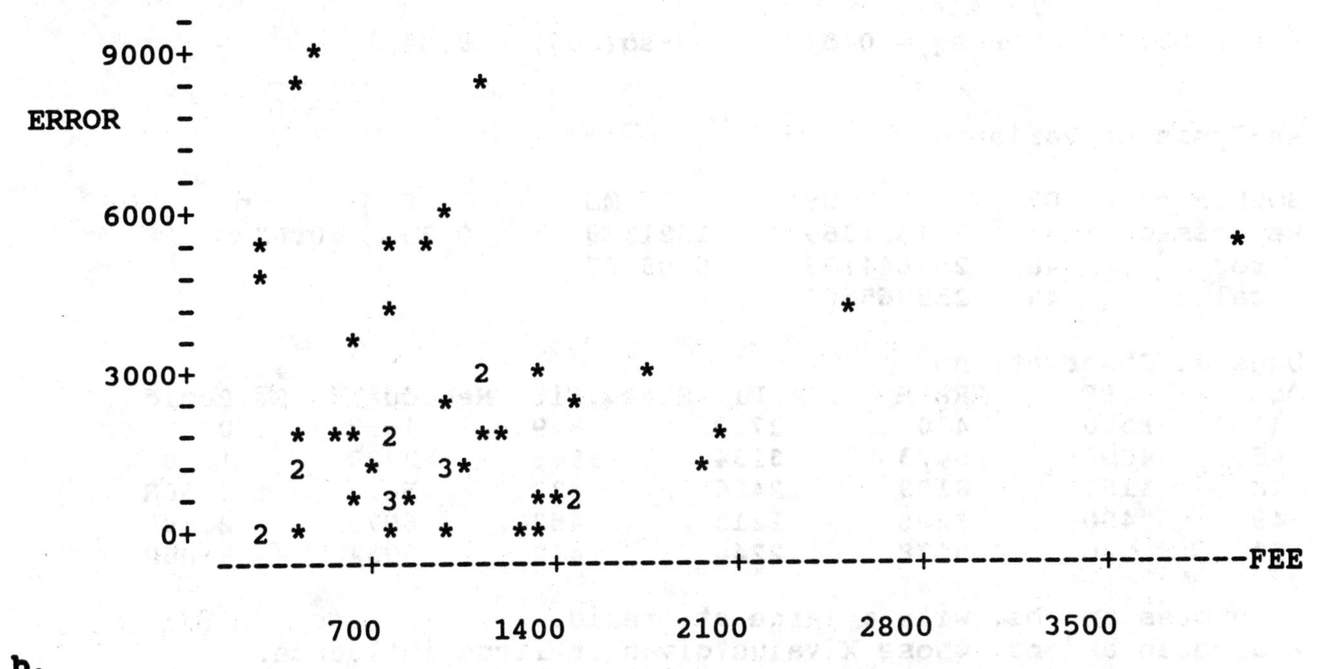

b.

```
MTB > # Part b
MTB > # FEE and ERROR do not appear to be correlated
MTB > CORR C1 C2
```

```
Correlation of FEE and ERROR = 0.072

MTB > LET K1 = (0.072)**2
MTB > PRINT K1 #Coefficient of Determination
K1      0.00518400
MTB > # This suggests that almost none (0.52%) of the variation in y
MTB > # is explained by the model.
c.
MTB > # Part c
MTB > REGR C2 1 C1

The regression equation is
ERROR = 2111 + 0.256 FEE

Predictor        Coef       Stdev    t-ratio        p
Constant       2110.8       612.4       3.45    0.001
FEE            0.2558      0.5126       0.50    0.620

s = 2303         R-sq = 0.5%      R-sq(adj) = 0.0%


Analysis of Variance

SOURCE       DF          SS          MS         F        p
Regression    1     1321369     1321369      0.25    0.620
Error        48   254644192     5305087
Total        49   255965568

Unusual Observations
Obs.      FEE       ERROR         Fit Stdev.Fit  Residual    St.Resid
 41      2500        4307        2750       829      1557      0.72 X
 45      4000        5473        3134      1566      2339      1.38 X
 48      1150        8103        2405       333      5698      2.50R
 49       400        8288        2213       452      6075      2.69R
 50       520        9178        2244       412      6934      3.06R

R denotes an obs. with a large st. resid.
X denotes an obs. whose X value gives it large influence.

MTB > # p-value for slope is 0.62 > 0.05; so we fail to reject Ho.
MTB > # Cannot conclude that x and y are linearly correlated.
```

REVIEW EXERCISES
CHAPTER 14

14.52 **a.** The scatter diagram for the data follows:

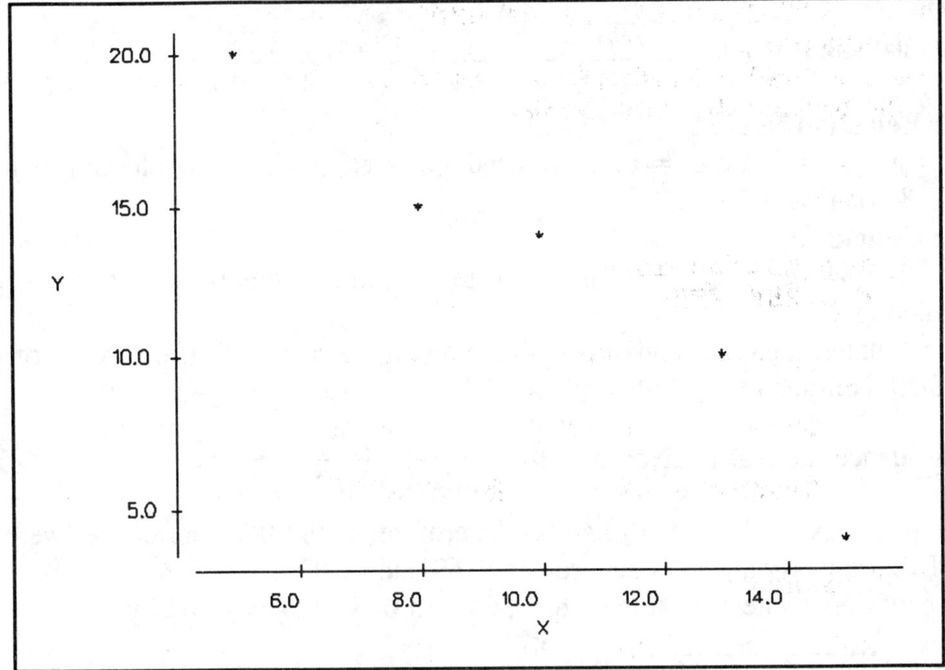

Exercise 14.52

b. We calculate: $\Sigma x = 51$; $\Sigma y = 63$; $\Sigma x^2 = 583$; $\Sigma y^2 = 937$; $\Sigma xy = 550$. Therefore:

$$SS(x) = \Sigma x^2 - \frac{(\Sigma x)^2}{n} = 583 - \frac{(51)^2}{5} = 62.8,$$

$$SS(y) = \Sigma y^2 - \frac{(\Sigma y)^2}{n} = 937 - \frac{(63)^2}{5} = 143.2, \text{ and}$$

$$SS(xy) = \Sigma xy - \frac{(\Sigma x)(\Sigma y)}{n} = 550 - \frac{(51)(63)}{5} = -92.6.$$

The slope of the least squares line is: $\hat{\beta}_1 = \frac{SS(xy)}{SS(x)} = \frac{-92.6}{62.8} = -1.4745$,

and the y-intercept is $\hat{\beta}_0 = \bar{y} - \hat{\beta}_1 \bar{x} = \frac{63}{5} - (-1.4745)\frac{51}{5} = 27.640.$

The equation of the least squares line is: $\hat{y} = 27.640 - 1.4745x$.

c. The correlation coefficient is:

$$r = \frac{SS(xy)}{\sqrt{SS(x)\,SS(y)}} = \frac{-92.6}{\sqrt{(62.8)(143.2)}} = -0.976,$$

and the coefficient of determination is $r^2 = 0.953$.

d. **Step 1: Hypotheses**

H_0: $\beta_1 = 0$

H_a: $\beta_1 \neq 0$

Step 2: Significance Level

We are given that $\alpha = 0.05$.

Step 3: Calculations

We first calculate s:

$$SSE = SS(y) - \hat{\beta}_1\, SS(xy) = 143.2 - (-1.4745)(-92.6) = 6.661.$$

$$s^2 = \frac{SSE}{n-2} = \frac{6.661}{5-2} = 2.220 \text{ so } s = \sqrt{s^2} = \sqrt{2.220} = 1.490.$$

The test statistic is $t = \dfrac{\hat{\beta}_1\sqrt{SS(x)}}{s} = \dfrac{(-1.4745)\sqrt{62.8}}{1.490} = -7.84.$

Step 4: Rejection Region

For df = n - 2 = 3 and $\alpha = 0.05$, we find $t_{\alpha/2} = t_{0.025} = 3.182$. Therefore, we reject H_0 if $t < -3.182$ or $t > 3.182$.

Step 5: Conclusion

Since t = -7.84, we reject H_0. This indicates that knowledge of x contributes information for predicting y.

e. The estimated standard deviation of the model is s = 1.490. (See Step 3 of part d.)

f. For 90% confidence, and df = n - 2 = 3 we use $t_{\alpha/2} = t_{0.05} = 2.353$.

The confidence interval is given by $\hat{\beta}_1 \pm t_{\frac{\alpha}{2}}\dfrac{s}{\sqrt{SS(x)}} \rightarrow -1.4745 \pm 2.353\dfrac{1.490}{\sqrt{62.8}}$

which simplifies to -1.4745 ± 0.4424. Therefore, with 90% confidence, we estimate the slope of the regression line is between -1.9169 and -1.0321.

g. The confidence interval estimate for the mean of y at x_0 is given by

$\hat{y} \pm t_{\frac{\alpha}{2}} s\sqrt{\dfrac{1}{n} + \dfrac{(x_0-\bar{x})^2}{SS(x)}}$, where for df = n - 2 = 3 and 95% confidence, we have

$t_{\alpha/2} = t_{0.025} = 3.182$.

When $x_0 = 11$, we have $\hat{y} = 27.640 - 1.4745x_0 = 27.640 - (1.4745)\cdot(11) = 11.420$. From above, we have s = 1.490, $\bar{x} = 51/5 = 10.2$, and SS(x) = 62.8. We calculate

$11.420 \pm (3.182)\cdot(1.490)\sqrt{\dfrac{1}{5} + \dfrac{(11-10.2)^2}{62.8}}$, or 11.420 ± 2.174. Therefore,

with 95% confidence, we estimate the mean of y for x = 11 is between 9.246 and 13.594.

h. The prediction interval for predicting y when $x = x_0$ is given by

$\hat{y} \pm t_{\frac{\alpha}{2}} s\sqrt{1 + \dfrac{1}{n} + \dfrac{(x_0-\bar{x})^2}{SS(x)}}$, where the quantities are as above. We calculate

$11.420 \pm (3.182)\cdot(1.490)\sqrt{1 + \dfrac{1}{5} + \dfrac{(11-10.2)^2}{62.8}}$, or 11.420 ± 5.216.

Therefore, with 95% probability, we predict that when x = 11, y is between 6.204 and 16.636.

14.53 **a.** The equation for the model is $\hat{y} = 135 + 75x$, where x denotes the number of cars sold in a week, and y the salesperson's salary in dollars.

b. The model is deterministic since there is no random component; for a given x, we always get the same y-value.

14.54 **a.** We first find SST = SS(y):

$$SS(y) = \Sigma y^2 - \frac{(\Sigma y)^2}{n} = 12560 - \frac{(680)^2}{40} = 1000.$$ We are given $r^2 = 0.76$, so

we have: $$r^2 = 1 - \frac{SSE}{SS(y)} = 1 - \frac{SSE}{1000} = 0.76.$$

Solving for SSE, we obtain SSE = 240.

b. The estimated variance of the random error components is:

$$s^2 = \frac{SSE}{n-2} = \frac{240}{40-2} = 6.3158.$$

c. The estimated standard deviation of the model is $s = \sqrt{s^2} = \sqrt{6.3158} = 2.5131.$

14.55 **a.** The scatter diagram follows:

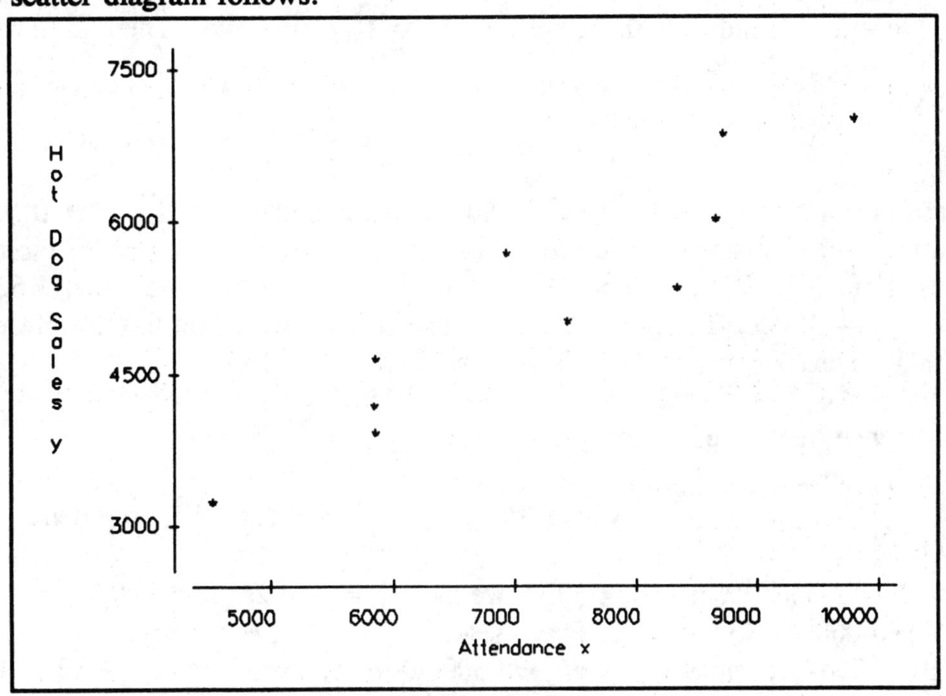

Exercise 14.55

b. We are given: $\Sigma x = 72{,}149$; $\Sigma y = 51{,}840$; SS(x) = 25,378,309, SS(y) = 13,893,508, and SS(xy) = 17,604,101. The slope of the least squares line is:

$$\beta_1 = \frac{SS(xy)}{SS(x)} = \frac{17{,}604{,}101}{25{,}378{,}309} = 0.693667,$$

and the y-intercept is $\beta_0 = \bar{y} - \beta_1 \bar{x} = \frac{51840}{10} - (0.693667)\frac{72149}{10} = 179.26.$

The equation of the least squares line is: $\hat{y} = 179.26 + 0.69367x$.

c. The correlation coefficient is:

$$r = \frac{SS(xy)}{\sqrt{SS(x)\,SS(y)}} = \frac{17{,}604{,}101}{\sqrt{(25{,}378{,}309)(13{,}893{,}508)}} = 0.9375,$$

and the coefficient of determination is $r^2 = 0.879$. This means that the least squares line accounts for 87.9% of the variation in the y values.

14.56 **Step 1: Hypotheses**
H_0: $\beta_1 = 0$
H_a: $\beta_1 \neq 0$
Step 2: Significance Level
We are given that $\alpha = 0.05$.
Step 3: Calculations
We first calculate s:

$$SSE = SS(y) - \hat{\beta}_1 SS(xy) = 13{,}893{,}508 - (0.693667)(17{,}604{,}101) = 1682124.$$

$$s^2 = \frac{SSE}{n-2} = \frac{1682124}{10-2} = 210266 \text{ so } s = \sqrt{s^2} = \sqrt{210266} = 458.5.$$

The test statistic is $t = \frac{\hat{\beta}_1\sqrt{SS(x)}}{s} = \frac{(0.693667)\sqrt{25{,}378{,}309}}{458.5} = 7.62.$

Step 4: Rejection Region
For df = n - 2 = 8 and $\alpha = 0.05$, we find $t_{\alpha/2} = t_{0.025} = 2.306$. Therefore, we reject H_0 if $t < -2.306$ or $t > 2.306$.
Step 5: Conclusion
Since t = 7.62, we reject H_0. This indicates that knowledge of x contributes information for predicting y.
p-value: Since a t-test is used, where the alternative hypothesis is H_a: $\beta_1 \neq 0$, the p-value is the area under the t distribution to the left of t = -7.62 plus the area under the t distribution to the right of 7.62. We use Table IV with df = n - 2 = 8. We note that 7.62 lies to the right of $t_{0.005} = 3.355$. Therefore, the area under the t distribution to the right of 7.62 is less than 0.005. Thus ½(p-value) < 0.005 so that p-value < 0.01.

14.57 The confidence interval estimate for the mean of y at x_0 is given by

$\hat{y} \pm t_{\frac{\alpha}{2}} s\sqrt{\frac{1}{n} + \frac{(x_0 - \bar{x})^2}{SS(x)}}$, where for df = n - 2 = 8 and 95% confidence, we have

$t_{\alpha/2} = t_{0.025} = 2.306$. When $x_0 = 6000$, we have $\hat{y} = 179.26 + 0.69367x_0 = 179.26 + (0.69367)\cdot(6000) = 4341$. From Exercises 14.55 and 14.56, we have s = 458.5, $\bar{x}$ = 72149/10 = 7214.9, and SS(x) = 25,378,309. We calculate

$$4341 \pm (2.306)\cdot(458.5)\sqrt{\frac{1}{10} + \frac{(6000 - 7214.9)^2}{25{,}378{,}309}},$$ or 4341 ± 420.

Therefore, with 95% confidence, we estimate the mean hot dog sales for days when the attendance is 6000 to be between 3921 and 4761.

14.58 The prediction interval for predicting y when $x = x_0$ is given by

$\hat{y} \pm t_{\frac{\alpha}{2}} s\sqrt{1 + \frac{1}{n} + \frac{(x_0 - \bar{x})^2}{SS(x)}}$, where for df = n - 2 = 8 and 95% probability, we have

$t_{\alpha/2} = t_{0.025} = 2.306$. When $x_0 = 6000$, we have $\hat{y} = 179.26 + 0.69367x_0 = 179.26 + (0.69367)\cdot(6000) = 4341$. From Exercise 14.57, we have s = 458.5, $\bar{x}$ = 72149/10 = 7214.9, and SS(x) = 25,378,309. We calculate

$$4341 \pm (2.306)\cdot(458.5)\sqrt{1+\frac{1}{10}+\frac{(6000-7214.9)^2}{25,378,309}}, \quad \text{or } 4341 \pm 1138.$$

Therefore, with 95% probability, we predict that for a day when attendance is 6000, hot dog sales will be between 3203 and 5479.

14.59 **a.** The scatter diagram is shown below.

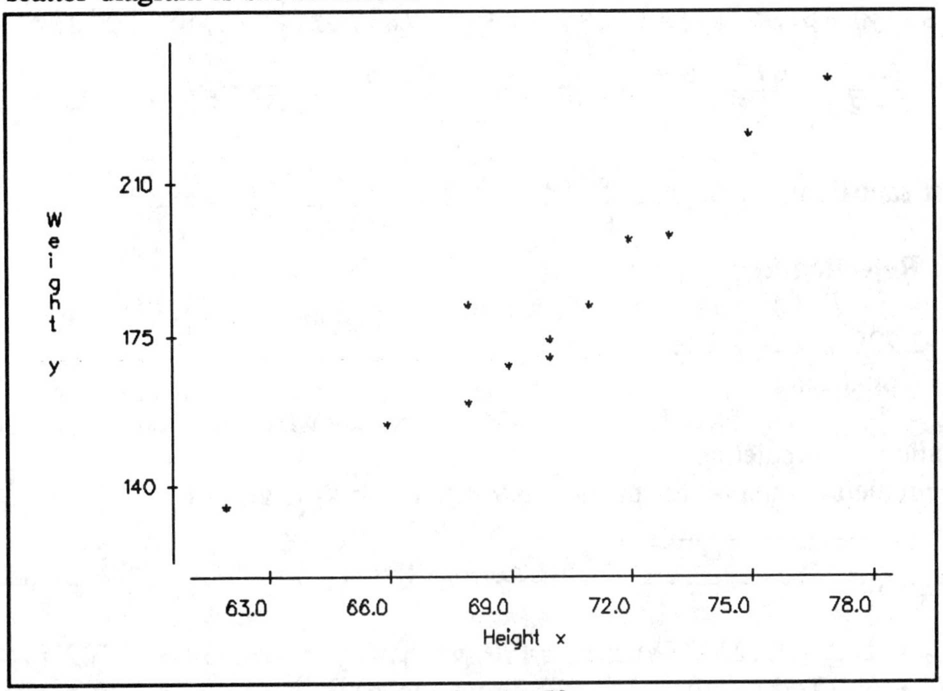

Exercise 14.59

b. We are given: $\Sigma x = 841$; $\Sigma y = 2,174$; $\Sigma x^2 = 59,117$; $\Sigma y^2 = 402,480$; $\Sigma xy = 153,547$. Therefore:

$$SS(x) = \Sigma x^2 - \frac{(\Sigma x)^2}{n} = 59,117 - \frac{(841)^2}{12} = 176.916667,$$

$$SS(y) = \Sigma y^2 - \frac{(\Sigma y)^2}{n} = 402,480 - \frac{(2,174)^2}{12} = 8623.666667, \text{ and}$$

$$SS(xy) = \Sigma xy - \frac{(\Sigma x)(\Sigma y)}{n} = 153,547 - \frac{(841)(2,174)}{12} = 1185.833334.$$

The slope of the least squares line is: $\beta_1 = \frac{SS(xy)}{SS(x)} = \frac{1185.833334}{176.916667} = 6.70278,$

and the y-intercept is $\beta_0 = \bar{y} - \beta_1\bar{x} = \frac{2174}{12} - (6.70278)\frac{841}{12} = -288.59.$

The equation of the least squares line is: $\hat{y} = -288.59 + 6.70278x$.

c. The correlation coefficient is:

$$r = \frac{SS(xy)}{\sqrt{SS(x)\,SS(y)}} = \frac{1185.833334}{\sqrt{(176.916667)(8623.666667)}} = 0.9600,$$

and the coefficient of determination is $r^2 = 0.922$. This means that the least squares line accounts for 92.2% of the variation in the y values.

d. Step 1: Hypotheses
H_0: $\beta_1 = 0$
H_a: $\beta_1 \neq 0$
Step 2: Significance Level
We are given that $\alpha = 0.05$.
Step 3: Calculations
We first calculate s:

$$SSE = SS(y) - \hat{\beta}_1 SS(xy) = 8623.666667 - (6.70278)(1185.833334) = 675.29.$$

$$s^2 = \frac{SSE}{n-2} = \frac{675.29}{12-2} = 67.53 \text{ so } s = \sqrt{s^2} = \sqrt{67.53} = 8.218.$$

The test statistic is $t = \dfrac{\hat{\beta}_1\sqrt{SS(x)}}{s} = \dfrac{(6.70278)\sqrt{176.916667}}{8.218} = 10.85.$

Step 4: Rejection Region
For df = n - 2 = 10 and $\alpha = 0.05$, we find $t_{\alpha/2} = t_{0.025} = 2.228$. Therefore, we reject H_0 if $t < -2.228$ or $t > 2.228$.
Step 5: Conclusion
Since t = 10.85, we reject H_0. This indicates that knowledge of height contributes information for predicting weight.
e. The prediction interval for predicting y when $x = x_0$ is given by

$$\hat{y} \pm t_{\frac{\alpha}{2}} s\sqrt{1 + \frac{1}{n} + \frac{(x - \bar{x}}{SS\ x}}$$ where for df = n - 2 = 10 and 95% probability, we

have $t_{\alpha/2} = t_{0.025} = 2.228$. When $x_0 = 70$, we have $\hat{y} = -288.59 + 6.70278x_0 = -288.59 + (6.70278)\cdot(70) = 180.61$. From above, we have s = 8.218, $\bar{x} = 841/12 = 70.0833$, and SS(x) = 176.916667. We calculate

$$180.61 \pm (2.228)\cdot(8.218)\sqrt{1 + \frac{1}{12} + \frac{(70-70.0833)^2}{176.916667}},$$ or 180.61 ± 19.06.

Therefore, with 95% probability, we predict that when the height of a player is 70, his weight will be between 161.55 and 199.67.

14.60 **a.** The scatter diagram follows:

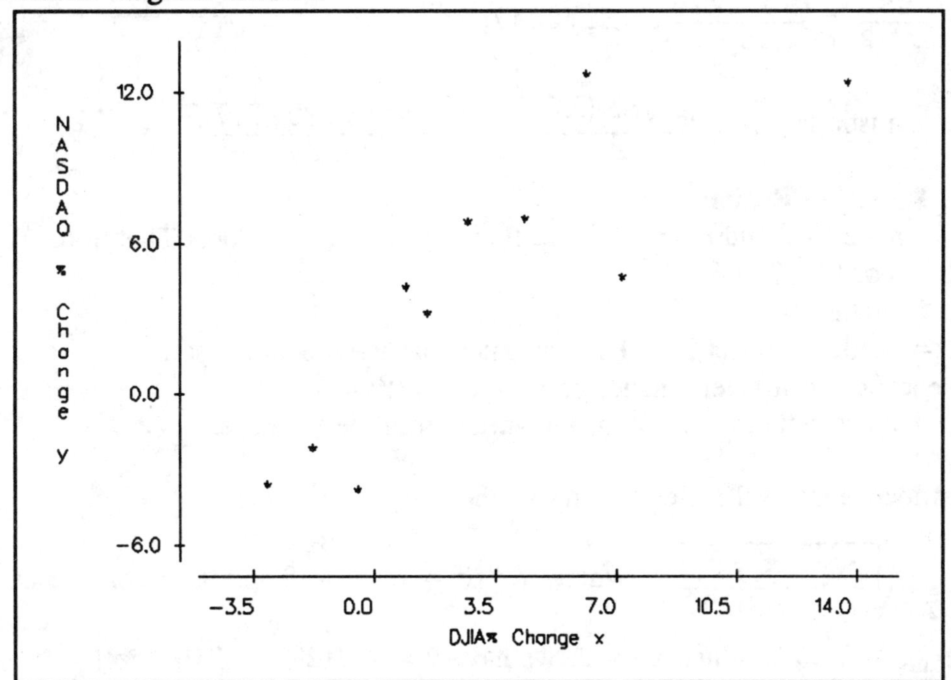

Exercise 14.60

b. From the data, we calculate: $\Sigma x = 31.9$; $\Sigma y = 41.7$; $\Sigma x^2 = 323.53$; $\Sigma y^2 = 495.37$; $\Sigma xy = 359.46$. Therefore:

$$SS(x) = \Sigma x^2 - \frac{(\Sigma x)^2}{n} = 323.53 - \frac{(31.9)^2}{10} = 221.769,$$

$$SS(y) = \Sigma y^2 - \frac{(\Sigma y)^2}{n} = 495.37 - \frac{(41.7)^2}{10} = 321.481 \text{, and}$$

$$SS(xy) = \Sigma xy - \frac{(\Sigma x)(\Sigma y)}{n} = 359.46 - \frac{(31.9)(41.7)}{10} = 226.437.$$

The slope of the least squares line is: $\beta_1 = \frac{SS(xy)}{SS(x)} = \frac{226.437}{221.769} = 1.02105$, and the y-intercept is $\beta_0 = \bar{y} - \beta_1\bar{x} = \frac{41.7}{10} - (1.02105)\frac{31.9}{10} = 0.91285$. The equation of the least squares line is: $\hat{y} = 0.91285 + 1.02105x$.

c. The correlation coefficient is:

$$r = \frac{SS(xy)}{\sqrt{SS(x)\,SS(y)}} = \frac{226.437}{\sqrt{(221.769)(321.481)}} = 0.848.$$

d. **Step 1: Hypotheses**

H_0: $\beta_1 = 0$

H_a: $\beta_1 \neq 0$

Step 2: Significance Level

We are given that $\alpha = 0.05$.

Step 3: Calculations

We first calculate s:

$$SSE = SS(y) - \beta_1 SS(xy) = 321.481 - (1.02105)(226.437) = 90.2775.$$

$$s^2 = \frac{SSE}{n-2} = \frac{90.2775}{10-2} = 11.2847 \text{ so } s = \sqrt{s^2} = \sqrt{11.2847} = 3.359.$$

The test statistic is $t = \frac{\hat{\beta}_1\sqrt{SS(x)}}{s} = \frac{(1.02105)\sqrt{221.769}}{3.359} = 4.53.$

Step 4: Rejection Region
For df = n - 2 = 8 and $\alpha = 0.05$, we find $t_{\alpha/2} = t_{0.025} = 2.306$. Therefore, we reject H_0 if $t < -2.306$ or $t > 2.306$.
Step 5: Conclusion
Since t = 4.53, we reject H_0. This indicates that the model is useful.
e. The coefficient of determination is $r^2 = (0.848)^2 = 0.719$. This means that the least squares line accounts for 71.9% of the variation in the y values.

14.61 The confidence interval for estimating y when $x = x_0$ is given by

$\hat{y} \pm t_{\frac{\alpha}{2}} s\sqrt{\frac{1}{n} + \frac{(x_0 - \bar{x})^2}{SS(x)}}$, where, for df = n - 2 = 8 and 95% confidence, we have

$t_{\alpha/2} = t_{0.025} = 2.306$. When $x_0 = 2$, we have $\hat{y} = 0.91285 + 1.02105x_0 =$ $0.91285 + (1.02105)\cdot(2) = 2.95$. From above, we have s = 3.359, $\bar{x} = 31.9/10 = 3.19$,

and SS(x) = 221.769. We calculate $2.95 \pm (2.306)\cdot(3.359)\sqrt{\frac{1}{10} + \frac{(2-3.19)^2}{221.769}}$,

or 2.95 ± 2.53. Therefore, with 95% confidence, we estimate that for Januarys in which the DJIA increases by 2 percent, the average percentage change in the NASDAQ composite will be between 0.42% and 5.48%.

14.62 The prediction interval for predicting y when $x = x_0$ is given by

$\hat{y} \pm t_{\frac{\alpha}{2}} s\sqrt{1 + \frac{1}{n} + \frac{(x_0 - \bar{x})^2}{SS(x)}}$, where for df = n - 2 = 8 and 95% probability, we have

$t_{\alpha/2} = t_{0.025} = 2.306$. When $x_0 = 2$, we have $\hat{y} = 0.91285 + 1.02105x_0 =$ $0.91285 + (1.02105)\cdot(2) = 2.95$. From Exercise 14.61, we have s = 3.359, $\bar{x} = 31.9/10$ = 3.19, and SS(x) = 221.769. We calculate

$2.95 \pm (2.306)\cdot(3.359)\sqrt{1 + \frac{1}{10} + \frac{(2-3.19)^2}{221.769}}$, or 2.95 ± 8.15. Therefore, with

95% probability, we predict that for a January in which the DJIA increases by 2 percent the percentage change in the NASDAQ composite will be between -5.2% and 11.1%.

MINITAB LAB ASSIGNMENTS

14.63(M)

```
MTB > # The data were read into C1 and C2
MTB > NAME C1'BA' C2 'WINS'
```

a.

```
MTB > # Part a
MTB > PLOT C2 C1

 WINS     -                                  *
          -
          -
       96+
          -                                        *
          -
          -                                                            *
          -                                                  *
       84+                                          *
          -
          -                                         **
          -            *                            *           *
          -                                   *              *  *
       72+
          -
          -      *
          -
          --+---------+---------+---------+---------+---------+---BA
         0.2400    0.2460    0.2520    0.2580    0.2640    0.2700
```

b.

```
MTB > # Part b
MTB > REGR C2 1 C1

The regression equation is
WINS = 19.8 + 236 BA

Predictor        Coef        Stdev    t-ratio        p
Constant        19.85        80.54       0.25    0.810
BA              235.8        310.7       0.76    0.463

s = 9.449        R-sq = 4.6%       R-sq(adj) = 0.0%

Analysis of Variance

SOURCE        DF           SS           MS          F         p
Regression     1        51.42        51.42       0.58     0.463
Error         12      1071.51        89.29
Total         13      1122.93

Unusual Observations
Obs.       BA       WINS         Fit Stdev.Fit   Residual    St.Resid
  1     0.254     103.00       79.73      2.98      23.27       2.59R

R denotes an obs. with a large st. resid.
MTB > # The least squares line is y = 19.8 + 236x
```

c.

```
MTB > # Part c: R-sq = 4.6%, so the model does not appear to be
MTB > # useful.
```

d.

```
MTB > # Part d: p-value for slope = 0.463 > 0.05. We fail to reject
MTB > # the null hypothesis that the slope is 0.
MTB > # There is insufficient evidence that the model has utility.
```

14.64(M) a. The estimated model for relating sales to the DJIA is $\hat{y}$ = -2.39 + 0.0270x.

b. The coefficient of determination is R-sq = r^2 =92.1%. The least squares line accounts for 92.1% of the total variation in sales.

14.65(M) **Step 1: Hypotheses**

H_0: $\beta_1 = 0$

H_a: $\beta_1 \neq 0$

Step 2: Significance Level

We are given that $\alpha = 0.01$.

Step 3: Calculations

We use the t-ratio for predictor X, so the test statistic is t = 12.75.

Step 4: Rejection Region

We reject H_0 if p-value $\leq \alpha = 0.01$.

Step 5: Conclusion

Since the p-value is 0.000, we reject H_0. This indicates that the model is useful.

14.66(M) For weeks when the average DJIA is 2300, we are 95% confident that the long run average weekly sales is between $57,441 and $62,099.

CHAPTER 15
NONPARAMETRIC TESTS
EXERCISES for Sections 15.1 & 15.2

The One-Sample Sign Test
The Two-Sample Sign Test: Paired Samples

15.1

Pair	I	II	Difference	Sign
1	390	395	-5	-
2	256	258	-2	-
3	190	199	-9	-
4	189	189	0	*
5	395	391	4	+
6	121	126	-5	-
7	254	253	1	+
8	189	197	-8	-
9	243	254	-11	-

Step 1: Hypotheses
Let $\tilde{\mu}_D$ denote the median of the hypothetical population of all possible differences in the pairs from populations I and II. We wish to show that population I tends to yield smaller values than population I, so we test:
H_0: $\tilde{\mu}_D = 0$
H_a: $\tilde{\mu}_D < 0$
Step 2: Significance Level
We are given that $\alpha = 0.05$.
Step 3: Calculations
The value of the test statistic is x = 2, since there are 2 plus signs. We note that n = 8, since one of the differences is 0.
Step 4: Rejection Region
If H_0 were true, x would have a binomial distribution with p = 0.5 and n = 8. Since H_a involves the < relation, the rejection region is left-tailed so we reject H_0 if x is too small. We use Table I to find the "left-tail" area which is closest to $\alpha = 0.05$. The rejection region may be obtained from the following diagram.

Probability 0.035

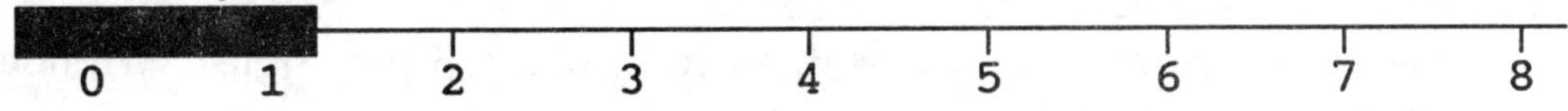

The diagram indicates that we should reject H_0 for x = 0 or 1 and that the level of significance is $\alpha = 0.035$ for this rejection region.
Step 5: Conclusion
Since x = 2, we are unable to reject H_0. We do not find sufficient evidence at the 0.035 level to conclude that population one tends to yield smaller values than population two.

15.3 **Step 1: Hypotheses**

Let $\tilde{\mu}$ denote the average age of blood donors during the last 12 months. In order to show that the median age has fallen below the previous value of 40 years, we test:

H_0: $\tilde{\mu} = 40$

H_a: $\tilde{\mu} < 40$

Step 2: Significance Level

We are given that $\alpha = 0.05$.

Step 3: Calculations

We replace with a plus sign (+) those data values which exceed 40, and with a minus sign (-) those data values which are below 40. We obtain - + + + - + - - - - - - - + - - - - - - - ; so the value of the test statistic is x = 5, since there are 5 plus signs. We note that n = 20, since none of the values is 40.

Step 4: Rejection Region

If H_0 were true, x would have a binomial distribution with p = 0.5 and n = 20. Since H_a involves the < relation, the rejection region is left-tailed so we reject H_0 if x is too small. We use Table I to find the "left-tail" area which is closest to $\alpha = 0.05$. The rejection region may be obtained from the following diagram.

Probability = 0.058

The diagram indicates that we should reject H_0 for x = 0, 1, 2, 3, 4, 5, or 6 and that the level of significance is $\alpha = 0.058$ for this rejection region.

Step 5: Conclusion

Since x = 5, we reject H_0. We find sufficient evidence at the 0.058 level to conclude that the average age of blood donors during the last 12 months has fallen below the previous value of 40 years.

15.5 **Step 1: Hypotheses**

Let $\tilde{\mu}$ denote the average weight of the bags. In order to show that the bags tend to be underweight, we test:

H_0: $\tilde{\mu} = 15$

H_a: $\tilde{\mu} < 15$

Step 2: Significance Level

We are given that $\alpha = 0.05$.

Step 3: Calculations

We replace with a plus sign (+) those data values which exceed 15, and with a minus sign (-) those data values which are below 15. We obtain - - - + - - + - + - - - - - - ; so the value of the test statistic is x = 3, since there are 3 plus signs. We note that n = 15, since none of the values is 15.

Step 4: Rejection Region

If H_0 were true, x would have a binomial distribution with p = 0.5 and n = 15. Since H_a involves the < relation, the rejection region is left-tailed so we reject H_0 if x is too small. We use Table I to find the "left-tail" area which is closest to $\alpha = 0.05$. The rejection region may be obtained from the following diagram.

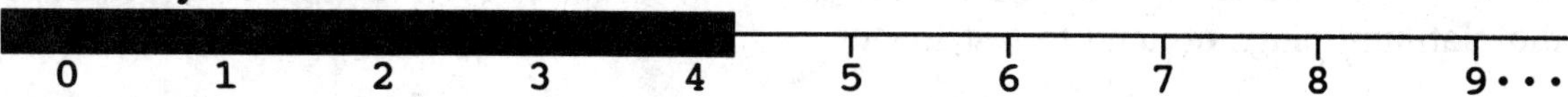

The diagram indicates that we should reject H_0 for $x = 0, 1, 2, 3$, or 4 and that the level of significance is $\alpha = 0.059$ for this rejection region.
Step 5: Conclusion
Since $x = 3$, we reject H_0. We find sufficient evidence at the 0.059 level to conclude that the bags tend to be underweight.

15.7 The p-value for the test in Exercise 15.6 is $P(x \geq 6) = 1 - P(x \leq 5) = 1 - 0.623 = 0.377$, where x is a binomial random variable with $n = 10$ and $p = 0.5$.

15.9

Engine	Platinum Plug	Conventional Plug	Difference	Sign
1	640	470	170	+
2	570	370	200	+
3	530	460	70	+
4	410	490	-80	-
5	600	380	220	+
6	580	410	170	+

Step 1: Hypotheses
Let $\tilde{\mu}_D$ denote the median of the population of all possible differences. In order to demonstrate that the platinum plugs tend to last longer, we test:
H_0: $\tilde{\mu}_D = 0$
H_a: $\tilde{\mu}_D > 0$
Step 2: Significance Level
We are given that $\alpha = 0.10$.
Step 3: Calculations
The value of the test statistic is $x = 5$, since there are 5 plus signs. We note that $n = 6$, since none of the differences is 0.
Step 4: Rejection Region
If H_0 were true, x would have a binomial distribution with $p = 0.5$ and $n = 6$. Since H_a involves the > relation, the rejection region is right-tailed so we reject H_0 if x is too large. We use Table I to find the "right-tail" area which is closest to $\alpha = 0.10$. The rejection region may be obtained from the following diagram.

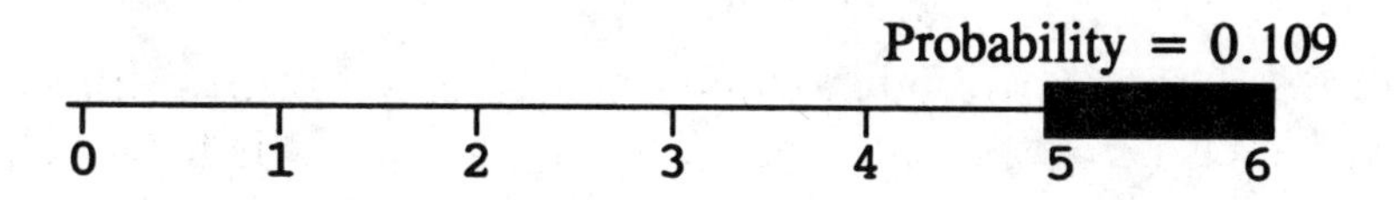

The diagram indicates that we should reject H_0 for $x = 5$ or 6 and that the level of significance is $\alpha = 0.109$ for this rejection region.

Step 5: Conclusion
Since x = 5, we reject H_0. We find sufficient evidence at the 0.109 level to conclude that the platinum plugs do tend to last longer.

15.11 The p-value for the test in Exercise 15.10 is $P(z > 2.54) = 1 - 0.9945 = 0.0055$.

15.13 The p-value for the test in Exercise 15.12 is $2 \cdot P(z < -2.26) = 2(0.0119) = 0.0238$.

15.15 The p-value for the test in Exercise 15.14 is $P(x \geq 6) = 1 - P(x \leq 5) = 1 - 0.937 = 0.063$, where x is a binomial random variable with n = 7 and p = 0.5.

MINITAB LAB ASSIGNMENTS

15.17(M)

```
MTB > # Exercise 15.17(M)
MTB > SET C1
DATA> 84 79 67 86 75 89 76 91 83 74 87 78 86 90 84 79 73 88
DATA> 87 79 73 76 87 86 80 82 74 85 87 69 84 83 74 81 85 79
DATA> END
MTB > STEST 80 C1;
SUBC> ALTERNATIVE = +1.

SIGN TEST OF MEDIAN = 80.00 VERSUS  G.T.  80.00

                N   BELOW   EQUAL   ABOVE    P-VALUE      MEDIAN
C1             36      15       1      20     0.2498       82.50
MTB > # Since p-value = 0.2498 > 0.05, we fail to reject Ho.
MTB > # Insufficient evidence that median exceeds 80.
```

15.19(M)

```
MTB > # Exercise 15.19(M)
MTB > # The data were read into C1 and C2
MTB > LET C3 = C1 - C2
MTB > STEST 0 C3;
SUBC> ALTERNATIVE = +1.

SIGN TEST OF MEDIAN = 0.00000 VERSUS  G.T.  0.00000

                N   BELOW   EQUAL   ABOVE    P-VALUE      MEDIAN
C3             36      10       1      25     0.0083       2.000
MTB > # Since p-value = 0.0083 < 0.01, we reject Ho.
MTB > # We find sufficient evidence that the herb is effective.
```

EXERCISES for Section 15.3

The Mann-Whitney U Test: Independent Samples

15.21 We let $\tilde{\mu}_1$ denote the median of Population One and $\tilde{\mu}_2$ the median of Population Two.

Step 1: Hypotheses
H_0: $\tilde{\mu}_1 - \tilde{\mu}_2 = 0$
H_a: $\tilde{\mu}_1 - \tilde{\mu}_2 \neq 0$

Step 2: Significance Level
We are given that $\alpha = 0.05$.

Step 3: Calculations
We construct a table which shows the ranks of the combined samples:

Sample 1	Overall Rank	Sample 2	Overall Rank
56	10	45	1
47	3	63	16.5
52	9	51	7
67	20	48	4
58	13	49	5
59	14	51	7
51	7	69	21
62	15	57	11.5
63	16.5	46	2
64	18	71	23
70	22	65	19
57	11.5		

The rank sum for the second sample is $R_2 = 117$. We calculate the value of U_1:
$U_1 = n_1 \cdot n_2 + n_2(n_2 + 1)/2 - R_2 = 12 \cdot 11 + 11(11 + 1)/2 - 117 = 81$.
The z-value for U_1 is used as the test statistic.

$$z = \frac{U_1 - \frac{n_1 n_2}{2}}{\sqrt{\frac{n_1 n_2 (n_1 + n_2 + 1)}{12}}} = \frac{81 - \frac{(12)(11)}{2}}{\sqrt{\frac{(12)(11)(12 + 11 + 1)}{12}}} = \frac{15}{16.248} = 0.92.$$

Step 4: Rejection Region
Since H_a involves the $\neq$ relation, the rejection region is two-tailed, so we reject H_0 if $z < -z_{\alpha/2} = -z_{0.025} = -1.96$ or if $z > z_{\alpha/2} = z_{0.025} = 1.96$.

Step 5: Conclusion
Since $z = 0.92$, we fail to reject H_0, so we do not find sufficient evidence at the 0.05 level to conclude that the medians of the sampled populations differ.

15.23 The p-value for the test in Exercise 15.22 is $2 \cdot P(z < -2.57) = 2(0.0051) = 0.0102$.

15.25 The p-value for the test in Exercise 15.24 is $2 \cdot P(z > 2.33) = 2(1 - 0.9901) = 2(0.0099) = 0.0198$.

15.27 The p-value for the test in Exercise 15.26 is $P(z > 1.98) = 1 - 0.9761 = 0.0239$.

15.29 The p-value for the test in Exercise 15.28 is $P(z > 2.77) = 1 - 0.9972 = 0.0028$.

MINITAB LAB ASSIGNMENTS

15.31(M)

```
MTB > # Exercise 15.31(M)
MTB > SET C1
DATA> 27.3 25.7 29.2 28.2 22.9 26.2 28.4 29.2 31.9 28.0 24.0
DATA> END
MTB > SET C2
DATA> 21.6 23.9 22.8 21.9 22.1 22.2 23.3 20.9 22.7 23.0 22.5
DATA> 22.4 20.9 21.1
DATA> END
MTB > MANN C1 C2;
SUBC> ALTERNATIVE 0.

Mann-Whitney Confidence Interval and Test

C1          N =  11     Median =        28.000
C2          N =  14     Median =        22.300
Point estimate for ETA1-ETA2 is          5.350
95.4 pct c.i. for ETA1-ETA2 is (3.501,6.699)
W = 217.0
Test of ETA1 = ETA2  vs.  ETA1 n.e. ETA2 is significant at 0.0001
The test is significant at 0.0001 (adjusted for ties)
MTB > # Since p-value = 0.0001 < 0.05, reject Ho.
MTB > # Sufficient evidence that the medians differ.
```

EXERCISES for Section 15.4

The Kruskal-Wallis H Test for a Completely Randomized Design

15.33 We let $\tilde{\mu}_i$ denote the median of population i, i = 1, 2, 3.

Step 1: Hypotheses
H_0: $\tilde{\mu}_1 = \tilde{\mu}_2 = \tilde{\mu}_3$
H_a: At least two of the medians differ

Step 2: Significance Level
We are given that $\alpha = 0.10$.

Step 3: Calculations
We construct a table which shows the ranks of the combined samples:

Sample 1	Overall Rank	Sample 2	Overall Rank	Sample 3	Overall Rank
69	13.0	63	8.0	51	1.0
73	15.0	64	9.0	56	2.5
70	14.0	60	5.5	59	4.0
68	12.0	61	7.0	56	2.5
74	16.0	65	10.0	60	5.5
		67	11.0		
$R_1 =$	70	$R_2 =$	50.5	$R_3 =$	15.5
$n_1 =$	5	$n_2 =$	6	$n_3 =$	5

Noting that $n = n_1 + n_2 + n_3 = 16$, the test statistic H equals

$$H = \frac{12}{n(n+1)} \sum_{i=1}^{k} \frac{R_i^2}{n_i} - 3(n+1)$$

$$= \frac{12}{(16)(17)} \left(\frac{(70)^2}{5} + \frac{(50.5)^2}{6} + \frac{(15.5)^2}{5} \right) - 3(17) = 13.11.$$

Step 4: Rejection Region
H has approximately a chi-square distribution with df = k - 1 = 2. We reject H_0 if $H > X^2_\alpha = X^2_{0.10} = 4.61$.

Step 5: Conclusion
Since H = 13.11, we reject H_0 and find sufficient evidence at the 0.10 level to conclude that the sampled populations have medians that are not the same.

15.35 We let $\tilde{\mu}_i$ denote the median of brand i, i = 1, 2.

Step 1: Hypotheses

H_0: $\tilde{\mu}_1 = \tilde{\mu}_2$

H_a: $\tilde{\mu}_1 \neq \tilde{\mu}_2$

Step 2: Significance Level

We are given that $\alpha = 0.05$.

Step 3: Calculations

We construct a table which shows the ranks of the combined samples:

Brand 1	Overall Rank	Brand 1 (cont.)	Overall Rank	Brand 2	Overall Rank	Brand 2 (cont.)	Overall Rank
27.3	19.0	28.4	22.0	21.6	4.0	20.9	1.5
25.7	17.0	29.2	23.5	23.9	15.0	22.7	10.0
29.2	23.5	31.9	25.0	22.8	11.0	23.0	13.0
28.2	21.0	28.0	20.0	21.9	5.0	22.5	9.0
22.9	12.0	24.0	16.0	22.1	6.0	22.4	8.0
26.2	18.0			22.2	7.0	20.9	1.5
				23.3	14.0	21.1	3.0
		R_1 =	217			R_2 =	108
		n_1 =	11			n_2 =	14

Noting that $n = n_1 + n_2 = 25$, the test statistic H equals

$$H = \frac{12}{n(n+1)} \sum_{i=1}^{k} \frac{R_i^2}{n_i} - 3(n+1)$$

$$= \frac{12}{(25)(26)}\left(\frac{(217)^2}{11} + \frac{(108)^2}{14}\right) - 3(26) = 16.41.$$

Step 4: Rejection Region

H has approximately a chi-square distribution with df = k - 1 = 1. We reject H_0 if $H > X^2_{\alpha} = X^2_{0.05} = 3.84$.

Step 5: Conclusion

Since H = 16.41, we reject H_0 and find sufficient evidence at the 0.05 level to conclude that a difference exists in the median protein content of the two brands.

15.37 For df = 2, we find $X^2_{0.05} = 5.99$ and $X^2_{0.10} = 4.61$. For the test value H = 5.28, we have $0.05 < \text{p-value} < 0.10$.

15.39 We let $\tilde{\mu}_i$ denote the median of programs i, where i = 1, 2, 3, 4.

Step 1: Hypotheses

H_0: $\tilde{\mu}_1 = \tilde{\mu}_2 = \tilde{\mu}_3 = \tilde{\mu}_4$

H_a: At least two of the medians differ

Step 2: Significance Level

We are given that $\alpha = 0.05$.

Step 3: Calculations

We construct a table which shows the ranks of the combined samples:

Prog. 1	Overall Rank	Prog. 2	Overall Rank	Prog. 3	Overall Rank	Prog. 4	Overall Rank
24	10.5	29	19.5	24	10.5	25	14.5
22	5.5	27	17.0	38	21.0	23	7.5
21	3.0	21	3.0	25	14.5	24	10.5
20	1.0	25	14.5	24	10.5	25	14.5
21	3.0	22	5.5	29	19.5	23	7.5
		28	18.0				
$R_1 =$ 23		$R_2 =$ 77.5		$R_3 =$ 76		$R_4 =$ 54.5	
$n_1 =$ 5		$n_2 =$ 6		$n_3 =$ 5		$n_4 =$ 5	

Noting that $n = n_1 + n_2 + n_3 + n_4 = 21$, the test statistic H equals

$$H = \frac{12}{n(n+1)} \sum_{i=1}^{k} \frac{R_i^2}{n_i} - 3(n+1)$$

$$= \frac{12}{(21)(22)}\left(\frac{(23)^2}{5} + \frac{(77.5)^2}{6} + \frac{(76)^2}{5} + \frac{(54.5)^2}{5}\right) - 3(22) = 8.18.$$

Step 4: Rejection Region

H has approximately a chi-square distribution with df = k - 1 = 3. We reject H_0 if $H > X^2_\alpha = X^2_{0.05} = 7.81$.

Step 5: Conclusion

Since H = 8.18, we reject H_0 and find sufficient evidence at the 0.05 level to indicate differences in performance ratings for at least two of the four word processing programs.

15.41 We let $\tilde{\mu}_i$ where i = 1, 2, 3, denote the median cost in cents for a quart of milk in the western, central, and eastern parts of the state, respectively.

Step 1: Hypotheses

H_0: $\tilde{\mu}_1 = \tilde{\mu}_2 = \tilde{\mu}_3$

H_a: At least two of the medians differ

Step 2: Significance Level

We are given that $\alpha = 0.05$.

Step 3: Calculations

We construct a table which shows the ranks of the combined samples:

Western	Overall Rank	Central	Overall Rank	Eastern	Overall Rank
63	7.5	67	15.0	68	17.5
64	9.5	62	5.0	68	17.5
63	7.5	69	21.0	73	24.0
60	2.5	68	17.5	64	9.5
55	1.0	65	12.0	69	21.0
62	5.0	65	12.0	72	23.0
60	2.5	65	12.0	69	21.0
62	5.0	66	14.0	68	17.5
$R_1 = 40.5$		$R_2 = 108.5$		$R_3 = 151$	
$n_1 = 8$		$n_2 = 8$		$n_3 = 8$	

Noting that $n = n_1 + n_2 + n_3 = 24$, the test statistic H equals

$$H = \frac{12}{n(n+1)} \sum_{i=1}^{k} \frac{R_i^2}{n_i} - 3(n+1)$$

$$= \frac{12}{(24)(25)} \left(\frac{(40.5)^2}{8} + \frac{(108.5)^2}{8} + \frac{(151)^2}{8} \right) - 3(25) = 15.53.$$

Step 4: Rejection Region
H has approximately a chi-square distribution with df = k - 1 = 2. We reject H_0 if $H > X^2_\alpha = X^2_{0.05} = 5.99$.
Step 5: Conclusion
Since H = 15.53 , we reject H_0 and find sufficient evidence at the 0.05 level to conclude that a difference exists in the median cost, in cents, for a quart of milk in the three regions.

15.43 For df = 3, we find $X^2_{0.005} = 12.84$. For the test value H = 18.44, we have p-value < 0.005.

MINITAB LAB ASSIGNMENTS

15.45(M)
```
MTB > # Exercise 15.45(M)
MTB > SET C1
DATA> 78 69 75 89 76 78 65 78 74 89 71 80 88 84 82 91
DATA> 73 65 78 76 73 65 69 74 62 80 76 72 56 69 61 71
DATA> 79 87 84 98 87 92 91 87 91 90 87 92 85 78 86 92
DATA> END
MTB > SET C2
DATA> (1:3)16
DATA> END
MTB > KRUS C1 C2 # Samples are in C1 and indices in C2
```

```
LEVEL      NOBS     MEDIAN  AVE. RANK    Z VALUE
    1        16      78.00       24.4      -0.02
    2        16      71.50       11.6      -4.52
    3        16      87.00       37.5       4.54
OVERALL      48                  24.5

H = 27.33  d.f. = 2  p = 0.000
H = 27.41  d.f. = 2  p = 0.000 (adj. for ties)
MTB > #We reject Ho; so there is a difference in the quality ratings.
```

EXERCISES for Section 15.5

The Runs Test for Randomness

15.47 The following sequence of letters contains 11 runs, each of which is underlined:

A B B B A A B A B B B B B A A B A B A A

15.49 The median of the following sample of data is 34. The sample values are classified as above (A) or below (B) the median. Those values which equal the median are discarded:

```
15  76  45  34  26  54  21  67  49  18  34
 B   A   A   -   B   A   B   A   A   B   -
```

The resulting sequence of letters has 7 runs which means that the number of runs above or below the median for the sequence of 11 numbers is also 7.

B A A B A B A A B

15.51 The following sequence of 13 numbers, for which the median is 25, has 5 runs above or below the median (other answers possible):

```
19  20  21  26  27  28  22  23  29  30  31  24  25
B   B   B   A   A   A   B   B   A   A   A   B   -
```

15.53 **Step 1: Hypotheses**

H_0: The sequence of heads and tails is random

H_a: The sequence of heads and tails is not random

Step 2: Significance Level

We are given that $\alpha = 0.05$.

Step 3: Calculations

H T T H H T T T H T T T T H H T H T H T H H H T H

The sequence has $x = 15$ runs, consisting of $n_1 = 12$ heads and $n_2 = 13$ tails.

Step 4: Rejection Region

From Table VII, with $n_1 = 12$ and $n_2 = 13$, and a tail area of $\alpha/2 = 0.025$, we obtain $L_{\alpha/2} = L_{0.025} = 8$ and $R_{\alpha/2} = R_{0.025} = 19$. Therefore, we reject H_0 if $x \leq 8$ or $x \geq 19$.

Step 5: Conclusion

Since $x = 15$ is not in the rejection region, we fail to reject H_0, so there is insufficient evidence to indicate that the sequence is not random.

15.55 **Step 1: Hypotheses**

H_0: The sequence of males (M) and females (F) is random

H_a: The sequence is not random

Step 2: Significance Level

We are given that $\alpha = 0.01$.

Step 3: Calculations

M M M M M F M M M F M M M M F F M M M

M M F M M F M M M M F F F M M M M M F

The sequence has $x = 14$ runs, consisting of $n_1 = 28$ males and $n_2 = 10$ females.

$$\mu = \frac{2n_1n_2}{n_1 + n_2} + 1 = \frac{2(28)(10)}{28 + 10} + 1 = 15.737$$

$$\sigma = \sqrt{\frac{2n_1n_2(2n_1n_2 - n_1 - n_2)}{(n_1 + n_2)^2(n_1 + n_2 - 1)}}$$

$$= \sqrt{\frac{2(28)(10)[2(28)(10) - 28 - 10]}{(28 + 10)^2(28 + 10 - 1)}} = 2.339$$

$$z = \frac{x - \mu}{\sigma} = \frac{14 - 15.737}{2.339} = -0.74.$$

Step 4: Rejection Region

We reject H_0 if $z < -z_{\alpha/2} = -z_{0.005} = -2.576$ or if $z > z_{\alpha/2} = z_{0.005} = 2.576$.

Step 5: Conclusion

Since $z = -0.74$ is not in the rejection region, we fail to reject H_0, so there is insufficient evidence to indicate that the sequence is not random.

15.57 **Step 1: Hypotheses**

H_0: The sequence of observations is random

H_a: The sequence is not random

Step 2: Significance Level

We are given that $\alpha = 0.05$.

Step 3: Calculations

A A A B A A B B B B B B A A A A B B B B

A A A A B A A A B B B A B B B B B A A A

The sequence has $x = 13$ runs, consisting of $n_1 = 20$ for A and $n_2 = 20$ for B.

$$\mu = \frac{2n_1n_2}{n_1 + n_2} + 1 = \frac{2(20)(20)}{20 + 20} + 1 = 21$$

$$\sigma = \sqrt{\frac{2n_1n_2(2n_1n_2 - n_1 - n_2)}{(n_1 + n_2)^2(n_1 + n_2 - 1)}}$$

$$= \sqrt{\frac{2(20)(20)[2(20)(20) - 20 - 20]}{(20 + 20)^2(20 + 20 - 1)}} = 3.121$$

$$z = \frac{x - \mu}{\sigma} = \frac{13 - 21}{3.121} = -2.56.$$

Step 4: Rejection Region
We reject H_0 if $z < -z_{\alpha/2} = -z_{0.025} = -1.96$ or if $z > z_{\alpha/2} = z_{0.025} = 1.96$.
Step 5: Conclusion
Since $z = -2.56$ is in the rejection region, we reject H_0, so at level of significance $\alpha = 0.05$, there is sufficient evidence to indicate that the sequence is not random.

15.59 **Step 1: Hypotheses**
H_0: The sequence of observations is random
H_a: The sequence is not random
Step 2: Significance Level
We are given that $\alpha = 0.05$.
Step 3: Calculations
In the following display, those values below the median (20.55) are classified as B and those above the median as A.

22.6	23.1	20.8	20.9	22.7	23.6	20.4	19.9	22.7	20.5
A	A	A	A	A	A	B	B	A	B

20.0	18.7	21.6	20.9	21.5	21.8	20.2	19.7	18.9	19.5
B	B	A	A	A	A	B	B	B	B

19.3	21.2	18.4	21.0	21.6	20.6	20.7	21.9	20.1	17.1
B	A	B	A	A	A	A	A	B	B

18.1	21.1	19.3	21.5	20.1	16.5	18.9	17.4	20.8	18.5
B	A	B	A	B	B	B	B	A	B

21.6	23.1	20.5	22.0	20.6	17.5	16.1	20.1	21.8	19.4
A	A	B	A	A	B	B	B	A	B

The sequence has $x = 22$ runs, consisting of $n_1 = 25$ A's and $n_2 = 25$ B's.

$$\mu = \frac{2n_1n_2}{n_1 + n_2} + 1 = \frac{2(25)(25)}{25 + 25} + 1 = 26$$

$$\sigma = \sqrt{\frac{2n_1n_2(2n_1n_2 - n_1 - n_2)}{(n_1 + n_2)^2(n_1 + n_2 - 1)}}$$

$$= \sqrt{\frac{2(25)(25)[2(25)(25) - 25 - 25]}{(25 + 25)^2(25 + 25 - 1)}} = 3.499$$

$$z = \frac{x - \mu}{\sigma} = \frac{22 - 26}{3.499} = -1.14.$$

Step 4: Rejection Region
We reject H_0 if $z < -z_{\alpha/2} = -z_{0.025} = -1.96$ or if $z > z_{\alpha/2} = z_{0.025} = 1.96$.
Step 5: Conclusion
Since $z = -1.14$ is not in the rejection region, we fail to reject H_0, so there is insufficient evidence to indicate that the sequence is not random.

MINITAB LAB ASSIGNMENTS

15.61(M)

```
MTB > # Exercise 15.61(M)
MTB > SET C1
DATA> 9 5 6 7 2 9 5 7 0 2 4 6 3 5
DATA> END
MTB > MEDIAN C1
   MEDIAN =        5.0000
MTB > RUNS 5 C1

    C1

    K =      5.0000


    THE OBSERVED NO. OF RUNS =  10
    THE EXPECTED NO. OF RUNS =    7.8571
     6 OBSERVATIONS ABOVE K     8 BELOW
  * N SMALL--FOLLOWING APPROX. MAY BE INVALID
                 THE TEST IS SIGNIFICANT AT  0.2231
                 CANNOT REJECT AT ALPHA = 0.05
```

EXERCISES for Section 15.6
Spearman's Rank Correlation Coefficient

15.63 Using Table VIII for n = 24 and $\alpha = 0.05$, we reject H_0: $\rho_s = 0$ if $r_s < -r_{\alpha/2} = -r_{0.025} = -0.409$ or if $r_s > r_{\alpha/2} = r_{0.025} = 0.409$.

15.65 For 13 pairs of data, the rejection region from Table VIII for detecting a negative rank correlation at the significance level $\alpha = 0.05$ is $r_s < -r_\alpha = -r_{0.05} = -0.475$.

15.67 The data, their relative ranks, differences in ranks, and the squares of the differences are displayed in the following table.

Area	1990 Population (in thousands)	1990 Population rank	1980 Population rank	d	d^2
LA	8771	1	2	-1	1
NY	8625	2	1	1	1
Chi	6308	3	3	0	0
Phil	4973	4	4	0	0
Det	4409	5	5	0	0
DC	3710	6	7	-1	1
Hos	3509	7	8	-1	1
Bos	2837	8	6	2	4
				$\Sigma d^2 =$	8

$$r_s = 1 - \frac{6\Sigma d^2}{n(n^2 - 1)} = 1 - \frac{6(8)}{8(8^2 - 1)} = 0.905.$$

15.69 The data, their relative ranks, differences in ranks, and the squares of the differences are displayed in the following table.

Sample	miles	tread	rank (x)	rank (y)	d	d^2
1	39	10	8	2.5	5.5	30.25
2	37	15	6	6	0	0
3	35	16	4	7	-3	9
4	15	30	1	9	-8	64
5	25	21	3	8	-5	25
6	38	14	7	4.5	2.5	6.25
7	18	34	2	10	-8	64
8	60	3	10	1	9	81
9	36	10	5	2.5	2.5	6.25
10	40	14	9	4.5	4.5	20.25
					$\Sigma d^2 =$	306

We calculate: $$r_s = 1 - \frac{6\Sigma d^2}{n(n^2 - 1)} = 1 - \frac{6(306)}{10(10^2 - 1)} = -0.855.$$

NOTE: Since there are ties in the values of the "tread" data, the above result, which is based on the shortcut formula, will differ slightly from that obtained using the exact formula. For the ranks, we calculate $\Sigma x = 55$; $\Sigma y = 55$; $\Sigma x^2 = 385$; $\Sigma y^2 = 384$; $\Sigma xy = 231.5$. Therefore, using the exact formula, we obtain:

$$r_s = \frac{\Sigma xy - \frac{(\Sigma x)(\Sigma y)}{n}}{\sqrt{\left[\Sigma x^2 - \frac{(\Sigma x)^2}{n}\right]\left[\Sigma y^2 - \frac{(\Sigma y)^2}{n}\right]}}$$

$$= \frac{231.5 - \frac{(55)(55)}{10}}{\sqrt{\left[385 - \frac{(55)^2}{10}\right]\left[384 - \frac{(55)^2}{10}\right]}} = -0.866.$$

15.71 **Step 1: Hypotheses**

H_0: $\rho_s = 0$ (no correlation between the rankings of the two magazines)

H_a: $\rho_s > 0$ (positive correlation between the rankings of the two magazines)

Step 2: Significance Level

We are given that $\alpha = 0.05$.

Step 3: Calculations

The data, their relative ranks, differences in ranks, and the squares of the differences are displayed in the following table.

Computer System	First Magazine's Ratings (x)	Second Magazine's Ratings (y)	d	d^2
1	6	7	-1	1
2	2	3	-1	1
3	8	9	-1	1
4	3	2	1	1
5	1	1	0	0
6	4	5	-1	1
7	9	8	1	1
8	5	4	1	1
9	7	6	1	1
			$\Sigma d^2 =$	8

$$r_s = 1 - \frac{6\Sigma d^2}{n(n^2 - 1)} = 1 - \frac{6(8)}{9(9^2 - 1)} = 0.933.$$

Step 4: Rejection Region

For a one-tailed alternative with $n = 9$ and $\alpha = 0.05$, we find, using Table VIII, $r_\alpha = r_{0.05} = 0.600$. Therefore, we reject H_0 if $r_s > 0.600$.

Step 5: Conclusion

Since $r_s = 0.933$ is in the rejection region, we reject H_0 and find sufficient evidence of a positive correlation between the rankings of the two magazines.

15.73 Step 1: Hypotheses

H_0: $\rho_s = 0$ (exam score and time required to complete the exam are not associated)
H_a: $\rho_s \neq 0$ (exam score and time required to complete the exam are associated)

Step 2: Significance Level

We are given that $\alpha = 0.01$.

Step 3: Calculations

The data, their relative ranks, differences in ranks, and the squares of the differences are displayed in the following table.

Student	Exam Score	Time	rank (x)	rank (y)	d	d^2
1	86	108	12	4	8	64
2	93	100	13.5	3	10.5	110.25
3	73	115	6	11	-5	25
4	78	113	8.5	10	-1.5	2.25
5	54	118	2	14	-12	144
6	93	99	13.5	2	11.5	132.25
7	69	110	4	6.5	-2.5	6.25
8	78	109	8.5	5	3.5	12.25
9	84	111	11	8	3	9
10	82	117	10	13	-3	9
11	41	120	1	15	-14	196
12	67	116	3	12	-9	81
13	98	89	15	1	14	196
14	74	112	7	9	-2	4
15	71	110	5	6.5	-1.5	2.25
					$\Sigma d^2 =$	993.5

$$r_s = 1 - \frac{6\Sigma d^2}{n(n^2 - 1)} = 1 - \frac{6(993.5)}{15(15^2 - 1)} = -0.774.$$

NOTE: Since there are ties in the values of the data, the above result, which is based on the shortcut formula, will differ slightly from that obtained using the exact formula. For the ranks, we calculate $\Sigma x = 120$; $\Sigma y = 120$; $\Sigma x^2 = 1239$; $\Sigma y^2 = 1239.5$; $\Sigma xy = 742.5$. Therefore, using the exact formula, we obtain:

$$r_s = \frac{\Sigma xy - \frac{(\Sigma x)(\Sigma y)}{n}}{\sqrt{\left[\Sigma x^2 - \frac{(\Sigma x)^2}{n}\right]\left[\Sigma y^2 - \frac{(\Sigma y)^2}{n}\right]}}$$

$$= \frac{742.5 - \frac{(120)(120)}{15}}{\sqrt{\left[1239 - \frac{(120)^2}{15}\right]\left[1239.5 - \frac{(120)^2}{15}\right]}} = -0.779.$$

Step 4: Rejection Region

For a two-tailed alternative with $n = 15$ and $\alpha = 0.01$, we find, using Table VIII, $r_{\alpha/2} = r_{0.005} = 0.689$. Therefore, we reject H_0 for $r_s < -0.689$ or for $r_s > 0.689$.

Step 5: Conclusion
Since $r_s = -0.779$ is in the rejection region, we reject H_0 and find sufficient evidence of an association between exam score and time required to complete the exam.

MINITAB LAB ASSIGNMENTS

15.75(M)

```
MTB > # Exercise 15.75(M)
MTB > # The exam score data were read into C1 and time into C2
MTB > RANK C1 C3
MTB > RANK C2 C4
MTB > PRINT C1-C4

 ROW    C1     C2      C3      C4

   1    86    108    12.0     4.0
   2    93    100    13.5     3.0
   3    73    115     6.0    11.0
   4    78    113     8.5    10.0
   5    54    118     2.0    14.0
   6    93     99    13.5     2.0
   7    69    110     4.0     6.5
   8    78    109     8.5     5.0
   9    84    111    11.0     8.0
  10    82    117    10.0    13.0
  11    41    120     1.0    15.0
  12    67    116     3.0    12.0
  13    98     89    15.0     1.0
  14    74    112     7.0     9.0
  15    71    110     5.0     6.5

MTB > CORRELATION C3 C4

Correlation of C3 and C4 = -0.779

MTB > # The rank correlation coefficient is r = -0.779.
```

REVIEW EXERCISES
CHAPTER 15

15.77 We let $\tilde{\mu}_1$ denote the median age for the 7:00 p.m. showing and $\tilde{\mu}_2$ the median age for the 9:30 p.m. showing.

Step 1: Hypotheses
H_0: $\tilde{\mu}_1 - \tilde{\mu}_2 = 0$
H_a: $\tilde{\mu}_1 - \tilde{\mu}_2 \neq 0$

Step 2: Significance Level
We are given that $\alpha = 0.05$.

Step 3: Calculations
We construct a table which shows the ranks of the combined samples:

7:00 p.m. Showing	Overall Rank	9:30 p.m.Showing	Overall Rank
57	26	23	5
49	22.5	20	2
41	19	32	12
56	25	43	20
39	16	21	3.5
40	17.5	27	10
37	14	25	6.5
26	8.5	52	24
19	1	40	17.5
68	27	26	8.5
49	22.5	34	13
73	28	38	15
28	11	25	6.5
44	21	21	3.5

The rank sum for the second sample is $R_2 = 147$. We calculate the value of U_1:
$U_1 = n_1 \cdot n_2 + n_2 (n_2 + 1) /2 - R_2 = 14 \cdot 14 + 14(14 + 1)/2 - 147 = 154.$
The z-value for U_1 is used as the test statistic.

$$z = \frac{U_1 - \frac{n_1 n_2}{2}}{\sqrt{\frac{n_1 n_2 (n_1 + n_2 + 1)}{12}}} = \frac{154 - \frac{(14)(14)}{2}}{\sqrt{\frac{(14)(14)(14 + 14 + 1)}{12}}} = \frac{56}{21.764} = 2.57.$$

Step 4: Rejection Region
Since H_a involves the $\neq$ relation, the rejection region is two-tailed, so we reject H_0 if $z < -z_{\alpha/2} = -z_{0.025} = -1.96$ or if $z > z_{\alpha/2} = z_{0.025} = 1.96$.

Step 5: Conclusion
Since $z = 2.57$, we reject H_0 and find sufficient evidence at the 0.05 level to conclude that there is a difference in the median age of patrons for the two showings.

15.78 **Step 1: Hypotheses**
H_0: The sex of the callers is random
H_a: The sex of the callers is not random
Step 2: Significance Level
We are given that $\alpha = 0.05$.
Step 3: Calculations

F MM FFFF M F M FF MMMMM F M FFFF MMMMM

The sequence has $x = 12$ runs, consisting of $n_1 = 13$ female and $n_2 = 15$ male callers.
Step 4: Rejection Region
From Table VII, with $n_1 = 13$ and $n_2 = 15$, and a tail area of $\alpha/2 = 0.025$, we obtain $L_{\alpha/2} = L_{0.025} = 9$ and $R_{\alpha/2} = R_{0.025} = 21$. Therefore, we reject H_0 if $x \leq 9$ or $x \geq 21$.
Step 5: Conclusion
Since $x = 12$ is not in the rejection region, we fail to reject H_0, so there is insufficient evidence to indicate that the sequence is not random.

15.79 **Step 1: Hypotheses**
H_0: The sex of the callers is random
H_a: The sex of the callers is not random
Step 2: Significance Level
We are given that $\alpha = 0.05$.
Step 3: Calculations
F MM FFFF M F M FF MMMMM F M FFFF MMMMM
The sequence has $x = 12$ runs, consisting of $n_1 = 13$ female and $n_2 = 15$ male callers.

$$\mu = \frac{2n_1n_2}{n_1 + n_2} + 1 = \frac{2(13)(15)}{13 + 15} + 1 = 14.929$$

$$\sigma = \sqrt{\frac{2n_1n_2(2n_1n_2 - n_1 - n_2)}{(n_1 + n_2)^2(n_1 + n_2 - 1)}}$$

$$= \sqrt{\frac{2(13)(15)[2(13)(15) - 13 - 15]}{(13 + 15)^2(13 + 15 - 1)}} = 2.583$$

$$z = \frac{x - \mu}{\sigma} = \frac{12 - 14.929}{2.583} = -1.13.$$

Step 4: Rejection Region
We reject H_0 if $z < -z_{\alpha/2} = -z_{0.025} = -1.96$ or if $z > z_{\alpha/2} = z_{0.025} = 1.96$.
Step 5: Conclusion
Since $z = -1.13$ is not in the rejection region, we fail to reject H_0, so there is insufficient evidence to indicate that the sequence is not random.

15.80

Secretary	Speed before the program	Speed after the program	Difference	Sign
1	59	64	5	+
2	43	51	8	+
3	58	55	-3	-
4	47	59	12	+
5	37	48	11	+
6	62	62	0	*
7	41	45	4	+
8	50	52	2	+
9	43	53	10	+
10	49	52	3	+
11	63	65	2	+

Step 1: Hypotheses
Let $\tilde{\mu}_D$ denote the median change in typing speed as a result of participating in a one day workshop. In order to determine whether the workshop increases typing speed, we test:
H_0: $\tilde{\mu}_D = 0$
H_a: $\tilde{\mu}_D > 0$
Step 2: Significance Level
We are given that $\alpha = 0.05$.
Step 3: Calculations
The value of the test statistic is x = 9, since there are 9 plus signs. We note that n = 10, since one of the differences is 0.
Step 4: Rejection Region
If H_0 were true, x would have a binomial distribution with p = 0.5 and n = 10. Since H_a involves the > relation, the rejection region is right-tailed so we reject H_0 if x is too large. We use Table I to find the "right-tail" area which is closest to $\alpha = 0.05$. The rejection region may be obtained from the following diagram.

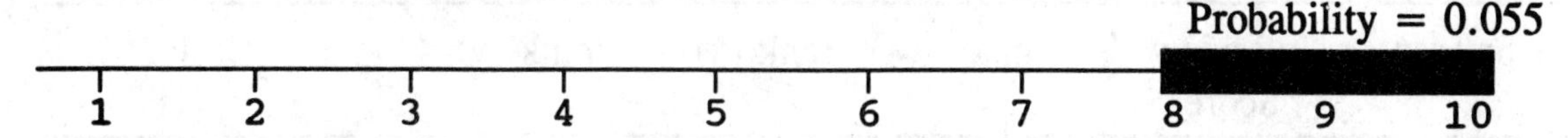

The diagram indicates that we should reject H_0 for x = 8, 9, or 10 and that the level of significance is $\alpha = 0.055$ for this rejection region.
Step 5: Conclusion
Since x = 9, we reject H_0. We find sufficient evidence at the 0.055 level to conclude that the course tends to increase one's typing speed.

15.81 The p-value for the test in Exercise 15.80 is $P(x \geq 9) = 1 - P(x \leq 8) = 1 - 0.989 = 0.011$, where x is a binomial random variable with n = 10 and p = 0.5.

15.82 **Step 1: Hypotheses**
Let $\tilde{\mu}$ denote the average (median) length of time of a ferris wheel ride. In order to determine whether the median duration of the ride is less than five minutes (300 seconds), we test:

H_0: $\tilde{\mu} = 300$
H_a: $\tilde{\mu} < 300$
Step 2: Significance Level
We are given that $\alpha = 0.01$.
Step 3: Calculations
We replace with a plus sign (+) those data values which exceed 300, with a minus sign (-) those data values which are below 300, and with as asterisk (*) those data values which equal 300. We obtain - - - + - - + - - - + + - - - - - * - + - - - - - - - + - - - + - + - - - - ; so the value of the test statistic is x = 8, since there are 8 plus signs. We note that n = 35, since one of the differences is 0. We use the normal approximation to the binomial:

$$z = \frac{x - 0.5n}{0.5\sqrt{n}} = \frac{8 - 0.5(35)}{0.5\sqrt{35}} = -3.21.$$

Step 4: Rejection Region
Since H_a involves the < relation, the rejection region is left-tailed so we reject H_0 if $z < -z_\alpha = -z_{0.01} = -2.326$.
Step 5: Conclusion
Since x = -3.21, we reject H_0 and find sufficient evidence at the 0.01 level to conclude that the median duration of the ride is less than 5 minutes (300 seconds).

15.83 The p-value for the test in Exercise 15.82 is $P(z \leq -3.21) = 0.0007$.

15.84 **Step 1: Hypotheses**
H_0: $\rho_s = 0$ (no correlation between the interview score and grade-point-average)
H_a: $\rho_s > 0$ (positive correlation between the interview score and grade-point-average)
Step 2: Significance Level
We are given that $\alpha = 0.01$.
Step 3: Calculations
The data, their relative ranks, differences in ranks, and the squares of the differences are displayed in the following table.

Applicant	Interview Score	gpa	rank (x)	rank (y)	d	d^2
1	47	3.12	10	6	4	16
2	43	3.59	7	9	-2	4
3	37	2.98	5	4	1	1
4	46	3.71	9	10	-1	1
5	30	2.76	1	1	0	0
6	31	3.21	2	7	-5	25
7	44	3.34	8	8	0	0
8	34	2.95	3	3	0	0
9	40	3.06	6	5	1	1
10	35	2.88	4	2	2	4
					$\Sigma d^2 =$	52

$$r_s = 1 - \frac{6\Sigma d^2}{n(n^2 - 1)} = 1 - \frac{6(52)}{10(10^2 - 1)} = 0.685.$$

Step 4: Rejection Region
For a one-tailed alternative with n = 10 and $\alpha = 0.01$, we find, using Table VIII, $r_\alpha = r_{0.01} = 0.745$. Therefore, we reject H_0 if $r_s > 0.745$.
Step 5: Conclusion
Since $r_s = 0.685$ is not in the rejection region, we fail to reject H_0 and do not find sufficient evidence of a positive correlation between the the interview score and grade-point-average.

15.85 For n = 10, we find in Table VIII, $r_{0.025} = 0.648$ and $r_{0.01} = 0.745$. Since the test value found in Exercise 15.84 was $r_s = 0.685$, we have $0.01 < \text{p-value} < 0.025$.

15.86 We let $\tilde{\mu}_i$ denote the median of Population i, where i = 1, 2, 3, 4.
Step 1: Hypotheses
H_0: $\tilde{\mu}_1 = \tilde{\mu}_2 = \tilde{\mu}_3 = \tilde{\mu}_4$
H_a: At least two of the medians differ
Step 2: Significance Level
We are given that $\alpha = 0.01$.
Step 3: Calculations
We construct a table which shows the ranks of the combined samples:

Sample 1	Overall Rank	Sample 2	Overall Rank	Sample 3	Overall Rank	Sample 4	Overall Rank
65	15.0	59	8.5	72	19.0	56	5.0
76	22.0	60	10.0	75	21.0	54	3.0
68	17.0	62	12.0	67	16.0	53	2.0
63	13.0	58	6.5	74	20.0	51	1.0
64	14.0	61	11.0	71	18.0	55	4.0
59	8.5					58	6.5
$R_1 =$	89.5	$R_2 =$	48	$R_3 =$	94	$R_4 =$	21.5
$n_1 =$	6	$n_2 =$	5	$n_3 =$	5	$n_4 =$	6

Noting that $n = n_1 + n_2 + n_3 + n_4 = 22$, the test statistic H equals

$$H = \frac{12}{n(n+1)} \sum_{i=1}^{k} \frac{R_i^2}{n_i} - 3(n+1)$$

$$= \frac{12}{(22)(23)}\left(\frac{(89.5)^2}{6} + \frac{(48)^2}{5} + \frac{(94)^2}{5} + \frac{(21.5)^2}{6}\right) - 3(23) = 17.33.$$

Step 4: Rejection Region
H has approximately a chi-square distribution with df = k - 1 = 3. We reject H_0 if $H > X^2_\alpha = X^2_{0.01} = 11.34$.
Step 5: Conclusion
Since H = 17.33, we reject H_0 and find sufficient evidence at the 0.01 level to indicate differences in location.

15.87 **Step 1: Hypotheses**
H_0: The sequence of observations is random
H_a: The sequence is not random
Step 2: Significance Level
We are given that $\alpha = 0.05$.
Step 3: Calculations
In the following display, those values below the median (485) are classified as B and those above the median as A.

```
711 356 189 347 982 597 735 916 069 901 735 914 036 300 918
 A   B   B   B   A   A   A   A   B   A   A   A   B   B   A

131 016 733 868 971 350 071 415 555 783 027 271 244 612 085
 B   B   A   A   A   B   B   B   A   A   B   B   B   A   B
```

The sequence has $x = 14$ runs, consisting of $n_1 = 15$ above (A) and $n_2 = 15$ below(B) the median.
Step 4: Rejection Region
From Table VII, with $n_1 = 15$ and $n_2 = 15$, and a tail area of $\alpha/2 = 0.025$, we obtain $L_{\alpha/2} = L_{0.025} = 10$ and $R_{\alpha/2} = R_{0.025} = 22$. Therefore, we reject H_0 if $x \leq 10$ or $x \geq 22$.
Step 5: Conclusion
Since $x = 14$ is not in the rejection region, we fail to reject H_0, so there is insufficient evidence to indicate that the sequence is not random.

15.88 **Step 1: Hypotheses**
H_0: The sequence of observations is random
H_a: The sequence is not random
Step 2: Significance Level
We are given that $\alpha = 0.05$.
Step 3: Calculations
In the following display, those values below the median (485) are classified as B and those above the median as A.

```
711 356 189 347 982 597 735 916 069 901 735 914 036 300 918
 A   B   B   B   A   A   A   A   B   A   A   A   B   B   A

131 016 733 868 971 350 071 415 555 783 027 271 244 612 085
 B   B   A   A   A   B   B   B   A   A   B   B   B   A   B
```

The sequence has $x = 14$ runs, consisting of $n_1 = 15$ above (A) and $n_2 = 15$ below(B) the median.

$$\mu = \frac{2n_1n_2}{n_1 + n_2} + 1 = \frac{2(15)(15)}{15 + 15} + 1 = 16$$

$$\sigma = \sqrt{\frac{2n_1n_2(2n_1n_2 - n_1 - n_2)}{(n_1 + n_2)^2(n_1 + n_2 - 1)}}$$

$$= \sqrt{\frac{2(15)(15)[2(15)(15) - 15 - 15]}{(15 + 15)^2(15 + 15 - 1)}} = 2.691$$

$$z = \frac{x - \mu}{\sigma} = \frac{14 - 16}{2.691} = -0.74.$$

Step 4: Rejection Region

We reject H_0 if $z < -z_{\alpha/2} = -z_{0.025} = -1.96$ or if $z > z_{\alpha/2} = z_{0.025} = 1.96$.

Step 5: Conclusion

Since $z = -0.74$ is not in the rejection region, we fail to reject H_0, so there is insufficient evidence to indicate that the sequence is not random.

15.89 We let $\tilde{\mu}_i$ denote the median levels of DDT in Lake i, where i = 1, 2, 3, 4.

Step 1: Hypotheses

H_0: $\tilde{\mu}_1 = \tilde{\mu}_2 = \tilde{\mu}_3 = \tilde{\mu}_4$

H_a: At least two of the medians differ

Step 2: Significance Level

We are given that $\alpha = 0.05$.

Step 3: Calculations

We construct a table which shows the ranks of the combined samples:

Lake One	Overall Rank	Lake Two	Overall Rank	Lake Three	Overall Rank	Lake Four	Overall Rank
1.7	11.0	0.3	2.0	2.7	19.0	1.2	8.0
1.4	9.0	0.7	4.0	1.9	14.0	3.1	21.0
1.9	14.0	0.5	3.0	2.0	16.0	1.9	14.0
1.1	6.5	0.1	1.0	1.5	10.0	3.7	23.0
2.1	17.0	1.1	6.5	2.6	18.0	2.8	20.0
1.8	12.0	0.9	5.0			3.5	22.0
$R_1 =$	69.5	$R_2 =$	21.5	$R_3 =$	77	$R_4 =$	108
$n_1 =$	6	$n_2 =$	6	$n_3 =$	5	$n_4 =$	6

Noting that $n = n_1 + n_2 + n_3 + n_4 = 23$, the test statistic H equals

$$H = \frac{12}{n(n + 1)}\sum_{i=1}^{k}\frac{R_i^2}{n_i} - 3(n + 1)$$

$$= \frac{12}{(23)(24)}\left(\frac{(69.5)^2}{6} + \frac{(21.5)^2}{6} + \frac{(77)^2}{5} + \frac{(108)^2}{6}\right) - 3(24) = 15.21.$$

Step 4: Rejection Region

H has approximately a chi-square distribution with df = k - 1 = 3. We reject H_0 if $H > X^2_\alpha = X^2_{0.05} = 7.81$.

Step 5: Conclusion

Since $H = 15.21$, we reject H_0 and find sufficient evidence at the 0.05 level to indicate differences in the median DDT levels at the four lakes.

15.90 Using Table V for df = 3, we find $X^2_{0.005} = 12.84$, so for the test value H = 15.21, we have p-value < 0.005.

MINITAB LAB ASSIGNMENTS

15.91(M)

```
MTB > # Exercise 15.91(M)
MTB > SET C1
DATA> 283 274 296 301 294 288 302 275 297 291 306 316 285 296 289 295
DATA> 300 291 305 289 298 287 281 295 291 290 316 275 296 284 303 295
DATA> 303 290 278 299
DATA> END
MTB > STEST 300 C1;
SUBC> ALTERNATIVE = -1.

SIGN TEST OF MEDIAN = 300.0 VERSUS  L.T.  300.0

                  N  BELOW  EQUAL  ABOVE   P-VALUE    MEDIAN
C1               36     27      1      8    0.0009     294.5
MTB > # Since p-value = 0.0009 < 0.05, we reject Ho.
MTB > # Sufficient evidence that median is less than 5 minutes.
```

15.92(M)

```
MTB > # Exercise 15.92(M)
MTB > # The data were read into C1 and C2
MTB > RANK C1 C3
MTB > RANK C2 C4
MTB > PRINT C1 - C4

 ROW    C1      C2     C3     C4

   1    47    3.12     10      6
   2    43    3.59      7      9
   3    37    2.98      5      4
   4    46    3.71      9     10
   5    30    2.76      1      1
   6    31    3.21      2      7
   7    44    3.34      8      8
   8    34    2.95      3      3
   9    40    3.06      6      5
  10    35    2.88      4      2

MTB > CORRELATION C3 C4

Correlation of C3 and C4 = 0.685
```

15.93(M)

```
MTB > # Exercise 15.93(M)
MTB > # The data were read into C1
MTB > MEDIAN C1
   MEDIAN =      485.00
MTB > RUNS 485 C1

    C1

    K =   485.0000


    THE OBSERVED NO. OF RUNS =  14
    THE EXPECTED NO. OF RUNS =  16.0000
    15 OBSERVATIONS ABOVE K   15 BELOW
              THE TEST IS SIGNIFICANT AT  0.4575
              CANNOT REJECT AT ALPHA = 0.05
```

15.94(M)

```
MTB > # Exercise 15.94(M)
MTB > SET C1
DATA> 80.6 81.3 82.8 81.5 80.4 79.7 82.3 81.7 80.6 81.5
DATA> 91.6 83.5 83.4 88.6 96.7 84.8 88.4 89.5 84.4 85.1
DATA> 90.5 98.5 97.5 99.9 96.9 90.5 96.7 93.8 97.8 96.8
DATA> 86.7 75.4 79.7 76.5 75.7 84.7 74.5 83.3 84.2 75.3
DATA> END
MTB > SET C2
DATA> (1:4)10
DATA> END
MTB > KRUS C1 C2

LEVEL     NOBS    MEDIAN  AVE. RANK   Z VALUE
    1       10     81.40       11.4     -2.83
    2       10     86.75       25.0      1.42
    3       10     96.85       35.0      4.51
    4       10     78.10       10.6     -3.11
OVERALL     40                 20.5

H = 30.03  d.f. = 3  p = 0.000
H = 30.04  d.f. = 3  p = 0.000 (adj. for ties)

MTB > # Since p-value = 0.00 < 0.05, we reject Ho
MTB > # Sufficient evidence that the medians are not all equal.
```

15.95(M)

```
MTB > # Exercise 15.95(M)
MTB > SET C1
DATA> 80.6 81.3 82.8 81.5 80.4 79.7 82.3 81.7 80.6 81.5
DATA> END
MTB > SET C3
DATA> 90.5 98.5 97.5 99.9 96.9 90.5 96.7 93.8 97.8 96.8
MTB > MANN C1 C3

Mann-Whitney Confidence Interval and Test

C1          N =  10      Median =        81.400
C3          N =  10      Median =        96.850
Point estimate for ETA1-ETA2 is      -15.500
95.5 pct c.i. for ETA1-ETA2 is (-17.000,-12.301)
W = 55.0
Test of ETA1 = ETA2  vs.  ETA1 n.e. ETA2 is significant at 0.0002
The test is significant at 0.0002 (adjusted for ties)

MTB > # Since p-value = 0.0002 < 0.01, we reject Ho
MTB > # Sufficient evidence that the medians 1 and 3 differ.
```

15.96(M)

```
MTB > # Exercise 15.96(M)
MTB > # The data were read into C1 and C2
MTB > LET C3 = C1 - C2
MTB > STEST 0 C3;
SUBC> ALTERNATIVE 0.

SIGN TEST OF MEDIAN = 0.00000 VERSUS  N.E.  0.00000

                N  BELOW  EQUAL  ABOVE   P-VALUE     MEDIAN
C3             33     4      3     26    0.0001    0.06000
MTB > # Since p-value = 0.0001 < 0.01, we reject Ho
MTB > # Sufficient evidence that the average prices differ.
```